Yamaha YZF-R1
Service and Repair Manual

by Matthew Coombs

(3754 - 8AB2 - 272)

Models covered

YZF-R1. 998cc. 1998 to 2001 (Europe and US)

© Haynes Publishing 2002

A book in the Haynes Service and Repair Manual Series

All rights reserved. No part of this book may be reproduced or transmitted in any form or by any means, electronic or mechanical, including photocopying, recording or by any information storage or retrieval system, without permission in writing from the copyright holder.

ISBN 1 85960 754 3

British Library Cataloguing in Publication Data
A catalogue record for this book is available from the British Library

Library of Congress Control Number 00-134616

ABCDE
FGHIJ
KLMNO
PQR

Printed in the USA

Haynes Publishing
Sparkford, Yeovil, Somerset BA22 7JJ, England

Haynes North America, Inc
861 Lawrence Drive, Newbury Park, California 91320, USA

Editions Haynes
4, Rue de l'Abreuvoir
92415 COURBEVOIE CEDEX, France

Haynes Publishing Nordiska AB
Box 1504, 751 45 UPPSALA, Sweden

Contents

LIVING WITH YOUR YAMAHA R1

Introduction

Yamaha – Musical instruments to Motorcycles	Page	0•4
Acknowledgements	Page	0•8
About this manual	Page	0•8
Identification numbers	Page	0•9
Buying spare parts	Page	0•9
Performance data and Bike spec	Page	0•10
Safety first!	Page	0•12

Daily (pre-ride) checks

Engine/transmission oil level check	Page	0•13
Brake fluid level checks	Page	0•14
Coolant level check	Page	0•15
Tyre checks	Page	0•16
Suspension, steering and final drive checks	Page	0•16
Legal and safety checks	Page	0•16

MAINTENANCE

Routine maintenance and servicing

Specifications	Page	1•2
Recommended lubricants and fluids	Page	1•2
Maintenance schedule	Page	1•3
Component locations	Page	1•4
Maintenance procedures	Page	1•6

Contents

REPAIRS AND OVERHAUL

Engine, transmission and associated systems

Engine, clutch and transmission	Page	**2•1**
Cooling system	Page	**3•1**
Fuel and exhaust systems	Page	**4•1**
Ignition system	Page	**5•1**

Chassis and bodywork components

Frame and suspension	Page	**6•1**
Final drive	Page	**6•18**
Brakes	Page	**7•1**
Wheels	Page	**7•13**
Tyres	Page	**7•18**
Fairing and bodywork	Page	**8•1**

Electrical system

Page **9•1**

Wiring diagrams

Page **9•26**

REFERENCE

Tools and Workshop Tips	Page	**REF•2**
Security	Page	**REF•20**
Lubricants and fluids	Page	**REF•23**
Conversion Factors	Page	**REF•26**
MOT test checks	Page	**REF•27**
Storage	Page	**REF•32**
Fault finding	Page	**REF•35**
Fault finding equipment	Page	**REF•45**
Technical terms explained	Page	**REF•49**

Index

Page **REF•53**

Yamaha Musical instruments to motorcycles

**The FS1E -
first bike of many sixteen year olds in the UK**

The Yamaha Motor Company

The Yamaha name can be traced back to 1889, when Torakusu Yamaha founded the Yamaha Organ Manufacturing Company. Such was the success of the company, that in 1897 it became Nippon Gakki Limited and manufactured a wide range of reed organs and pianos.

During World War II, Nippon Gakki's manufacturing base was utilised by the Japanese authorities to produce propellers and fuel tanks for their aviation industry. The end of the war brought about a huge public demand for low cost transport and many firms decided to utilise their obsolete aircraft tooling for the production of motorcycles. Nippon Gakki's first motorcycle went on sale in February 1955 and was named the 125 YA-1 Red Dragonfly. This machine was a copy of the German DKW RT125 motorcycle, featuring a single cylinder two-stroke engine with a four-speed gearbox. Due to the outstanding success of this model the motorcycle operation was separated from Nippon Gakki in July 1955 and the Yamaha Motor Company was formed.

The YA-1 also received acclaim by winning two of Japan's biggest road races, the Mount Fuji Climbing race and the Asama Volcano race. The high level of public demand for the YA-1 led to the development of a whole series of two-stroke singles and twins.

Having made a large impact on their home market, Yamahas were exported to the USA in 1958 and to the UK in 1962. In the UK the signing of an Anglo-Japanese trade agreement during 1962 enabled the sale of Japanese lightweight motorcycles and scooters in Britain. At that time, competition between the many motorcycle producers in Japan had reduced numbers significantly and by the end of the sixties, only the big-four which are familiar with today remained.

Yamaha Europe was founded in 1968 and based in Holland. Although originally set up to market marine products, the Dutch base is now the official European Headquarters and distribution centre. Yamaha motorcycles are built at factories in Holland, Denmark, Norway, Italy, France, Spain and Portugal. Yamahas are imported into the UK by Yamaha Motor UK Ltd, formerly Mitsui Machinery Sales (UK) Ltd. Mitsui and Co. were originally a trading house, handling the shipping, distribution and marketing of Japanese products into western countries. Ultimately Mitsui Machinery Sales was formed to handle Yamaha motorcycles and outboard motors.

Based on the technology derived from its motorcycle operation, Yamaha have produced many other products, such as automobile and lightweight aircraft engines, marine engines and boats, generators, pumps, ATVs, snowmobiles, golf cars, industrial robots, lawnmowers, swimming pools and archery equipment.

Two-strokes first

Part of Yamaha's success was a whole string of innovations in the two-stroke world. Autolube engine lubrication, torque induction, multi-ported engines, reed valves and power valves kept their two-strokes at the forefront of technology. Many advances were achieved with the use of racing as a development laboratory. They went to the USA in the late 1950s with an air-cooled 250cc twin but didn't hit the GPs until the early 1960s when Fumio Ito scored a hat-trick of sixth places in the Isle of Man TT, the Dutch TT and the Belgian GP. This experiment gave rise to the idea of the over-the-counter racer, an idea that became reality in the TD1, the first in an unmatched series of two-stroke racers that were the standard issue for privateers at national and international level for years and helped Yamaha develop their road engines. While privateers raced the twins, Yamaha built the outrageously complicated vee-four 250 for Phil Read and followed it with a vee-four 125 that Bill Ivy lapped the Isle of Man on at over 100mph! When the FIM regulations were changed to limit the smaller GP classes to two cylinders, these exotic bikes died but set the scene for an unparalleled dynasty of mass-produced racers based on the same technology as the road bikes.

In the 1960s and 70s the two-stroke engined YAS3 125, YDS1 to YDS7 250 and YR5 350 formed the core of Yamaha's range. By the mid-70s they had been superseded by the RD (Race-Developed) 125, 250, and 350 range of two-stroke twins, featuring improved 7-port engines with reed valve induction. Braking was improved by the use of an hydraulic brake on the front wheel of DX models, instead of the drum arrangement used previously, and cast alloy wheels were available as an option on later RD models. The RD350 was replaced by the RD400 in 1976.

Running parallel with the RD twins was a range of single-cylinder two-strokes. Used in a variety of chassis types, the engine was used in the popular 50 cc FS1-E moped, the V50 to 90 step-thrus, RS100 and 125, YB100 and the DT trail range.

The TD racers got water-cooling in 1973 to become the TZs, the most successful and numerous over-the-counter racers ever built. That same year, Jarno Saarinen became the first rider to win a 500cc GP on a four-cylinder two-stroke on the new in-line four which was effectively a pair of TZs side-by-side. TZs won everywhere – including the Daytona 200 and 500 races when overbored to 351cc. A 700cc TZ also appeared, one year later taken out to 750cc. Steve Baker won the first Formula 750 world title – one of the precursors of Superbike – on one in 1977. The following year Kenny Roberts won Yamaha's first world 500 title and would be succeeded by Wayne Rainey and Eddie Lawson before Mick Doohan and the NSR500 took over.

The air-cooled single and twin cylinder RD road bikes were eventually replaced by the LC series in 1980, featuring liquid-cooled engines, radical new styling, spiral pattern cast wheels and cantilever rear suspension (Yamaha's Monoshock). Of all the LC models, the RD350LC, or RD350R as it was later known, has made the most impact in the market. Later models had YPVS (Yamaha Power Valve System) engines, another first for Yamaha – this was essentially a valve located in the exhaust ports which was electronically operated to alter port timing to achieve maximum power output. The RD500LC was the largest two-stroke made by Yamaha and differed from the other LCs by the use of its vee-four cylinder engine.

With the exception of the RD350R, now manufactured in Brazil, the LC range has been discontinued. Two-stroke engined models have given way to environmental pressure, and thus with a few exceptions, such as the TZR125 and TZR250, are used only in scooters and small capacity bikes.

The distinctive paintwork and trim of the RD models

Introduction

The Four-strokes

Yamaha concentrated solely on two-stroke models until 1970 when the XS1 was produced, their first four-stroke motorcycle. It was perhaps Yamaha's success with two-strokes that postponed an earlier move into the four-stroke motorcycle market, although their work with Toyota during the 1960s had given them a sound base in four-stroke technology.

The XS1 had a 650 cc twin-cylinder SOHC engine and was later to become known as the XS650, appearing also in the popular SE custom form. Yamaha introduced a three cylinder 750 cc engine in 1976, fitted in a sport-tourer frame and called the XS750, TX750 in the USA. The XS750 established itself well in the sport tourer class and remained in production with very few changes until uprated to 850 cc in 1980.

Other four-strokes followed in 1976, with the introduction of the XS250/360/400 series twins. The XS range was strengthened in 1978 by the four-cylinder XS1100.

The 1980s saw a new family of four-strokes, the XJ550, 650, 750 and 900 Fours. Improvements over the XS range amounted to a slimmer DOHC engine unit due to the relocation of the alternator behind the cylinders, electronic ignition and uprated braking and suspension systems. Models were available mainly in standard trim, although custom-styled Maxims were produced especially for the US market. The XJ650T was the first model from Yamaha to have a turbo-charged engine. Although these early XJ models have now been discontinued, their roots live on in the XJ600S and XJ900S Diversion (Seca II) models.

The XS650 led the way for Yamaha's four-stroke range

The FZR prefix encompasses the pure sports Yamaha models. With the exception of the 16-valve FZR400 and FZR600 models, the FZ/FZR750 and FZR1000 used 20-valve engines, two exhaust valves and three inlet

Yamaha's XS750 was produced from 1976 to 1982 and then uprated to 850 cc

valves per cylinder. This concept was called Genesis and gave improved gas flow to the combustion chambers. Other features of the new engine were the use of down-draught carburetors and the engine's inclined angle in the frame, plus the change to liquid-cooling. Lightweight Deltabox design aluminium frames and uprated suspension improved the bikes's handling. The Genesis engine lives on in the YZF750 and 1000 models.

The Genesis concept was the basis of Yamaha's foray into four-stroke racing, first with a bike known simply as 'The Genesis', an FZ750 motor in a TT Formula 1 bike with which the factory attempted to steal the Honda RVF750's thunder at important events like the Suzuka 8 Hours and the Bol d'Or although they never fielded it for a whole World Championship season. That had to wait for the advent of the World Superbike Championship, although there was no full works team until 1995, instead it was left to individual importers to support teams. It was the Australian Dealer Team Yamaha which scored the factory's first World Superbike win in the series debut year of 1988. The rider? Mick Doohan. Slightly, embarrassingly, it was the steel framed FZ750 rather than the FZR homologation special that won races. The OW01 was a race winner, mainly in the hands of Fabrizio Pirovano, the factory's most successful Superbike racer with ten victories, but national success in the UK, Japan, and in the Daytona 200 has not been translated into World Championships for any of Yamaha's 750s.

The vee-twin engine has been the mainstay of the XV Virago range. Since 1981 XVs have been produced in 535, 700, 750, 920, 1000 and 1100 engine sizes, all using the same basic air-cooled sohc vee-twin engine. Other uses of vee engines have been in the XZ550 of the early 1980s, the XVZ12 Venture and the mighty VMX-12 V-Max.

Yamaha has always been a sporting-orientated company whose motto could be 'Racing Improves the Breed', so it's no surprise that the latest generation of lightweight sportsters are at the cutting edge of performance on and off the track. The R6 won more races than any other machine in the

A new family of four-strokes was released in 1980 with the introduction of the XJ range

inaugural year of the World Supersports Championship, the R7 won a race in its debut year in World Superbike in the hands of the mercurial Noriyuki Haga, and the mighty 1000cc R1 ended Honda's domination of the Isle of Man F1 TT when David Jefferies won three races in a week in 1999.

In Grand Prix racing, the factory took several years to get over the shock of Wayne Rainey's crippling accident. and first 500cc win since the American's enforced retirement didn't come

The XV535 Virago vee-twin

Introduction

until 1998 when Simon Crafar won at Donington Park. For 1999, Yamaha refocussed their ambitions and signed Italian superstar Max Biaggi plus Spanish trier Carlos Checa for the works team, while dashing young Frenchman Regis Laconi and tough little Aussie Gary McCoy rode for the WCM satellite team. Both teams got a win in the '99 season and with a new TZ250 being developed for 2000 it looks as if Yamaha's spirit of competition will go on unabated into the new Millenium.

Yamaha YZF-R1 – The Blade Blunter

Very few motorcycles can be said to have rewritten the rule book but that's what the R1 did on its appearance in 1998. For six years the undisputed king of the big-bore sportsbikes had been the Honda FireBlade but as soon as the bike press got its hands on the first R1 it was obvious there was a new king. The first comparison tests confirmed the verdict: the R1 was smaller, faster, more aggressive, handled better and was easier to ride than the rest. Yamaha had beaten the 'Blade at its own game by packing more power into a smaller package than had ever been managed before. How?

The motor was an all-new unit with the crankcase and cylinder block cast in one piece and the gearbox layshaft positioned directly above the mainshaft rather than behind it. The first feature allowed the engine to be used as a heavily stressed member of the chassis and the gearbox layout made the new unit 81 mm shorter than the old YZF1000R Thunderace motor. This enabled the designers to achieve a wheel-base of 1395 mm, ten millimetres shorter than a FireBlade, without restorting to an over-steep head angle. At 24 degrees, the R1's steering head angle is well within normal bounds for a sportsbike. The most noticeable feature of the chassis is the swinging arm, or rather its length. It's long, well braced, and keeps weight over the front wheel for well-nigh perfect 50-50 weight distribution front and rear.

The first update happened for the 2000 model year with over 250 minor modifications, most of which you'd be hard pushed to spot. However, together they made the new R1 even easier to ride thanks mainly to the slightly revised riding position. There really wasn't much need to do more than fine-tune the R1, its crown as king of the supersports bikes was still firmly in place and, just as significantly, it was making a strong case on the track as well. Yamaha's claim that the chassis was developed with the help of their Grand Prix experience was obviously no idle boast.

At the 1999 Isle of Man TT David Jefferies won the F1 race, his first Island victory but the significance of the win was more than just personal. He was riding a Pirelli-slick shod R1 prepared by tuning house V&M, essentially a privateer on a modified production bike. It was the first time a Honda had failed to win the F1 TT since 1981. The end of the 750 cc limit helped of course, but Jerreries and the R1 had to beat full factory Superbikes and 500 V-twin Grand Prix two-strokes. Just to show it was no fluke, Jefferies then won both the Production and Senior TTs. It was the first time a mass-production bike had won three TTs in the same year. Jefferies also won the Ulster GP and the North West 200 to underline both his and the R1's domination of the British road-racing scene. In 2000, Jefferies and the R1 took the absolute lap-record for the Isle of Man Mountain Circuit over 125 mph, beating the mark set by Jim Moodie on a full factory Honda superbike.

All this on what was basically a tuned streetbike which features neither fuel-injection nor ram-air induction. One of the facts about the R1 that can pass you by is that the designers are quite happy with the 150 bhp of the original bike. Weight saving, aerodynamic efficiency and sheer useability are given far higher priorities. Oh, and it looks as sharp as it goes, which helps too.

Acknowledgements

Our thanks are due to Bransons Motorcycles of Yeovil who supplied the machine featured in the illustrations throughout this manual. We would also like to thank Mitsui Machinery Sales (UK) Ltd for permission to reproduce certain illustrations used in this manual and for supplying some of the cover photographs, also NGK Spark Plugs (UK) Ltd for supplying the colour spark plug condition photographs and the Avon Rubber Company for supplying information on tyre fitting.

Thanks are also due to Kel Edge who supplied the colour transparency of the racing R1 on the rear cover.

About this manual

The aim of this manual is to help you get the best value from your motorcycle. It can do so in several ways. It can help you decide what work must be done, even if you choose to have it done by a dealer; it provides information and procedures for routine maintenance and servicing; and it offers diagnostic and repair procedures to follow when trouble occurs.

We hope you use the manual to tackle the work yourself. For many simpler jobs, doing it yourself may be quicker than arranging an appointment to get the motorcycle into a dealer and making the trips to leave it and pick it up. More importantly, a lot of money can be saved by avoiding the expense the shop must pass on to you to cover its labour and overhead costs. An added benefit is the sense of satisfaction and accomplishment that you feel after doing the job yourself.

References to the left or right side of the motorcycle assume you are sitting on the seat, facing forward.

We take great pride in the accuracy of information given in this manual, but motorcycle manufacturers make alterations and design changes during the production run of a particular motorcycle of which they do not inform us. No liability can be accepted by the authors or publishers for loss, damage or injury caused by any errors in, or omissions from, the information given.

The Yamaha YZF-R1

Identification numbers 0•9

Frame and engine numbers

The frame serial number is stamped into the right-hand side of the steering head. The engine number is stamped into the top of the crankcase on the right-hand side of the engine. The model code label is on the top of the frame cross-piece under the passenger seat. These numbers should be recorded and kept in a safe place so they can be furnished to law enforcement officials in the event of a theft. There is also a carburettor identification number on the intake side of each carburettor body.

The frame serial number, engine serial number, carburettor identification number and model code should be recorded and kept in a handy place (such as with your driver's licence) so that they are always available when purchasing or ordering parts for your machine.

UK/Europe models	Year	Code
YZF-R1	1998	4XV1 (4XV2 – Finland/Sweden)
YZF-R1	1999	4XV7 (4XV8 – Finland/Sweden)
YZF-R1	2000	5JJ1
YZF-R1	2001	5JJ8
US models (US)	Year	Code
YZF-R1 K – 49-state	1998	4XV4
YZF-R1 KC – California	1998	4XV5
YZF-R1 L – 49-state	1999	4XVA
YZF-R1 LC – California	1999	4XVB
YZF-R1 M – 49-state	2000	5JJ4
YZF-R1 MC – California	2000	5JJ5

The procedures in this manual identify the bikes by year and model (eg 1998 YZF1000R). The model codes for all years and models covered are tabled above. Where available, the initial frame and engine numbers are also given.

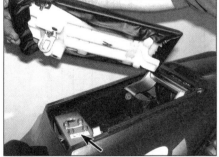

The model code label is stuck to the frame (arrowed)

The engine number is stamped into the crankcase on the right-hand side of the engine

The frame number is stamped into the right-hand side of the steering head

Buying spare parts

Once you have found all the identification numbers, record them for reference when buying parts. Since the manufacturers change specifications, parts and vendors (companies that manufacture various components on the machine), providing the ID numbers is the only way to be reasonably sure that you are buying the correct parts.

Whenever possible, take the worn part to the dealer so direct comparison with the new component can be made. Along the trail from the manufacturer to the parts shelf, there are numerous places that the part can end up with the wrong number or be listed incorrectly.

The two places to purchase new parts for your motorcycle – the accessory store and the franchised dealer – differ in the type of parts they carry. While dealers can obtain virtually every part for your motorcycle, the accessory dealer is usually limited to normal high wear items such as shock absorbers, tune-up parts, various engine gaskets, cables, chains, brake parts, etc. Rarely will an accessory outlet have major suspension components, cylinders, transmission gears, or cases.

Used parts can be obtained for roughly half the price of new ones, but you can't always be sure of what you're getting. Once again, take your worn part to the breaker's yard for direct comparison.

Whether buying new, used or rebuilt parts, the best course is to deal directly with someone who specialises in parts for your particular make.

Performance data and Bike spec

Model development

When Yamaha introduced the R1 in 1998, it instantly become the must-have big-bore sportsbike. The opinions of all the motorcycle press road testers were unanimous, the R1 was king of the sportsbikes, making the Honda Fireblade seem sedate by comparison.

The R1 continued unchanged for the 1999 model year before undergoing a host of revisions for 2000.

For the 2000 model year, Yamaha made over 250 modifications to improve the overall package. Although most of these changes were subtle ones, almost unnoticeable to the untrained eye, the list was extensive and major changes included those shown in the table below.

All the modifications resulted in a total weight-saving of approximately 2 kg, but also changed the weight distribution of the motorcycle slightly; the new model had a weight bias of 51% towards the front compared to 50.5% on the earlier model. These figures may sound trivial, but the improvements they made were dramatic. The new power delivery of the engine complemented the revised chassis perfectly, resulting in a motorcycle which was far more refined and easier to ride hard. It was official, the best had just got better!

Summary of changes on the 2000 R1

- *The bodywork and fuel tank were reshaped to improve rider comfort and aerodynamics.*
- *The frame was strengthened to improve the handling characteristics of the bike.*
- *The ignition timing and EXUP valve were remapped and the carburettors rejetted to smooth the power delivery of the engine. The exhaust system was also modified to complement the changes.*
- *The gearchange linkage was modified to improve the quality of the gearchange action. The transmission internals were also modified to suit.*
- *The front fork internals and rear shock absorber spring were modified to improve handling.*
- *Front and rear brakes were modified to reduce unsprung weight.*

Performance data

Maximum power
- 1998/9 models ....140 bhp (104 kW) @ 9900 rpm
- 2000 model ....147.9 bhp (110 kW) @ 10,000 rpm

Maximum torque
- 1998/9 models ....79 lbf ft (107 Nm) @ 7300 rpm
- 2000 model ....79.5 lbf ft (108 Nm) @ 8500 rpm

Power-to-weight ratio (approximate)
- 1998/9 models ....790 bhp per tonne (0.59 kW per kg)
- 2000 model ....850 bhp per tonne (0.63 kW per kg)

Top speed
- 1998/9 models ....174 mph (280 km/h)
- 2000 model ....176 mph (283 km/h)

Acceleration
- Time taken to cover a ¼ mile from a standing start ...10.3 seconds
- Terminal speed after ¼ mile ....144 mph (232 km/h)

Average fuel consumption *
- * (miles per Imp gal, miles per litre, litres per 100 km)
- 1998/9 models ....32 mpg, 7.0 mpl, 8.8 l/100 km
- 2000 model ....34 mpg, 7.5 mpl, 8.3 l/100 km

Fuel tank capacity ....18 litres (4 Imp gal, 4.75 US gal)

Fuel tank range ....130 miles (209 km)

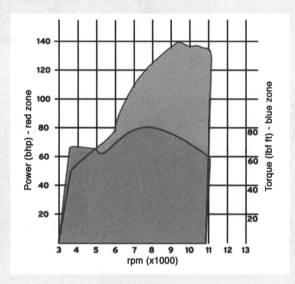

Performance data sourced from Motor Cycle News road test features. See the MCN website for up-to-date biking news.

MCN www.motorcyclenews.com

Performance data and Bike spec

Bike spec

Weights and dimensions

Wheelbase	1395 mm
Overall length	2035 mm
Overall height	
1998 and 1999 models	1095 mm
2000-on models	1105 mm
Overall width	695 mm
Seat height	815 mm
Ground clearance	140 mm
Dry weight*	
1998 and 1999 models	177 kg
2000-on models	175 kg
Wet weight (with fuel and oil)*	
1998 and 1999 models	198 kg
2000 models	194 kg

Add 1 kg for California models

Engine

Type	Liquid-cooled, 20V in-line four cylinder with EXUP (EXhaust Ultimate Powervalve) system
Capacity	998 cc
Bore	74 mm
Stroke	58 mm
Compression ratio	11.8:1
Camshafts	DOHC, chain-driven from the right-hand end of the crankshaft
Valves	5 per cylinder (3 intake and 2 exhaust)
Carburettors	4 x 40 mm Mikuni BDSR40
Clutch	Wet multi-plate, cable-operated
Transmission	Six-speed constant mesh
Final drive	
Chain	DAIDO 50ZVM (114 links)
Sprockets	16 tooth front/43 tooth rear

Chassis

Type	Aluminium alloy, twin spar Deltabox II
Rake	24°
Trail	92 mm
Front suspension	
Type	41 mm diameter upside-down hydraulic forks
Travel	135 mm
Adjustments	Spring preload, compression damping and rebound damping adjustment
Rear suspension	
Type	Monoshock with rising-rate linkage
Travel	130 mm
Adjustments	Spring preload, compression damping and rebound damping adjustment
Tyre sizes *	
Front	120/70 ZR 17 (58W)
Rear	190/50 ZR 17 (73W)

* *Refer to page 0•16 for tyre pressures*

Brakes	
Front	2 x 298 mm discs with Sumitumo four-piston calipers
Rear	245 mm disc with Sumitumo two-piston caliper

Safety first!

Professional mechanics are trained in safe working procedures. However enthusiastic you may be about getting on with the job at hand, take the time to ensure that your safety is not put at risk. A moment's lack of attention can result in an accident, as can failure to observe simple precautions.

There will always be new ways of having accidents, and the following is not a comprehensive list of all dangers; it is intended rather to make you aware of the risks and to encourage a safe approach to all work you carry out on your bike.

Asbestos

● Certain friction, insulating, sealing and other products - such as brake pads, clutch linings, gaskets, etc. - contain asbestos. Extreme care must be taken to avoid inhalation of dust from such products since it is hazardous to health. If in doubt, assume that they do contain asbestos.

Fire

● Remember at all times that petrol is highly flammable. Never smoke or have any kind of naked flame around, when working on the vehicle. But the risk does not end there - a spark caused by an electrical short-circuit, by two metal surfaces contacting each other, by careless use of tools, or even by static electricity built up in your body under certain conditions, can ignite petrol vapour, which in a confined space is highly explosive. Never use petrol as a cleaning solvent. Use an approved safety solvent.

● Always disconnect the battery earth terminal before working on any part of the fuel or electrical system, and never risk spilling fuel on to a hot engine or exhaust.
● It is recommended that a fire extinguisher of a type suitable for fuel and electrical fires is kept handy in the garage or workplace at all times. Never try to extinguish a fuel or electrical fire with water.

Fumes

● Certain fumes are highly toxic and can quickly cause unconsciousness and even death if inhaled to any extent. Petrol vapour comes into this category, as do the vapours from certain solvents such as trichloro-ethylene. Any draining or pouring of such volatile fluids should be done in a well ventilated area.
● When using cleaning fluids and solvents, read the instructions carefully. Never use materials from unmarked containers - they may give off poisonous vapours.
● Never run the engine of a motor vehicle in an enclosed space such as a garage. Exhaust fumes contain carbon monoxide which is extremely poisonous; if you need to run the engine, always do so in the open air or at least have the rear of the vehicle outside the workplace.

The battery

● Never cause a spark, or allow a naked light near the vehicle's battery. It will normally be giving off a certain amount of hydrogen gas, which is highly explosive.

● Always disconnect the battery ground (earth) terminal before working on the fuel or electrical systems (except where noted).
● If possible, loosen the filler plugs or cover when charging the battery from an external source. Do not charge at an excessive rate or the battery may burst.
● Take care when topping up, cleaning or carrying the battery. The acid electrolyte, evenwhen diluted, is very corrosive and should not be allowed to contact the eyes or skin. Always wear rubber gloves and goggles or a face shield. If you ever need to prepare electrolyte yourself, always add the acid slowly to the water; never add the water to the acid.

Electricity

● When using an electric power tool, inspection light etc., always ensure that the appliance is correctly connected to its plug and that, where necessary, it is properly grounded (earthed). Do not use such appliances in damp conditions and, again, beware of creating a spark or applying excessive heat in the vicinity of fuel or fuel vapour. Also ensure that the appliances meet national safety standards.
● A severe electric shock can result from touching certain parts of the electrical system, such as the spark plug wires (HT leads), when the engine is running or being cranked, particularly if components are damp or the insulation is defective. Where an electronic ignition system is used, the secondary (HT) voltage is much higher and could prove fatal.

Remember...

✗ **Don't** start the engine without first ascertaining that the transmission is in neutral.
✗ **Don't** suddenly remove the pressure cap from a hot cooling system - cover it with a cloth and release the pressure gradually first, or you may get scalded by escaping coolant.
✗ **Don't** attempt to drain oil until you are sure it has cooled sufficiently to avoid scalding you.
✗ **Don't** grasp any part of the engine or exhaust system without first ascertaining that it is cool enough not to burn you.
✗ **Don't** allow brake fluid or antifreeze to contact the machine's paintwork or plastic components.
✗ **Don't** siphon toxic liquids such as fuel, hydraulic fluid or antifreeze by mouth, or allow them to remain on your skin.
✗ **Don't** inhale dust - it may be injurious to health (see Asbestos heading).
✗ **Don't** allow any spilled oil or grease to remain on the floor - wipe it up right away, before someone slips on it.
✗ **Don't** use ill-fitting spanners or other tools which may slip and cause injury.
✗ **Don't** lift a heavy component which may be beyond your capability - get assistance.
✗ **Don't** rush to finish a job or take unverified short cuts.
✗ **Don't** allow children or animals in or around an unattended vehicle.
✗ **Don't** inflate a tyre above the recommended pressure. Apart from overstressing the carcass, in extreme cases the tyre may blow off forcibly.
✔ **Do** ensure that the machine is supported securely at all times. This is especially important when the machine is blocked up to aid wheel or fork removal.
✔ **Do** take care when attempting to loosen a stubborn nut or bolt. It is generally better to pull on a spanner, rather than push, so that if you slip, you fall away from the machine rather than onto it.
✔ **Do** wear eye protection when using power tools such as drill, sander, bench grinder etc.
✔ **Do** use a barrier cream on your hands prior to undertaking dirty jobs - it will protect your skin from infection as well as making the dirt easier to remove afterwards; but make sure your hands aren't left slippery. Note that long-term contact with used engine oil can be a health hazard.
✔ **Do** keep loose clothing (cuffs, ties etc. and long hair) well out of the way of moving mechanical parts.
✔ **Do** remove rings, wristwatch etc., before working on the vehicle - especially the electrical system.
✔ **Do** keep your work area tidy - it is only too easy to fall over articles left lying around.
✔ **Do** exercise caution when compressing springs for removal or installation. Ensure that the tension is applied and released in a controlled manner, using suitable tools which preclude the possibility of the spring escaping violently.
✔ **Do** ensure that any lifting tackle used has a safe working load rating adequate for the job.
✔ **Do** get someone to check periodically that all is well, when working alone on the vehicle.
✔ **Do** carry out work in a logical sequence and check that everything is correctly assembled and tightened afterwards.
✔ **Do** remember that your vehicle's safety affects that of yourself and others. If in doubt on any point, get professional advice.
● If in spite of following these precautions, you are unfortunate enough to injure yourself, seek medical attention as soon as possible.

Daily or (pre-ride) checks

Note: *The daily (pre-ride) checks outlined in the owner's manual covers those items which should be inspected on a daily basis.*

Engine/transmission oil level check

Before you start:
✔ Support the motorcycle in an upright position, using an auxiliary stand if required. Make sure it is on level ground.

✔ Start the engine and let it idle for several minutes to allow it to reach normal operating temperature.

Caution: Do not run the engine in an enclosed space such as a garage or workshop.

✔ Leave the motorcycle undisturbed for a few minutes to allow the oil level to stabilise.

The correct oil
● Modern, high-revving engines place great demands on their oil. It is very important that the correct oil for your bike is used.
● Always top up with a good quality oil of the specified type and viscosity and do not overfill the engine.
Caution: Do not use chemical additives or oils with a grade of CD or higher, or use oils labelled "ENERGY CONSERVING II". Such additives or oils could cause clutch slip.

Oil type	API grade SE, SF or SG
Oil viscosity*	
European models	SAE 10W30 or 10W40
US models	SAE 10W30 or 20W40

Refer to the viscosity table to select the oil best suited to your conditions.

Bike care:
● If you have to add oil frequently, you should check whether you have any oil leaks. If there is no sign of oil leakage from the joints and gaskets the engine could be burning oil (see *Fault Finding*).

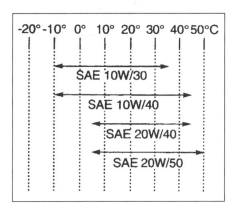

Oil viscosity table; select the oil best suited to the conditions

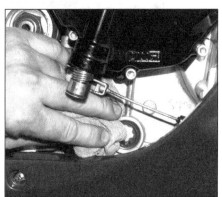

1 Wipe the oil level inspection window, located on the right-hand side of the engine, so that it is clean.

2 With the motorcycle vertical, the oil level should lie between the maximum and minimum levels on the window.

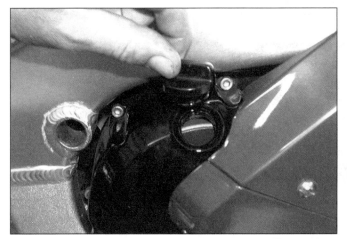

3 If the level is below the minimum line, remove the filler cap from the top of the clutch cover.

4 Top the engine up with the recommended grade and type of oil, to bring the level up to the maximum level on the window. Install the filler cap.

0•14 Daily or (pre-ride) checks

Brake fluid level checks

> **Warning:** Brake hydraulic fluid can harm your eyes and damage painted surfaces, so use extreme caution when handling and pouring it and cover surrounding surfaces with rag. Do not use fluid that has been standing open for some time, as it absorbs moisture from the air which can cause a dangerous loss of braking effectiveness.

Before you start:

✔ Support the motorcycle in an upright position, using an auxiliary stand if required. Turn the handlebars until the top of the front master cylinder is as level as possible. The rear master cylinder reservoir is located behind the right-hand side cover.

✔ Make sure you have the correct hydraulic fluid. DOT 4 is recommended.

✔ Wrap a rag around the reservoir being worked on to ensure that any spillage does not come into contact with painted surfaces.

Bike care:

● The fluid in the front and rear brake master cylinder reservoirs will drop slightly as the brake pads wear down.
● If any fluid reservoir requires repeated topping-up this is an indication of an hydraulic leak somewhere in the system, which should be investigated immediately.
● Check for signs of fluid leakage from the hydraulic hoses and brake system components – if found, rectify immediately (see Chapter 7).
● Check the operation of both brakes before taking the machine on the road; if there is evidence of air in the system (spongy feel to lever or pedal), it must be bled as described in Chapter 7.

FRONT BRAKE FLUID LEVEL

1 The front brake fluid level is visible through the reservoir body – it must be between the UPPER and LOWER level lines.

2 If the level is below the LOWER level line, undo the reservoir cap clamp screw (A) and remove the clamp, then unscrew the cap (B) and remove the diaphragm plate and the diaphragm.

3 Top up with new clean DOT 4 hydraulic fluid, until the level is above the LOWER level line. Take care to avoid spills (see **Warning** above).

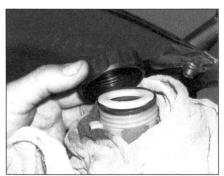

4 Ensure that the diaphragm is correctly seated before installing the plate and cap. Secure the cap with its clamp.

REAR BRAKE FLUID LEVEL

1 The rear brake fluid level is visible through the reservoir body – it must be above LOWER level line.

2 If the level is below the LOWER level line, remove the reservoir mounting screw (arrowed) and displace the reservoir to provide clearance. On 1998 and 1999 models, remove the cap clamp.

3 Support the reservoir upright, then unscrew the cap and remove the diaphragm plate and the diaphragm.

Daily or (pre-ride) checks

REAR BRAKE FLUID LEVEL (CONTINUED)

4 Top up with new clean DOT 4 hydraulic fluid, until the level is above the lower mark. Take care to avoid spills (see **Warning** on opposite page).

5 Ensure that the diaphragm is correctly seated before installing the plate and cap. Tighten the cap securely. On 1998 and 1999 models, fit the cap clamp. Install the reservoir and tighten the mounting screw.

Coolant level check

 Warning: DO NOT remove the cooling system pressure cap to add coolant. Topping up is done via the coolant reservoir tank filler. DO NOT leave open containers of coolant about, as it is poisonous.

Before you start:
✔ Make sure you have a supply of coolant available – a mixture of 50% distilled water and 50% corrosion inhibited ethylene glycol anti-freeze is needed. **Note:** *Yamaha specify that soft tap water can be used, but NOT hard water. If in doubt, boil the water first or use only distilled water.*

✔ Always check the coolant level when the engine is cold.
✔ Support the motorcycle in an upright position, using an auxiliary stand if required. Make sure it is on level ground.

Bike care:
● Use only the specified coolant mixture. It is important that anti-freeze is used in the system all year round, and not just in the winter. Do not top the system up using only water, as the system will become too diluted.
● Do not overfill the reservoir. If the coolant is significantly above the FULL level line at any time, the surplus should be siphoned or drained off to prevent the possibility of it being expelled out of the overflow hose.
● If the coolant level falls steadily, check the system for leaks (see Chapter 1). If no leaks are found and the level continues to fall, it is recommended that the machine be taken to a Yamaha dealer for a pressure test.

1 The reservoir is mounted on the right-hand end of the radiator, and is visible via the aperture in the right-hand fairing side panel. The coolant FULL and LOW level lines are marked on the reservoir.

2 If the coolant level does not lie between the FULL and LOW level lines, remove the right-hand cockpit trim panel (see Chapter 8). Remove the reservoir filler cap.

3 Top the coolant level up with the recommended coolant mixture. Fit the cap securely, then install the trim panel (see Chapter 8).

0•16 Daily or (pre-ride) checks

Tyre checks

The correct pressures:
● The tyres must be checked when **cold**, not immediately after riding. Note that low tyre pressures may cause the tyre to slip on the rim or come off. High tyre pressures will cause abnormal tread wear and unsafe handling.

● Use an accurate pressure gauge. Many garage forecourt gauges are wildly inaccurate. If you buy your own, spend as much as you can justify on a quality gauge.

● Correct air pressure will increase tyre life and provide maximum stability, handling capability and ride comfort.

Tyre care:
● Check the tyres carefully for cuts, tears, embedded nails or other sharp objects and excessive wear. Operation of the motorcycle with excessively worn tyres is extremely hazardous, as traction and handling are directly affected.
● Check the condition of the tyre valve and ensure the dust cap is in place.
● Pick out any stones or nails which may have become embedded in the tyre tread. If left, they will eventually penetrate through the casing and cause a puncture.
● If tyre damage is apparent, or unexplained loss of pressure is experienced, seek the advice of a tyre fitting specialist without delay.

Tyre tread depth:
● At the time of writing UK law requires that tread depth must be at least 1 mm over the entire tread breadth all the way around the tyre, with no bald patches. Many riders, however, consider 2 mm tread depth minimum to be a safer limit. Yamaha recommend a minimum of 1.6 mm.

● Many tyres now incorporate wear indicators in the tread. Identify the triangular pointer or TWI mark on the tyre sidewall to locate the indicator bar and renew the tyre if the tread has worn down to the bar.

Loading*/speed	Front	Rear
Up to 90 kg (198 lb) load	36 psi (2.50 bar)	36 psi (2.50 bar)
90 kg (198 lb) up to max. load of 196 kg (432 lb)	36 psi (2.50 bar)	42 psi (2.90 bar)
High speed riding	36 psi (2.50 bar)	36 psi (2.50 bar)

*Load is the total weight of the rider, passenger, luggage and any accessories

1 Check the tyre pressures when the tyres are **cold** and keep them properly inflated.

2 Measure tread depth at the centre of the tyre using a tread depth gauge.

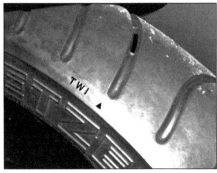

3 Tyre tread wear indicator bar and its location marking (usually either an arrow, a triangle or the letters TWI) on the sidewall.

Suspension, steering and final drive checks

Suspension and Steering:
● Check that the front and rear suspension operates smoothly without binding.

● Check that the suspension is adjusted as required.

● Check that the steering moves smoothly from lock-to-lock.

Final drive:
● Check that the drive chain slack isn't excessive, and adjust it if necessary (see Chapter 1).
● If the chain looks dry, lubricate it (see Chapter 1).

Legal and safety checks

Lighting and signalling:
● Take a minute to check that the headlight, tail light, brake light, instrument lights and turn signals all work correctly.
● Check that the horn sounds when the switch is operated.
● A working speedometer graduated in mph is a statutory requirement in the UK.

Safety:
● Check that the throttle grip rotates smoothly and snaps shut when released, in all steering positions. Also check for the correct amount of freeplay (see Chapter 1).
● Check that the engine shuts off when the kill switch is operated.
● Check that sidestand return spring holds the stand securely up when retracted.

Fuel:
● This may seem obvious, but check that you have enough fuel to complete your journey. If you notice signs of fuel leakage – rectify the cause immediately.
● Ensure you use the correct grade fuel – see Chapter 4 Specifications.

Chapter 1
Routine maintenance and Servicing

Contents

Air filter – cleaning and renewal 5	Engine/transmission – oil change 7
Battery – charging . . . . see Chapter 9	Engine/transmission oil level check . . . . . . .see *Daily (pre-ride) checks*
Battery – check 11	Front forks – oil change 34
Battery – removal, installation, inspection and	Fuel hoses – renewal 33
maintenance . . . . .see Chapter 9	Fuel system, EXUP system and air induction system (AIS) – check . 6
Brake fluid level check . . . . .see *Daily (pre-ride) checks*	Headlight aim – check and adjustment 30
Brake hoses – renewal 29	Idle speed – check and adjustment 3
Brake master cylinder and caliper seals – renewal 25	Nuts and bolts – tightness check 15
Brake pads – wear check 8	Sidestand – check 14
Brake system – check 9	Spark plugs – gap check and adjustment 2
Brakes– fluid change 26	Stand, lever pivots and cables – lubrication 18
Carburettors – synchronisation 4	Steering head bearings – freeplay check and adjustment 20
Clutch – check and adjustment 10	Steering head bearings – re-greasing 24
Coolant level check . . . . .see *Daily (pre-ride) checks*	Suspension – check 19
Cooling system – check 16	Swingarm and suspension linkage bearings – re-greasing 23
Cooling system – draining, flushing and refilling 22	Throttle and choke cables – check 17
Cylinder compression – check 31	Tyre pressure and tread depth . . . . .see *Daily (pre-ride) checks*
Drive chain and sprockets – check, adjustment and lubrication 1	Valve clearances – check and adjustment 28
Engine oil pressure – check 32	Wheel bearings – check 13
Engine/transmission – oil and filter change 21	Wheels and tyres – general check 12

Degrees of difficulty

| **Easy,** suitable for novice with little experience | | **Fairly easy,** suitable for beginner with some experience | | **Fairly difficult,** suitable for competent DIY mechanic | | **Difficult,** suitable for experienced DIY mechanic | | **Very difficult,** suitable for expert DIY or professional | 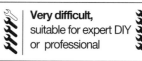 |

1•2 Servicing specifications

Engine
Spark plugs
    Type . . . . NGK CR9E or Nippondenso U27ESR-N
    Electrode gap . . . . 0.7 to 0.8 mm
Engine idle speed
    1998 and 1999 models . . . . 1050 to 1150 rpm
    2000-on models . . . . 1000 to 1100 rpm
Cylinder identification . . . . numbered 1 to 4 from left to right
Carburettor synchronisation – intake vacuum at idle . . . . 220 mmHg
Carburettor synchronisation – max. difference between carburettors . . 10 mmHg
Valve clearances (COLD engine)   **Intake valves**   **Exhaust valves**
    All 1998 and 1999 models, 2000-on US models . . . . 0.11 to 0.20 mm   0.21 to 0.30 mm
    2000-on Europe models . . . . 0.11 to 0.20 mm   0.21 to 0.25 mm
Cylinder compression
    Standard . . . . 206 psi (14.5 bars)
    Maximum . . . . 213 psi (15.0 bars)
    Minimum . . . . 192 psi (13.5 bars)
    Max. difference between cylinders . . . . 14.5 psi (1.0 bar)
Engine oil pressure . . . . 64 psi (4.5 bars) @ 1100 rpm

Cycle parts
Drive chain slack . . . . 40 to 50 mm
Drive chain stretch limit . . . . 150.1 mm
Rear brake pedal height . . . . 35 to 40 mm
Throttle cable freeplay . . . . 3 to 5 mm
EXUP cable freeplay . . . . 1.5 mm
Clutch cable freeplay . . . . 10 to 15 mm
Tyre pressures (cold) . . . . see *Daily (pre-ride) checks*

Recommended lubricants and fluids
Engine/transmission oil type . . . . see *Daily (pre-ride) checks*
Engine/transmission oil capacity
    Oil change . . . . 2.7 litres
    Oil and filter change . . . . 2.9 litres
    Following engine overhaul – dry engine, new filter . . . . 3.6 litres
Coolant type . . . . 50% distilled water, 50% corrosion inhibited ethylene glycol anti-freeze. **Note:** Yamaha specify that soft tap water can be used, but NOT hard water. If in doubt, boil the water first or use only distilled water.
Coolant capacity . . . . 2.55 litres
Brake fluid . . . . DOT 4
Drive chain . . . . Engine oil or chain lubricant suitable for O-ring chains
Steering head bearings . . . . Lithium-based multi-purpose grease
Swingarm pivot and bearings . . . . Lithium-based multi-purpose grease
Suspension linkage bearings . . . . Lithium-based multi-purpose grease
Bearing seal lips . . . . Lithium-based multi-purpose grease
Gearchange lever, clutch lever, front brake lever,
    rear brake pedal, sidestand pivots . . . . Lithium-based multi-purpose grease
Cables . . . . 10W30 motor oil or Yamaha cable lubricant
Throttle grip . . . . Multi-purpose grease or dry film lubricant

Torque wrench settings
Rear axle nut . . . . 150 Nm
Spark plugs . . . . 13 Nm
EXUP pulley cover bolts . . . . 10 Nm
Oil drain plug . . . . 43 Nm
Oil filter . . . . 17 Nm
Steering head bearing adjuster nut
    Initial setting . . . . 28 Nm
    Final setting . . . . 9 Nm
Steering stem nut . . . . 115 Nm
Fork clamp bolts (top yoke) . . . . 23 Nm
Handlebar positioning bolts . . . . 13 Nm
Handlebar clamp bolts . . . . 17 Nm
Cooling system drain plug . . . . 7 Nm
Timing rotor/pick-up coil cover bolts . . . . 12 Nm
Oil gallery bolt . . . . 20 Nm

Maintenance schedule

Note: *The daily (pre-ride) checks outlined in the owner's manual covers those items which should be inspected on a daily basis. Always perform the pre-ride inspection at every maintenance interval (in addition to the procedures listed). The intervals listed below are the intervals recommended by the manufacturer for each particular operation during the model years covered in this manual. Your owner's manual may have different intervals for your model.*

Daily (pre-ride)
- [ ] See *Daily (pre-ride) checks* at the beginning of this manual.

After the initial 600 miles (1000 km)
Note: *This check is usually performed by a Yamaha dealer after the first 600 miles (1000 km) from new. Thereafter, maintenance is carried out according to the following intervals of the schedule.*

Every 600 miles (1000 km)
- [ ] Check, adjust, clean and lubricate the drive chain (Section 1)

Every 4000 miles (6000 km) or 6 months – 1998/9 models
Every 6000 miles (10,000 km) – 2000-on models
- [ ] Check the spark plug gaps (Section 2)
- [ ] Check and adjust the idle speed (Section 3)
- [ ] Check/adjust the carburettor synchronisation (Section 4)
- [ ] Clean and check the air filter element (Section 5)
- [ ] Check the fuel system and EXUP system (Section 6)
- [ ] Check the air induction system (AIS) components – 2000-on models (Section 6)
- [ ] Change the engine/transmission oil (Section 7)
- [ ] Check the brake pads (Section 8)
- [ ] Check the brake system and brake light switch operation (Section 9)
- [ ] Check the clutch (Section 10)
- [ ] Check the battery (Section 11)
- [ ] Check the condition of the wheels and tyres (Section 12)
- [ ] Check the wheel bearings (Section 13)
- [ ] Check the sidestand (Section 14)
- [ ] Check the tightness of all nuts, bolts and fasteners (Section 15)
- [ ] Check the cooling system (Section 16)
- [ ] Check and adjust the throttle and choke cables (Section 17)
- [ ] Lubricate the clutch/gearchange/brake lever/brake pedal/sidestand pivots and the throttle/choke cables (Section 18)
- [ ] Check the suspension (Section 19)
- [ ] Check and adjust the steering head bearings (Section 20)

Every 8000 miles (12,000 km) or 12 months – 1998/9 models
Every 12,000 miles (20,000 km) – 2000-on models
Carry out all the items under the 4000/6000 mile (6000/10,000 km) check, plus the following
- [ ] Change the engine/transmission oil and filter (Section 21)
- [ ] Change the coolant (Section 22)

Every 16,000 miles (24,000 km) or two years
- [ ] Re-grease the swingarm and suspension linkage bearings (Section 23).
- [ ] Re-grease the steering head bearings (Section 24).

Every 20,000 miles (32,000 km)
- [ ] Renew the in-line fuel filter (Section 6)

Every 24,000 miles (40,000 km)
- [ ] Check and adjust the valve clearances (Section 28)

Every two years
- [ ] Renew the brake master cylinder and caliper seals (Section 25)
- [ ] Change the brake fluid (Section 26)

Every four years
- [ ] Renew the brake hoses (Section 29)

Non-scheduled maintenance
- [ ] Check and adjust the headlight aim (Section 30)
- [ ] Check the cylinder compression (Section 31)
- [ ] Check the engine oil pressure (Section 32)
- [ ] Renew the fuel hoses (Section 33)
- [ ] Change the front fork oil (Section 34)

1•4 Maintenance - component location

Component locations on left-hand side

1 Clutch cable upper adjuster
2 Steering head bearing adjuster
3 Air filter
4 Fuel in-line filter
5 Idle speed adjuster
6 Battery
7 Drive chain adjuster
8 EXUP pulley
9 Oil drain bolt
10 Oil filter
11 Coolant drain bolt
12 Fork seal
13 Oil pressure check bolt

Maintenance - component location

Component locations on right-hand side

1 Rear brake fluid reservoir
2 Oil filler cap
3 Front brake fluid reservoir
4 Throttle cable adjuster
5 Fork seal
6 Coolant reservoir tank
7 Clutch cable lower adjuster
8 Oil level inspection window
9 Rear brake light switch
10 Brake pedal height adjuster
11 Drive chain adjuster

Maintenance procedures

Introduction

1 This Chapter is designed to help the home mechanic maintain his/her motorcycle for safety, economy, long life and peak performance.

2 Deciding where to start or plug into the routine maintenance schedule depends on several factors. If the warranty period on your motorcycle has just expired, and if it has been maintained according to the warranty standards, you may want to pick up routine maintenance as it coincides with the next mileage or calendar interval. If you have owned the machine for some time but have never performed any maintenance on it, then you may want to start at the beginning and include all frequent procedures to ensure that nothing important is overlooked. If you have just had a major engine overhaul, then you should start the engine maintenance routines from the beginning. If you have a used machine and have no knowledge of its history or maintenance record, you should combine all the checks into one large initial service and then settle into the maintenance schedule prescribed.

3 Before beginning any maintenance or repair, the machine should be cleaned thoroughly, especially around the oil filter, spark plugs, valve cover, side panels, carburettors, etc. Cleaning will help ensure that dirt does not contaminate the engine and will allow you to detect wear and damage that could otherwise easily go unnoticed.

4 Certain maintenance information is sometimes printed on decals attached to the motorcycle. If any information on the decals differs from that included here, use the information on the decal.

Every 600 miles (1000 km)

1 Drive chain and sprockets – check, adjustment and lubrication

Check

1 As the chain stretches with wear, adjustment will periodically be necessary. A neglected drive chain won't last long and can quickly damage the sprockets. Routine chain adjustment and lubrication isn't difficult and will ensure maximum chain and sprocket life.

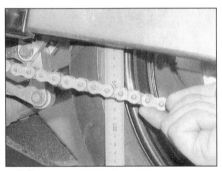

1.3 Push up on the chain and measure the slack

2 To check the chain, support the bike upright, but do not have someone sit on it to do this, and shift the transmission into neutral.

3 Push up on the bottom run of the chain and measure the slack midway between the two sprockets (see illustration), then compare your measurement to that listed in this Chapter's Specifications. Since the chain will rarely wear evenly, resulting in a tight spot, rotate the rear wheel so that another section of chain can be checked; do this several times to check the entire length of chain. Any adjustment should be based upon the measurement taken at the tightest point.

4 If the chain has reached the end of its adjustment, it has probably stretched beyond its service limit. This can be checked by measuring a section of the chain with the chain held taut, and comparing the measurement to the limit in the Specifications (see illustration). Ideally this should be done with the chain removed, although it is possible to take the measurement on the bike if the chainguard is removed and the measurement is taken on the chain's top run, midway between the sprockets. Take three measurements in different places on the chain. If the chain has stretched beyond the limit, replace it with a new one (see Chapter 6).

5 In some cases where lubrication has been neglected, corrosion and galling may cause the links to bind and kink, which effectively shortens the chain's length. Any such links should be thoroughly cleaned and worked free. If the chain is tight between the sprockets, rusty or kinked, it is time to replace it with a new one. If you find a tight area, mark it with felt pen or paint, and repeat the measurement after the bike has been ridden. If the chain is still tight in the same area, it may be damaged or worn. Because a tight or kinked chain can damage the transmission output shaft bearing, it is a good idea to replace it with a new one (see Chapter 6).

6 Check the entire length of the chain for damaged rollers, loose links and pins, and missing O-rings and replace with a new one it if damage is found. **Note:** *Never install a new chain on old sprockets, and never use the old chain if you install new sprockets – renew the chain and sprockets as a set.*

7 Remove the front sprocket cover (see Chapter 6). Check the teeth on the engine sprocket and the rear wheel sprocket for wear (see illustration).

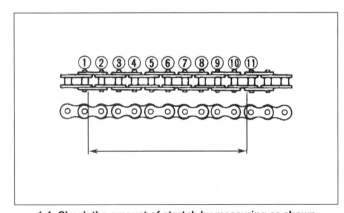

1.4 Check the amount of stretch by measuring as shown

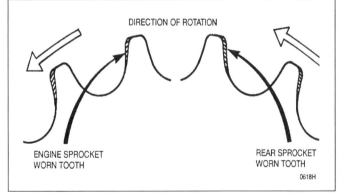

1.7 Check the sprockets in the areas indicated to see if they are worn excessively

Maintenance procedures 1•7

1.10 Slacken the rear axle nut (arrowed)

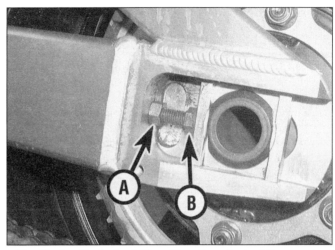

1.11a Slacken the locknut (A) and turn the adjuster (B) as required

8 Inspect the drive chain slider on the swingarm for excessive wear and renew it if worn (see Chapter 6, Section 14).

Adjustment

9 Rotate the rear wheel until the chain is positioned with the tightest point at the centre of its bottom run. Support the bike upright (but not by having someone sit on it).
10 Slacken the axle nut **(see illustration)**.
11 Slacken the adjuster locknut on each side of the swingarm, then turn the adjusters evenly until the amount of freeplay specified at the beginning of the Chapter is obtained at the centre of the bottom run of the chain **(see illustration)**. Following chain adjustment, check that the back edge of each chain adjustment marker is in the same position in relation to the marks on the swingarm **(see illustration)**. It is important that each adjuster aligns with the same mark; if not, the rear wheel will be out of alignment with the front. Also check that there is no clearance between the adjuster and the front of the adjustment marker – push or kick the wheel forwards to eliminate any freeplay.

> **HAYNES HiNT** *Refer to Chapter 7 for information on checking wheel alignment.*

12 If there is a discrepancy in the chain adjuster positions, adjust one of them so that its position is exactly the same as the other. Check the chain freeplay as described above and readjust if necessary.
13 Tighten the axle nut to the torque setting specified at the beginning of the Chapter, then tighten the adjuster locknuts securely **(see illustration)**. Recheck the adjustment.

Lubrication

14 If required, wash the chain in paraffin (kerosene), then wipe it off and allow it to dry, using compressed air if available. If the chain is excessively dirty it should be removed from the machine and allowed to soak in the paraffin (see Chapter 6).

Caution: Don't use petrol (gasoline), solvent or other cleaning fluids which might damage the internal sealing properties of the chain. Don't use high-pressure water. The entire process shouldn't take longer than ten minutes – if it does, the O-rings in the chain rollers could be damaged.

15 For routine lubrication, the best time to lubricate the chain is after the motorcycle has been ridden. When the chain is warm, the lubricant will penetrate the joints between the side plates better than when cold. **Note:** *Yamaha specifies engine oil or chain lube that is specifically for O-ring chains; do not use chain lube that is not specifically for O-ring chains, as it may contain solvents that could damage the O-rings.* Apply the lubricant to the area where the side plates overlap – not the middle of the rollers **(see illustration)**.

> **HAYNES HiNT** *Apply the lubricant to the top of the lower chain run, so centrifugal force will work it into the chain when the bike is moving. After applying the lubricant, let it soak in a few minutes before wiping off any excess.*

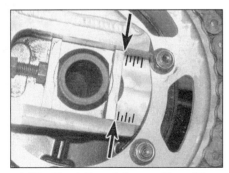

1.11b Check the relative position of the back edge of the adjustment marker (arrowed) with the marks on the swingarm

1.13 Tighten the axle nut to the specified torque

1.15 Apply the lubricant to the overlap between the sideplates

Maintenance procedures

2.3 Unscrewing the spark plug using the Yamaha tool

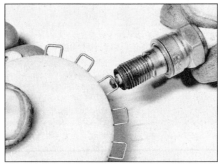

2.7a Using a wire type gauge to measure the spark plug electrode gap

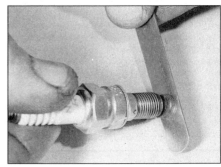

2.7b Using a feeler gauge to measure the spark plug electrode gap

Every 4000 miles (6000 km) – 1998/9 models
Every 6000 miles (10,000 km) – 2000-on models

2 Spark plugs – gap check and adjustment

1 Make sure your spark plug socket is the correct size before attempting to remove the plugs – a suitable one is supplied in the motorcycle's tool kit which is stored under the seat. Remove the air filter housing (see Chapter 4), and the ignition coils (see Chapter 5).

2 Using compressed air if available, clean the area around the base of the spark plugs to prevent any dirt falling into the engine when the plugs are removed.

3 Using either the plug removing tool supplied in the bike's toolkit or a deep socket type wrench, unscrew the plugs from the cylinder head (see illustration). Lay each plug out in relation to its cylinder so that, if any plug shows up a problem, it will be easy to identify the troublesome cylinder.

4 Inspect the electrodes for wear. Both the centre and side electrodes should have square edges and the side electrodes should be of uniform thickness. Look for excessive deposits and evidence of a cracked or chipped insulator around the centre electrode. Compare your spark plugs to the colour spark plug reading chart at the end of this manual. Check the threads, the washer and the ceramic insulator body for cracks and other damage.

5 If the electrodes are not excessively worn, and if the deposits can be easily removed with a wire brush, and there are no cracks or chips visible in the insulator, the plugs can be re-gapped and re-used. If in doubt concerning the condition of the plugs, replace them with new ones, as the expense is minimal. Note that the spark plugs should be renewed at every second service interval, i.e. every 8000 miles (1998 and 1999 models) or every 12,000 miles (2000-on models).

6 Cleaning spark plugs by sandblasting is permitted, provided you clean the plugs with a high flash-point solvent afterwards.

7 Before installing the plugs, make sure they are the correct type and heat range and check the gap between the electrodes (see illustrations). Compare the gap to that specified and adjust as necessary. If the gap must be adjusted, bend the side electrodes only and be very careful not to chip or crack the insulator nose (see illustration). Make sure the sealing washer is in place on the plug before installing it.

8 Since the cylinder head is made of aluminium, which is soft and easily damaged, thread the plugs into the heads and turn the tool by hand (see illustration). Once the plugs are finger-tight, the job can be finished with a spanner on the tool supplied or a socket drive (see illustration 2.3). If a torque wrench can be applied, tighten the spark plugs to the torque setting specified at the beginning of the Chapter. Otherwise tighten them by 1/4 to 1/2 turn after they have been fully hand tightened and have seated. Do not over-tighten them.

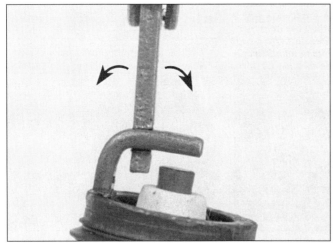

2.7c Adjust the electrode gap by bending the side electrode only

2.8 Thread the plug as far as possible turning the tool by hand

Maintenance procedures 1•9

 HAYNES HiNT *As the plugs are quite recessed, you can slip a short length of hose over the end of the plug to use as a tool to thread it into place. The hose will grip the plug well enough to turn it, but will start to slip if the plug begins to cross-thread in the hole – this will prevent damaged threads.*

9 Install the ignition coils and air filter housing (see Chapters 5 and 4).

 HAYNES HiNT *Stripped plug threads in the cylinder head can be repaired with a thread insert – see Section 2 of 'Tools and Workshop Tips' in the Reference section.*

3 Idle speed – check and adjustment

1 The idle speed should be checked and adjusted before and after the carburettors are synchronised (balanced), and when it is obviously too high or too low. Before adjusting the idle speed, make sure the valve clearances and spark plug gaps are correct, and the air filter is clean. Also, turn the handlebars back-and-forth and see if the idle speed changes as this is done. If it does, the throttle cables may not be adjusted or routed correctly, or may be worn out. This is a dangerous condition that can cause loss of control of the bike. Be sure to correct this problem before proceeding.

2 The engine should be at normal operating temperature, which is usually reached after 10 to 15 minutes of stop-and-go riding. Make sure the transmission is in neutral, and place the motorcycle on its stand.

3 The idle speed adjuster is located on the left-hand side of the motorcycle under the frame beam **(see illustration)**. With the engine idling, turn the adjuster until the speed listed in this Chapter's Specifications is obtained. Turn the knob clockwise to increase idle speed, and anti-clockwise to decrease it.

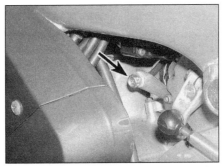

3.3 Idle speed adjuster knob (arrowed)

4 Snap the throttle open and shut a few times, then recheck the idle speed. If necessary, repeat the adjustment procedure.

5 If a smooth, steady idle cannot be achieved, the fuel/air mixture may be incorrect (check the pilot screw settings – see Chapter 4 Specifications) or the carburettors may need synchronising (see Section 4), or on 2000-on models, the AIS system may be faulty. Also check the intake manifold rubbers for cracks which will cause an air leak, resulting in a weak mixture.

4 Carburettors – synchronisation

⚠ **Warning:** *Petrol (gasoline) is extremely flammable, so take extra precautions when you work on any part of the fuel system. Don't smoke or allow open flames or bare light bulbs near the work area, and don't work in a garage where a natural gas-type appliance is present. If you spill any fuel on your skin, rinse it off immediately with soap and water. When you perform any kind of work on the fuel system, wear safety glasses and have a fire extinguisher suitable for a Class B type fire (flammable liquids) on hand.*

⚠ **Warning:** *Take great care not to burn your hand on the hot engine unit when accessing the gauge take-off points on the intake manifolds. Do not allow exhaust gases to build up in the work area; either perform the check outside or use an exhaust gas extraction system.*

1 Carburettor synchronisation is simply the process of adjusting the carburettors so that they pass the same amount of fuel/air mixture to each cylinder. This is done by measuring the vacuum produced in each cylinder. Carburettors that are out of synchronisation will result in decreased fuel mileage, increased engine temperature, less than ideal throttle response and higher vibration levels. Before synchronising the carburettors, make sure that the valve clearances and idle speed are properly set.

2 To properly synchronise the carburettors you will need a set of vacuum gauges or a manometer. These instruments measure engine vacuum, and can be obtained from motorcycle dealers or mail order parts suppliers. The equipment used should be suitable for a four cylinder engine and come complete with the necessary adapters and hoses to fit the take off points. **Note:** *Because of the nature of the synchronisation procedure and the need for special instruments, most owners leave the task to a Yamaha dealer.*

3 Start the engine and let it run until it reaches normal operating temperature, then shut it off. Remove the fuel tank (see Chapter 4).

4 On 1998 and 1999 models, unscrew the blanking bolts from the take-off points on the intake manifolds **(see illustration)**. Thread suitable adapters in their place and attach the gauge or manometer hoses **(see illustrations)**. Make sure the No. 1 gauge is attached to the hose from the No. 1 (left-hand) intake manifold, and so on.

5 On 2000-on models, release the clamp securing the AIS system hose to the take-off union on the intake manifold for cylinder No. 1 (left-hand cylinder) and attach the No. 1 gauge or manometer hose in its place. Unscrew the blanking bolts from the take-off points on the intake manifolds for cylinders Nos. 2, 3 and 4, then thread suitable adapters in their place and attach the other gauge or manometer hoses **(see illustrations 4.4b and c)**. Make sure the No. 1 gauge is attached to the hose from the No. 1 (left-hand) intake manifold, and so on.

6 Arrange a temporary fuel supply using an auxiliary tank and some hosing.

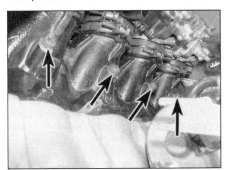

4.4a Remove the bolts (arrowed) ...

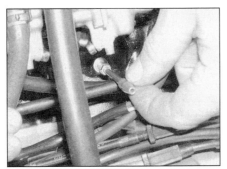

4.4b ... then install suitable adapters ...

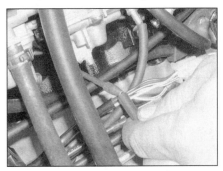

4.4c ... and attach the gauge hoses to them

1•10 Maintenance procedures

4.8 Carburettor synchronisation set-up

4.9 The screws are located in the throttle linkage between the carburettors (arrowed)

4.10 Install the blanking bolts with their copper washers

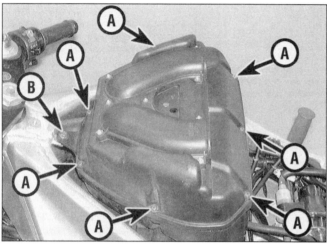

5.2a Remove the screws (A) and the bolt (B) . . .

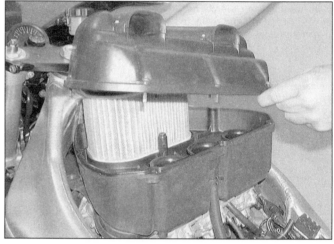

5.2b . . . then lift off the cover . . .

7 Start the engine and let it idle. If the gauges are fitted with damping adjustment, set this so that the needle flutter is just eliminated but so that they can still respond to small changes in pressure.

8 The vacuum readings for all cylinders should be the same **(see illustration)**. If the vacuum readings differ, proceed as follows.

9 The carburettors are balanced by turning the synchronising screws situated in between each carburettor, in the throttle linkage **(see illustration)**. **Note:** *Do not press on the screws whilst adjusting them, otherwise a false reading will be obtained.* First synchronise No. 1 carburettor to No. 2 using the left-hand synchronising screw until the readings are the same. Then synchronise No. 3 carburettor to No. 4 using the right-hand screw. Finally synchronise Nos. 1 and 2 carburettors to Nos. 3 and 4 using the centre screw. When all the carburettors are synchronised, open and close the throttle quickly to settle the linkage, and recheck the gauge readings, readjusting if necessary.

10 When the adjustment is complete, recheck the vacuum readings, then adjust the idle speed (see Section 3), and check the throttle cable freeplay (see Section 17). Remove the gauges and refit the blanking bolts with their copper washers **(see illustration)**. On 2000-on models attach the AIS system hose to the No. 1 cylinder union. Detach the temporary fuel supply and install the fuel tank (see Chapter 4).

5 Air filter – cleaning and renewal

1 Remove the fuel tank (see Chapter 4).
2 Remove the screws and bolt securing the air filter cover to the filter housing, then remove the cover and withdraw the filter element from the housing **(see illustrations)**.
3 Tap the element on a hard surface to dislodge any large particles of dirt. If compressed air is available, use it to clean the element, directing the air in the opposite direction of normal airflow **(see illustration)**.

> **HAYNES HiNT** *If using compressed air to clean the element, place either your hand, a rag, or a piece of card on the inside of the element to prevent any dust and debris being blown from one side of the element into the other.*

Caution: *If the machine is continually ridden in dusty conditions, the filter should be cleaned more frequently.*

5.2c . . . and withdraw the element

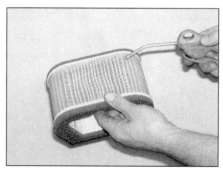

5.3 Direct the air in the opposite direction to normal airflow

Maintenance procedures 1•11

4 Check the element for signs of damage. If the element is torn or cannot be cleaned, or is obviously beyond further use, replace it with a new one.

5 Install the filter element, making sure it is properly seated **(see illustration 5.2c)**. Fit the air filter cover, making sure the rubber seal is in place in the rim of the housing **(see illustrations 5.2b and a)**. Install the fuel tank (see Chapter 4).

6 Check that the collector in the air filter housing drain hose has not become blocked, and drain it if necessary – the hose comes out of the rear left side of the housing.

7 Check the crankcase breather hose between the engine and the rear of the air filter housing for loose connections, cracks and deterioration, and replace it with a new one if necessary.

6 Fuel system, EXUP system and air induction system (AIS) – check

Warning: Petrol (gasoline) is extremely flammable, so take extra precautions when you work on any part of the fuel system. Don't smoke or allow open flames or bare light bulbs near the work area, and don't work in a garage where a natural gas-type appliance is present. If you spill any fuel on your skin, rinse it off immediately with soap and water. When you perform any kind of work on the fuel system, wear safety glasses and have a fire extinguisher suitable for a Class B type fire (flammable liquids) on hand.

Fuel system – check

1 Remove the fuel tank (see Chapter 4) and check the tank, fuel tap and fuel hoses, and the filter and fuel pump, for signs of leakage, deterioration or damage; in particular check that there is no leakage from the fuel hoses. Renew any hoses which are cracked or have deteriorated (see Section 33).

2 If the tap has been leaking from the face, tightening the assembly screws on the face may help. Slacken the screws a little first, then tighten them evenly and a little at a time to ensure the cover seats properly on the tap body. If leakage persists, remove the screws on the face of the tap and disassemble it, noting how the components fit. Inspect all components for wear or damage, and replace the seal and O-ring with new ones. None of the other components are available individually, so if they are damaged fit a new tap. If the tap has been leaking from the base, tightening the mounting screws may help. Otherwise, remove the tap and fit a new O-ring (see Chapter 4). Remove any corrosion or paint bubbles before installing the tap.

3 If the carburettor gaskets are leaking, the carburettors should be disassembled and rebuilt using new gaskets and seals (see Chapter 4).

4 On California models, check the EVAP system hoses for loose connections, cracks and deterioration and replace them with new ones if necessary.

Fuel strainer/filter cleaning and renewal

5 A fuel strainer is mounted in the tank and is integral with the fuel tap. Cleaning of the strainer is advised after a high mileage has been covered but not every 4000/6000 miles. It is also necessary if fuel starvation is suspected. Remove the fuel tank and the fuel tap (see Chapter 4). Clean the gauze strainer to remove all traces of dirt and fuel sediment. Check the gauze for holes. If any are found, a new tap should be fitted – the strainer is not available separately. If the strainer is dirty, check the condition of the inside of your tank – if there is evidence of rust, drain and clean the tank (see Chapter 4).

6 An in-line fuel filter is fitted in the hose from the fuel tap to the fuel pump **(see illustration)**. Note that Yamaha recommend that the in-line filter should be renewed every 20,000 miles (32,000 km). Remove the fuel tank (see Chapter 4). If the filter is dirty or clogged, replace it with a new one – this type of filter cannot be cleaned. Have a rag handy to soak up any residual fuel, then release the clamps and disconnect the hoses from the filter, noting which fits where. Release the filter from its holder and discard it. Install the new filter so that its arrow points in the direction of fuel flow (i.e. towards the pump). Fit the hoses onto the unions on the filter and secure them with the clamps. Start the engine and check that there are no leaks.

EXUP system – check

7 When the ignition is first switched ON, or while the engine is running, the EXUP system performs its own self-diagnosis. If a fault occurs, the tachometer will be seen to display zero rpm for 3 seconds, then 7000 rpm for 2.5 seconds, then the actual engine speed for 3 seconds, whereupon it will repeat the cycle until the engine is switched off. Note that the motorcycle can be ridden even though a fault has been diagnosed, though a difference in performance will be noticed.

8 Remove the lower fairing (see Chapter 8) and the fuel tank (see Chapter 4).

9 Unscrew the bolts securing the EXUP pulley cover on the exhaust and remove the cover **(see illustration)**.

10 Start the engine and increase the revs to 2000 rpm. The servo should start to turn at 2000 rpm and continue as the revs rise. As it turns, simultaneously check that the pulley on the exhaust is turning the valve. If the servo turns but the pulley does not, remove the throttle cables and check them. Lubricate them, or replace with new ones, as necessary (see Chapter 4). If the pulley operates, but the movement of the valve is sticky or not complete, first check the cables, and if they are good, disassemble the valve for cleaning and inspection (see Chapter 4). If the servo does not operate, refer to Chapter 4 for further tests.

6.6 In-line fuel filter (arrowed)

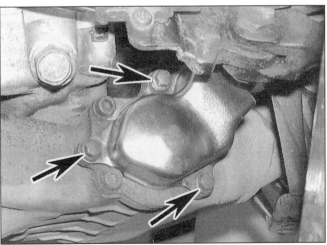

6.9 Unscrew the bolts (arrowed) and remove the cover

1•12 Maintenance procedures

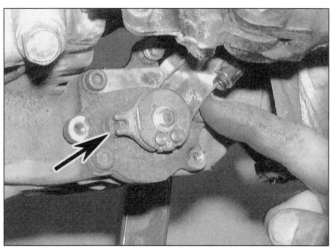

6.12a Check the freeplay in the cables. If it is incorrect, adjust it as described, inserting the pin into the hole (arrowed) so it locks the pulley

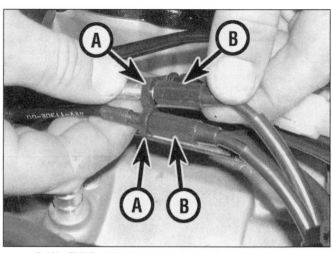

6.12b EXUP cable freeplay adjuster lockrings (A) and adjusters (B)

11 The EXUP pulley and valve are prone to corrosion, due to their location. The pulley cover and valve cover bolts can seize, the cable ends can seize in their sockets in the pulley, and the valve shaft can seize in its bushes. Keep the area well lubricated with WD40 or a suitable equivalent. It is highly advisable to disassemble the system from time to time (according to the conditions you ride in) and lubricate the shaft, bushes, cable ends and bolts with copper-based grease (see Chapter 4).

12 Measure the amount of freeplay in each EXUP cable as shown and check that it is as specified at the beginning of the Chapter **(see illustration)**. If not, slacken the lockring on the adjuster in each cable and turn the adjusters in to create freeplay in the cable **(see illustration)**. Now turn the cable pulley on the valve until the notch in the pulley aligns with the hole in the housing, then insert a 4 mm pin into the hole **(see illustration 6.12a)**. Turn each cable adjuster out until it becomes tight, then turn it in by 1/2 turn, then tighten the lockrings. Remove the pin from the hole and check that the freeplay is as specified and equal for each cable.

13 Apply some copper grease to the cover bolt threads. Install the cover and tighten the bolts to the torque setting specified at the beginning of the Chapter **(see illustration)**. Install the lower fairing (see Chapter 8) and the fuel tank (see Chapter 4).

Air induction system (AIS) – check (2000-on models)

14 If the valves clearances are all correct and the carburettors have been synchronised, and have no other faults, and the idle speed cannot be set properly, it is possible that the AIS is faulty. Further information on the function of the system is in Chapter 4.

15 Remove the fairing side panels (see Chapter 8). Check the air cut-off valve and the reed valve assembly on the front of the engine for signs of physical damage and replace it with a new one if necessary (see Chapter 4).

16 Check the AIS hoses and pipes for signs of deterioration or damage, and check that they are all securely connected with the hoses clamped at each end. Replace any hoses which are cracked or deteriorated with new ones (see Chapter 4).

17 Remove the valve assembly and disassemble it (see Chapter 4). Check the reeds for cracks, warpage and any other damage or deterioration **(see illustration)**. Also check the contact areas between the reeds and the reed holders, and the holders themselves, for any signs of damage or deterioration. Any carbon deposits or other foreign particles can be cleaned off using a high flash-point solvent. Check that the maximum amount of bend in each reed is such that the gap between the end of each reed and the holder is 0.4 mm maximum – you can remove the reed from the valve and place it on a surface plate to check this.

18 Any further testing of the air induction system requires the use of exhaust gas analysers and temperature sensors. If the system is thought to be faulty, take the bike to a Yamaha dealer for assessment. Make sure that the idle speed, valve clearances and carburettor synchronisation have all been checked before assuming that the AIS is faulty.

7 Engine/transmission – oil change

⚠️ **Warning:** Be careful when draining the oil, as the exhaust pipes, the engine, and the oil itself can cause severe burns.

1 Consistent routine oil and filter changes are the single most important maintenance procedure you can perform on a motorcycle. The oil not only lubricates the internal parts of the engine, transmission and clutch, but it also acts as a coolant, a cleaner, a sealant, and a protector. Because of these demands, the oil takes a terrific amount of abuse and should be changed often with new oil of the recommended grade and type. Saving a little money on the difference in cost between a good oil and a cheap oil won't pay off if the engine is damaged. The oil filter should be changed with every second oil change.

2 Before changing the oil, warm up the engine so the oil will drain easily. Remove the lower fairing (see Chapter 8).

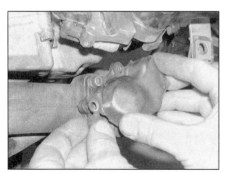

6.13 Install the cover and tighten the bolts

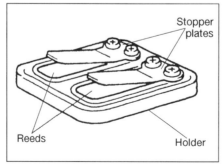

6.17 Reed valve components

Maintenance procedures 1•13

7.3 Remove the oil filler cap from the clutch cover

7.4a Unscrew the crankcase oil drain plug . . .

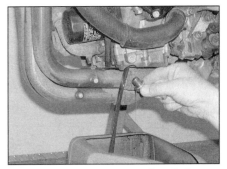

7.4b . . . and allow the oil to drain

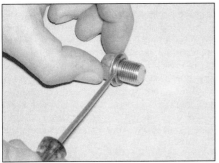

7.4c Replace the sealing washer with a new one if necessary

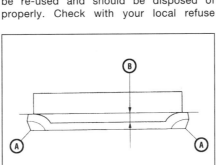

7.5a Install the drain plug, using a new sealing washer if necessary . . .

7.5b . . . and tighten it to the specified torque

3 Position a clean drain tray below the engine. Unscrew the oil filler cap from the clutch cover to vent it and to act as a reminder that there is no oil in the engine **(see illustration)**.

4 Unscrew the oil drain plug from the underside of the engine and allow the oil to flow into the drain tray **(see illustrations)**. Check the condition of the sealing washer on the drain plug and discard it if it is damaged or worn – it will probably be necessary to cut the old one off using pliers **(see illustration)**. It is always advisable to use a new one even if the old one looks all right.

5 When the oil has completely drained, fit the plug into the crankcase, using a new sealing washer if required, and tighten it to the torque setting specified at the beginning of the Chapter **(see illustrations)**. Avoid overtightening, as you will damage the sump.

6 Refill the engine to the proper level using the recommended type and amount of oil (see

> **HAYNES HiNT**
> Check the old oil carefully – if it is very metallic coloured, then the engine is experiencing wear from break-in (new engine) or from insufficient lubrication. If there are flakes or chips of metal in the oil, then something is drastically wrong internally and the engine will have to be disassembled for inspection and repair. If there are pieces of fibre-like material in the oil, the clutch is experiencing excessive wear and should be checked.

Daily (pre-ride) checks). With the motorcycle vertical, the oil level should lie between the maximum and minimum level lines on the inspection window (see *Daily (pre-ride) checks*). Install the filler cap **(see illustration 7.3)**. Start the engine and let it run for two or three minutes. Stop the engine, wait a few minutes, then check the oil level. If necessary, add more oil to bring the level up to the maximum level line on the window. Check that there are no leaks around the drain plug. If there are, and you didn't use a new sealing washer, drain the oil again and replace the washer with a new one.

7 Every so often, and especially as Yamaha do not fit an oil pressure switch and warning light (the system fitted uses an oil level sensor), it is advisable to perform an oil pressure check (see Section 32).

8 The old oil drained from the engine cannot be re-used and should be disposed of properly. Check with your local refuse disposal company, disposal facility or environmental agency to see whether they will accept the used oil for recycling. Don't pour used oil into drains or onto the ground.

9 Install the lower fairing (see Chapter 8).

Note: It is antisocial and illegal to dump oil down the drain. In the UK, call this number free to find the location of your local oil recycling bank. In the USA, note that any oil supplier must accept used oil for recycling.

8 Brake pads – wear check

1 Each brake pad has wear indicators that can be viewed without removing the pads from the caliper.

2 The turned-in corners of the brake pad backing material form the wear indicators – when they are almost contacting the disc itself the pads must be renewed **(see illustration)**. The indicators are visible by looking at the bottom corner of the pads **(see illustration)**.

Note: Some after-market pads may use different indicators (such as a groove cut into the friction material); the pad is worn when the groove is no longer visible.

Caution: *Do not allow the pads to wear to the extent that the indicators contact the disc itself, as the disc will be damaged.*

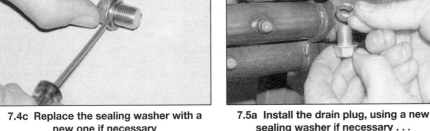

8.2a Brake pad wear indicators (A) and minimum thickness (B)

1•14 Maintenance procedures

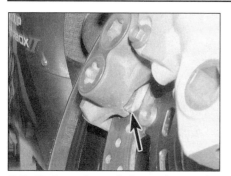

8.2b Brake pad wear indicator (arrowed) – front brake shown

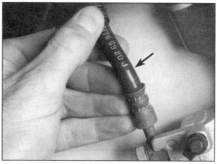

9.3 Flex the brake hose and check for cracks, bulges and leaking fluid

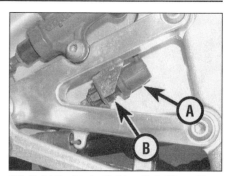

9.5 Hold the brake light switch (A) and turn the adjuster nut (B)

3 If the pads are worn to the indicators, new ones must be installed. If the pads are dirty or if you are in doubt as to the amount of friction material remaining, remove them for inspection (see Chapter 7). If required, measure the amount of friction material remaining – the minimum is 0.5 mm.

4 Refer to Chapter 7 for details of pad renewal.

9 Brake system – check

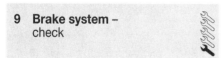

1 A routine general check of the brake system will ensure that any problems are discovered and remedied before the rider's safety is jeopardised.

2 Check the brake lever and pedal for looseness, improper or rough action, excessive play, bends, and other damage. Replace any damaged parts with new ones (see Chapter 7). Clean and lubricate the lever and pedal pivots if their action is stiff or rough (see Section 18).

3 Make sure all brake fasteners are tight. Check the brake pads for wear (see Section 8) and make sure the fluid level in the reservoirs is correct (see *Daily (pre-ride) checks*). Look for leaks at the hose connections and check for cracks in the hoses themselves **(see illustration)**. If the lever or pedal is spongy, bleed the brakes (see Chapter 7). The brake fluid should be changed every two years (see Section 26) and the hoses renewed if they

deteriorate, or every four years irrespective of their condition (see Section 29). The master cylinder and caliper seals should be renewed every two years, or if leakage from them is evident (see Section 25).

4 Make sure the brake light operates when the front brake lever is pulled in. The front brake light switch, mounted on the underside of the master cylinder, is not adjustable. If it fails to operate properly, check it (see Chapter 9).

5 Make sure the brake light is activated just before the rear brake takes effect. If adjustment is necessary, hold the switch and turn the adjuster nut on the switch body until the brake light is activated when required **(see illustration)**. If the brake light comes on too late, turn the nut clockwise. If the brake light comes on too soon or is permanently on, turn the nut anti-clockwise. If the switch doesn't operate the brake light, check it (see Chapter 9).

6 The front brake lever has a span adjuster which alters the distance of the lever from the handlebar. Each setting is identified by a number on the adjuster which aligns with the arrow on the lever bracket. Pull the lever away from the handlebar and turn the adjuster ring until the setting which best suits the rider is obtained **(see illustration)**. There are four settings – setting 1 gives the largest span, setting 4 the smallest. When making adjustment ensure that the pin set in the lever bracket is engaged in its detent in the adjuster.

7 Check the position of the brake pedal

(brake pedal height). Yamaha recommend the distance between the tip of the brake pedal and the top of the rider's footrest should be as specified at the beginning of the Chapter **(see illustration)**. If the pedal height is incorrect, or if the rider's preference is different, slacken the clevis locknut on the master cylinder pushrod, then turn the pushrod using a spanner on the hex at the top of the rod until the pedal is at the correct or desired height **(see illustration)**. After adjustment check that the pushrod end is still visible in the hole in the clevis. On completion tighten the locknut securely. Adjust the rear brake light switch after adjusting the pedal height (see Step 5).

10 Clutch – check and adjustment

1 Check that the clutch lever operates smoothly and easily.

2 If the lever action is heavy or stiff, remove the cable (see Chapter 2) and lubricate it (see Section 18). If the inner cable still does not run smoothly in the outer cable, replace it with a new one. Install the lubricated or new cable (see Chapter 2). If the action is still stiff, remove the lever and check for damage or distortion, or any other cause, and remedy as necessary. Clean and lubricate the pivot and contact areas (see Section 18). If the lever is good, refer to Chapter 2 and check the release mechanism in the cover and the clutch itself.

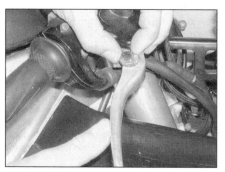

9.6 Adjusting the front brake lever span

9.7a Measure the distance between the top of the footpeg and the top of the brake pedal as shown

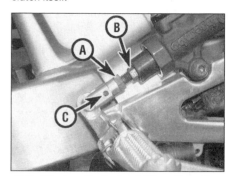

9.7b Slacken the locknut (A) and turn the pushrod using the hex (B) making sure the rod end is still visible in the hole (C)

Maintenance procedures 1•15

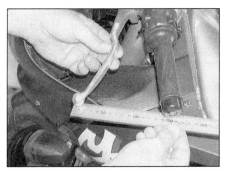

10.3 Check the amount of freeplay in the cable as shown

10.4 Turn the adjuster as described to set the correct freeplay

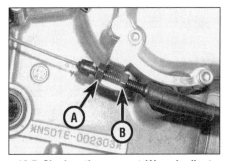

10.5 Slacken the rear nut (A) and adjust the freeplay using the front nut (B) as described

3 With the cable operating smoothly, check that the clutch lever is correctly adjusted. Periodic adjustment is necessary to compensate for wear in the clutch plates and stretch of the cable. Check that the amount of freeplay at the clutch lever end is within the specifications listed at the beginning of the Chapter **(see illustration)**.

4 If adjustment is required, turn the adjuster in or out until the required amount of freeplay is obtained **(see illustration)**. To increase freeplay, turn the adjuster clockwise (into the lever bracket). To reduce freeplay, turn the adjuster anti-clockwise (out of the lever bracket). Tighten the locking ring securely.

5 If all the adjustment has been taken up at the lever, reset the adjuster to give the maximum amount of freeplay, then set the correct amount of freeplay using the adjuster nuts on each end of the threaded section in the cable bracket on the right-hand side of the engine. Remove the lower fairing to access it (see Chapter 8). Slacken the rear nut until the front nut can be drawn free of its locking lug on the bracket, then turn the front nut as required to obtain the correct freeplay **(see illustration)**. To increase freeplay, thread the front nut up the threaded section of the cable. To reduce freeplay, thread the front nut down the threaded section. Locate the front nut against the lug on the bracket, then tighten the rear nut against the bracket. Subsequent adjustments can now be made using the lever adjuster only.

11 Battery – check

1 All models are fitted with a sealed (maintenance-free) battery which requires no maintenance. **Note:** *Do not attempt to remove the battery caps to check the electrolyte level or battery specific gravity. Removal will damage the caps, resulting in electrolyte leakage and battery damage.*

2 All that should be done is to check that the terminals are clean and tight and that the casing is not damaged or leaking. See Chapter 9 for further details.

Caution: Be extremely careful when handling or working around the battery.

The electrolyte is very caustic and an explosive gas (hydrogen) is given off when the battery is charging.

3 If the machine is not in regular use, disconnect the battery and give it a refresher charge every month to six weeks (see Chapter 9, Section 3).

12 Wheels and tyres – general check
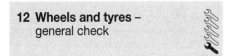

Tyres

1 Check the tyre condition and tread depth thoroughly – see *Daily (pre-ride) checks*.

Wheels

2 Cast wheels are virtually maintenance free, but they should be kept clean and checked periodically for cracks and other damage. Also check the wheel runout and alignment (see Chapter 7). Never attempt to repair damaged cast wheels; they must be replaced with new ones. Check the valve rubber for signs of damage or deterioration and have it replaced with a new one if necessary. Also, make sure the valve cap is in place and tight.

13 Wheel bearings – check
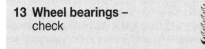

1 Wheel bearings will wear over a period of time and result in handling problems.

2 Support the motorcycle upright using an auxiliary stand. Check for any play in the bearings by pushing and pulling the wheel against the hub **(see illustration)**. Also rotate the wheel and check that it rotates smoothly.

3 If any play is detected in the front wheel hub, or if the wheel does not rotate smoothly (and this is not due to brake or transmission drag), the wheel bearings must be removed and inspected for wear or damage (see Chapter 7).

4 On the rear wheel there should be a small amount of freeplay because it is fitted with a needle roller bearing in one side, which by nature exhibits more freeplay than conventional caged ball bearings. However, if the freeplay is excessive, or if the wheel does not rotate smoothly (and this is not due to brake or transmission drag), inspect the bearings for wear and damage and replace them with new ones if necessary (see Chapter 7). Note that once a needle bearing has been removed, it cannot be reused.

14 Sidestand – check
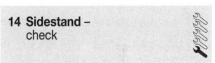

1 The stand return springs must be capable of retracting the stand fully and holding the stand retracted when the motorcycle is in use. If a spring is sagged or broken it must be replaced with a new one.

2 Lubricate the stand pivot regularly (see Section 18).

3 The sidestand switch prevents the motorcycle being started if it is in gear and the stand is down, and cuts the engine if the stand is put down while it is running and in gear. Check its operation by shifting the transmission into neutral, retracting the stand and starting the engine. Pull in the clutch lever and select a gear. Extend the sidestand. The engine should stop as the sidestand is extended. If the sidestand switch does not operate as described, check its circuit (see Chapter 9). The clutch switch is also part of the same circuit – to check it, retract the sidestand, then select a gear. With the clutch lever pulled in, start the engine – if the engine starts, the switch is good. Otherwise, check the circuit (see Chapter 9).

13.2 Checking for play in the wheel bearings

1

1•16 Maintenance procedures

15 Nuts and bolts – tightness check

1 Since vibration of the machine tends to loosen fasteners, all nuts, bolts, screws, etc. should be periodically checked for proper tightness.
2 Pay particular attention to the following:
  Spark plugs
  Engine oil drain plug
  Gearchange lever, brake and clutch lever, and brake pedal bolts
  Footrest and stand bolts
  Engine mounting bolts
  Shock absorber and suspension linkage bolts and swingarm pivot bolts
  Handlebar clamp bolts
  Front axle bolt and axle clamp bolts
  Front fork clamp bolts (top and bottom yoke)
  Rear axle nut
  Brake caliper mounting bolts
  Brake hose banjo bolts and caliper bleed valves
  Brake disc bolts
  Exhaust system bolts/nuts
3 If a torque wrench is available, use it along with the torque specifications at the beginning of this and other Chapters.

16 Cooling system – check

Warning: The engine must be cool before beginning this procedure.

1 Check the coolant level (see *Daily (pre-ride) checks*).
2 The entire cooling system should be checked for evidence of leakage. Remove the lower fairing and the fairing side panels (see Chapter 8). Examine each rubber coolant hose along its entire length. Look for cracks, abrasions and other damage. Squeeze each hose at various points. They should feel firm, yet pliable, and return to their original shape when released. If they are dried out or hard, replace them with new ones.
3 Check for evidence of leaks at each cooling system joint. If necessary, tighten the hose clips carefully to prevent future leaks.
4 To prevent leakage of water from the cooling system to the lubrication system and vice versa, two seals are fitted on the pump shaft. On the bottom of the sump there is a drain hole fed by a pipe coming from the pump **(see illustration)**. If either seal fails, the drain allows the coolant or oil to escape and prevents them mixing. The seal on the water pump side is of the mechanical type which bears on the rear face of the impeller. The second seal, which is mounted behind the mechanical seal, is of the normal feathered lip type. If on inspection the drain hole shows

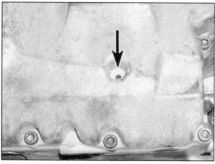

16.4 The drain hole (arrowed) is on the bottom of the sump

signs of coolant leakage, remove the pump and replace the mechanical seal with a new one. If it is oil that is leaking, or if the leakage is white and with the texture of emulsion, replace both seals with new ones (the mechanical seal has to be removed in order to remove the oil seal, and it cannot be reused). Refer to Chapter 3 for seal renewal.
5 Check the radiator for leaks and other damage. Leaks in the radiator leave tell-tale scale deposits or coolant stains on the outside of the core below the leak. If leaks are noted, remove the radiator (see Chapter 3) and have it repaired by a professional.
Caution: Do not use a liquid leak stopping compound to try to repair leaks.
6 Check the radiator fins for mud, dirt and insects, which may impede the flow of air through the radiator. If the fins are dirty, remove the radiator (see Chapter 3) and clean it, using water or low pressure compressed air directed through the fins from the back side. If the fins are bent or distorted, straighten them carefully with a screwdriver. If the air flow is restricted by bent or damaged fins over more than 30% of the radiator's surface area, install a new radiator.
7 Remove the right-hand cockpit trim panel (see Chapter 8). Remove the pressure cap from the radiator filler neck by turning it anti-clockwise until it reaches a stop **(see illustration)**. If you hear a hissing sound (indicating that there is still pressure in the system), wait until it stops. Now press down on the cap and continue turning the cap until it can be removed. Check the condition of the coolant in the system. If it is rust-coloured or if

16.7 Remove the pressure cap as described

accumulations of scale are visible, drain, flush and refill the system with new coolant (See Section 22). Check the cap seal for cracks and other damage. If in doubt about the pressure cap's condition, have it tested by a Yamaha dealer or replace it with a new one. Install the cap by turning it clockwise until it reaches the first stop, then push down on the cap and continue turning until it can turn no further.
8 Check the antifreeze content of the coolant with an antifreeze hydrometer. Sometimes coolant looks like it is in good condition, but it might be too weak to offer adequate protection. If the hydrometer indicates a weak mixture, drain, flush and refill the system (see Section 22).
9 Start the engine and let it reach normal operating temperature, then check for leaks again. As the coolant temperature increases beyond normal, the fan should come on automatically and the temperature should begin to drop. If it does not drop, refer to Chapter 3 and check the fan switch, fan motor and fan circuit carefully.
10 If the coolant level is consistently low, and no evidence of leaks can be found, have the entire system pressure-checked by a Yamaha dealer.

17 Throttle and choke cables – check

Throttle cables

1 Make sure the throttle grip rotates easily from fully closed to fully open with the front wheel turned at various angles. The grip should return automatically from fully open to fully closed when released.
2 If the throttle sticks, this is probably due to a cable fault. Remove the cables (see Chapter 4) and lubricate them (see Section 18). If the inner cables still do not run smoothly in the outer cables, replace them with new ones. With the cables removed, check that the twistgrip runs smoothly and freely around the handlebar – dirt and debris combined with a lack of lubrication can cause the action to be stiff. Install the lubricated or new cables, making sure they are correctly routed (see Chapter 4). If this fails to improve the operation of the throttle, the fault could lie in the carburettors. Remove them and check the action of the throttle linkage and butterflies (see Chapter 4).
3 With the throttle operating smoothly, check for a small amount of freeplay in the cables, measured in terms of the amount of twistgrip rotation before the throttle opens, and compare the amount to that listed in this Chapter's Specifications **(see illustration)**. If it is incorrect, adjust the cables to correct it.
4 Freeplay adjustments can be made at the throttle end of the cable. Pull back the rubber cover on the adjuster, then loosen the locknut

Maintenance procedures 1•17

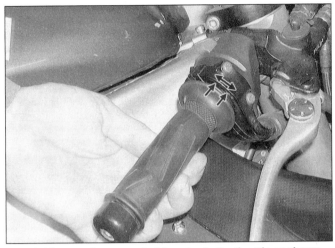

17.3 Measure the amount of freeplay in the throttle as shown

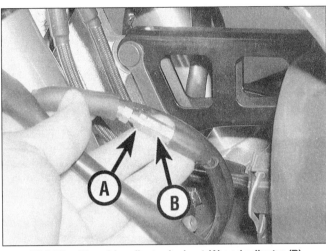

17.4 Throttle cable adjuster locknut (A) and adjuster (B)

(see illustration). Turn the adjuster until the specified amount of freeplay is obtained (see this Chapter's Specifications), then retighten the locknut. Turn the adjuster in to increase freeplay and out to reduce it. Refit the rubber boot on completion.

5 If the adjuster has reached its limit of adjustment, reset it so that the freeplay is at a maximum, then adjust the cable at the carburettor end as follows. Remove the fuel tank (see Chapter 4), the air filter housing (see Chapter 4), and the ignition coils (see Chapter 5). Release the trim clips securing the rubber cover and remove the cover, noting how it fits **(see illustrations)**. To release the clips, push the centre into the body, then draw the clip out **(see illustration)**.

6 Slacken the accelerator (opening) cable top nut and slide the cable down in the bracket until the bottom nut is clear of the lug, then thread the bottom nut up or down as required – thread it down to reduce freeplay, and thread it up to increase it **(see illustration)**. Draw the cable up into the bracket so the bottom nut becomes captive against the lug, then tighten the top nut down onto the bracket. Further adjustments can now be made at the throttle end. If the cable cannot be adjusted as specified, renew the cable (see Chapter 4).

⚠ *Warning: Turn the handlebars all the way through their travel with the engine idling. Idle speed should not change. If it does, the cable may be routed incorrectly. Correct this condition before riding the bike.*

7 Check that the throttle twistgrip operates smoothly and snaps shut quickly when released. Install all disturbed components and assemblies. To install the clips securing the rubber cover, first push the centre back out so that it protrudes from the top of the body. Fit the clip into its socket, then push the centre in so that it is flush with the top of the body **(see illustrations)**.

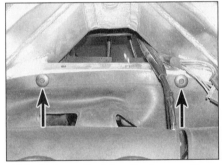

17.5a Release the trim clips (arrowed) . . .

17.5b . . . and remove the cover

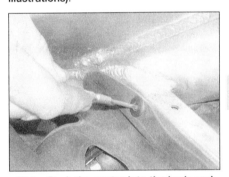

17.5c Push the centre into the body and remove the clip

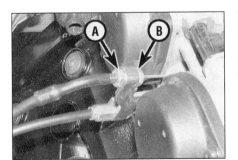

17.6 Slacken the top nut (A) and slide the cable in the bracket until the bottom nut (B) is clear of the lug, then thread it up or down as required

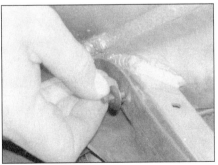

17.7a Fit the clip into its socket . . .

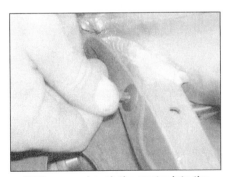

17.7b . . . and push the centre into the body to secure it

1•18 Maintenance procedures

17.10 Slacken the clamp screw (arrowed) and slide the cable into the clamp a little

Choke cable

8 If the choke does not operate smoothly this is probably due to a cable fault. Remove the cable (see Chapter 4) and lubricate it (see Section 18). If the inner cable still does not run smoothly in the outer cable, replace it with a new one. Install the cable, routing it so it takes the smoothest route possible.

9 If this fails to improve the operation of the choke, check that the lever is not binding in the switch housing. If the lever action is good, the fault could lie in the carburettors rather than the cable, necessitating their removal and inspection of the choke plungers (see Chapter 4).

10 Make sure there is a small amount of freeplay in the cable before the plungers move. If there isn't, check that the cable is seating correctly at the carburettor end – remove the air filter housing (see Chapter 4), the ignition coils (see Chapter 5), and the rubber cover for access. You can create some freeplay in the cable by slackening the outer cable clamp screw on the carburettor and

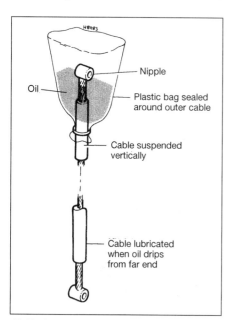

18.3b Lubricating a cable with a makeshift funnel and motor oil

18.3a Lubricating a cable with a pressure lubricator. Make sure the tool seals around the inner cable

sliding the cable further into the clamp **(see illustration)**. Otherwise, renew the cable.

18 Stand, lever pivots and cables – lubrication

1 Since the controls, cables and various other components of a motorcycle are exposed to the elements, they should be lubricated periodically to ensure safe and trouble-free operation.

2 The footrests, clutch and brake levers, brake pedal, gearchange lever linkage and sidestand pivots should be lubricated frequently. In order that the lubricant is applied where it will do the most good, the component should be disassembled. However, if chain and cable lubricant is being used, it can be applied to the pivot joint gaps and will usually work its way into the areas where friction occurs. If motor oil or light grease is being used, apply it sparingly as it may attract dirt (which could cause the controls to bind or wear at an accelerated rate). **Note:** *One of the best lubricants for the control lever pivots is a dry-film lubricant (available from many sources by different names).*

3 To lubricate the throttle and choke cables, disconnect the relevant cable at its upper end, then lubricate the cable with a pressure adapter or, if one is not available, using the set-up shown **(see illustrations)**. See Chapter 4 for the throttle and choke cable removal procedures, and Chapter 2 for the clutch cable.

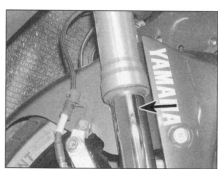

19.3a Check the top of the inner tube for scratches, corrosion, pitting and oil leakage

19 Suspension – check

1 The suspension components must be maintained in top operating condition to ensure rider safety. Loose, worn or damaged suspension parts decrease the motorcycle's stability and control.

Front suspension

2 While standing alongside the motorcycle, apply the front brake and push on the handlebars to compress the forks several times. See if they move up and down smoothly without binding. If binding is felt, the forks should be disassembled and inspected (see Chapter 6).

3 Inspect the top of the inner tube for signs of scratches, corrosion and pitting, and oil leakage, then carefully lever up the dust seal using a flat-bladed screwdriver and inspect the area around the fork seal **(see illustrations)**. Any scratches, corrosion and pitting will cause premature seal failure. If the damage is excessive, new tubes should be installed (see Chapter 6). If leakage is evident, new seals must be fitted (see Chapter 6).

4 Check the tightness of all suspension nuts and bolts to be sure none have worked loose, referring to the torque settings specified at the beginning of Chapter 6.

Rear suspension

5 Inspect the rear shock for fluid leakage and tightness of its mountings. If leakage is found, a new shock should be installed (see Chapter 6).

6 With the aid of an assistant to support the bike, compress the rear suspension several times. It should move up and down freely without binding. If any binding is felt, the worn or faulty component must be identified and renewed. The problem could be due to either the shock absorber, the suspension linkage components or the swingarm components.

7 Support the motorcycle using an auxiliary stand so that the rear wheel is off the ground. Grab the swingarm and rock it from side to side – there should be no discernible movement at the rear (Yamaha specify a

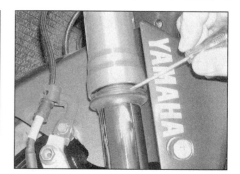

19.3b Lever off the dust seal and check for signs of oil leakage

maximum of 1 mm) **(see illustration)**. If there is a little movement or a slight clicking can be heard, inspect the tightness of all the rear suspension mounting bolts and nuts, referring to the torque settings specified at the beginning of Chapter 6, and re-check for movement. Next, grasp the top of the rear wheel and pull it upwards – there should be no discernible freeplay before the shock absorber begins to compress **(see illustration)**. Any freeplay felt in either check indicates worn bearings in the suspension linkage or swingarm, or worn shock absorber mountings. The worn components must be renewed (see Chapter 6).

8 To make an accurate assessment of the swingarm bearings, remove the rear wheel (see Chapter 7) and the bolt securing the suspension linkage rods to the swingarm (see Chapter 6). Grasp the rear of the swingarm with one hand and place your other hand at the junction of the swingarm and the frame. Try to move the rear of the swingarm from side to side. Any wear (play) in the bearings should be felt as movement between the swingarm and the frame at the front. If there is any play, the swingarm will be felt to move forward and backward at the front (not from side-to-side). Alternatively, measure the amount of freeplay at the swingarm end – Yamaha specify a maximum of 1 mm. Next, move the swingarm up and down through its full travel. It should move freely, without any binding or rough spots. If any play in the swingarm is noted or if the swingarm does not move freely, the bearings must be removed for inspection or renewal (see Chapter 6).

19.7a Checking for play in the swingarm bearings

20 Steering head bearings – freeplay check and adjustment

1 This motorcycle is equipped with caged ball steering head bearings which can become dented, rough or loose during normal use of the machine. In extreme cases, worn or loose steering head bearings can cause steering wobble – a condition that is potentially dangerous.

Check

2 Support the motorcycle in an upright position using an auxiliary stand. Raise the front wheel off the ground either by having an assistant push down on the rear, or by placing a support under the engine, in which case remove the lower fairing first (see Chapter 8).
3 Point the front wheel straight ahead, and

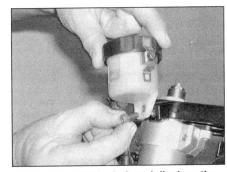

20.6 Unscrew the bolt and displace the reservoir

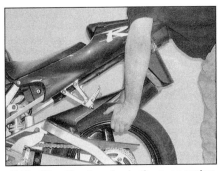

19.7b Checking for play in the suspension linkage bearings

slowly move the handlebars from side to side. Any dents or roughness in the bearing races will be felt and the bars will not move smoothly and freely.
4 Next, grasp the fork sliders and try to pull and push them forwards and backwards **(see illustration)**. Any looseness in the steering head bearings will be felt as front-to-rear movement of the forks. If play is felt in the bearings, adjust the steering head as follows.

> **HAYNES HiNT** *Freeplay in the fork due to worn fork bushes can be misinterpreted as steering head bearing play – do not confuse the two.*

Adjustment

5 Position the motorcycle in an upright position using an auxiliary stand. Remove the fuel tank (see Chapter 4) and the fairing (see Chapter 8). **Note:** *Although it is not strictly necessary to remove the fuel tank and fairing, doing so will prevent the possibility of damage, should a tool slip.*
6 Unscrew the bolt securing the front brake master cylinder to the top yoke and remove it **(see illustration)**. Keep the master cylinder upright to prevent fluid leakage.
7 Remove the blanking caps from the handlebar positioning bolts using a small flat-bladed screwdriver, then unscrew the bolts **(see illustrations)**. Slacken the handlebar clamp bolts, and the fork clamp bolts in the top yoke **(see illustration)**.

20.4 Checking for play in the steering head bearings

20.7a Remove the blanking caps . . .

20.7b . . . then unscrew the handlebar positioning bolts

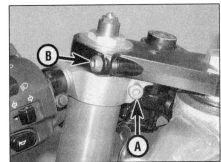

20.7c Slacken the handlebar clamp bolts (A) and the fork clamp bolts (B)

1•20 Maintenance procedures

20.8 Unscrew the steering stem nut and remove the washer

20.9 Ease the top yoke up off the steering stem and forks

20.10a Remove the tabbed lockwasher

20.10b The locknut should only be finger-tight . . .

20.10c . . . but use one of the specified tools if required . . .

20.10d . . . and remove the rubber washer

8 Unscrew the steering stem nut and remove it along with its washer **(see illustration)**.
9 Gently ease the top yoke upwards off the fork tubes and position it clear, using a rag to protect the tank or other components **(see illustration)**.
10 Remove the tabbed lockwasher, noting how it fits, then unscrew and remove the locknut, using either a C-spanner, a peg spanner or a drift located in one of the notches if required – though it should only be finger-tight **(see illustrations)**. Remove the rubber washer **(see illustration)**.
11 To adjust the bearings as specified by Yamaha, a special service tool (Pt. No. 90890-01403 for Europe, or YU-33975 for USA) and a torque wrench are required. If the tool is available, first slacken the adjuster nut, then tighten it to the initial torque setting specified at the beginning of the Chapter, making sure the torque wrench handle is at right-angles

(90°) to the line between the adjuster nut and the wrench socket in the special tool **(see illustration)**. Now slacken the nut so that it is loose, then tighten it to the final torque setting specified. Check that the steering is still able to move freely from side to side, but that all freeplay is eliminated.
12 If the Yamaha tool is not available, using either a C-spanner, a peg spanner or a drift located in one of the notches, slacken the adjuster nut slightly until pressure is just released, then tighten it until all freeplay is removed, then tighten it a little more **(see illustration)**. This pre-loads the bearings. Now slacken the nut, then tighten it again, setting it so that all freeplay is just removed, yet the steering is able to move freely from side to side. To do this tighten the nut only a little at a time, and after each tightening repeat the checks outlined above (Steps 2 to 4) until the bearings are correctly set. The

object is to set the adjuster nut so that the bearings are under a very light loading, just enough to remove any freeplay.
Caution: Take great care not to apply excessive pressure because this will cause premature failure of the bearings.
13 With the bearings correctly adjusted, install the rubber washer and the locknut **(see illustrations 20.10d and b)**. Tighten the locknut finger-tight, then tighten it further until its notches align with those in the adjuster nut. If necessary, counter-hold the adjuster nut and tighten the locknut using a C-spanner or drift until the notches align, but make sure the adjuster nut does not turn as well. Install the tabbed lockwasher so that the tabs fit into the notches in both the locknut and adjuster nut **(see illustration 20.10a)**.
14 Fit the top yoke onto the steering stem **(see illustration 20.9)**, then install the washer and steering stem nut and tighten it to the torque setting specified at the beginning of the Chapter **(see illustration 20.8)**. Now tighten both the fork clamp bolts to the specified torque setting **(see illustration)**.
15 Install the handlebar positioning bolts and tighten them to the specified torque setting, then tighten the handlebar clamp bolts to the specified torque **(see illustrations)**. Fit the blanking caps into the positioning bolts **(see illustration 20.7a)**.
16 Mount the front brake master cylinder on the top yoke and tighten the bolt securely **(see illustration 20.6)**.
17 Re-check the bearing adjustment as described above and re-adjust if necessary.

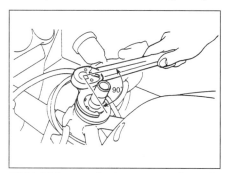

20.11 Make sure the torque wrench arm is at right angles (90°) to the tool

20.12 If the tool is not available, adjust the bearings as described

Maintenance procedures

20.14 Tighten the fork clamp bolts to the specified torque

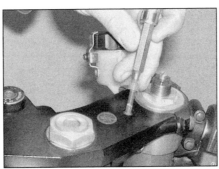

20.15a Align the handlebars, then install the positioning bolts . . .

20.15b . . . and tighten them and the handlebar clamp bolts to the specified torque

Every 8000 miles (12,000 km) – 1998/9 models
Every 12,000 miles (20,000 km) – 2000-on models

Carry out all the items under the 4000/6000 mile (6000/10,000 km) check, plus the following

21 Engine/transmission – oil and filter change

⚠️ **Warning: Be careful when draining the oil, as the exhaust pipes, the engine, and the oil itself can cause severe burns.**

1 Drain the engine oil as described in Section 7, Steps 1 to 5.

21.2 Unscrew the filter and drain it into the tray

2 Now place the drain tray below the oil filter, which is at the front of the engine. Unscrew the oil filter using a filter adaptor tool, or a strap wrench and tip any residual oil into the drain tray **(see illustration)**. Wipe any oil off the exhaust pipes to prevent too much smoke when you start it.

3 Smear clean engine oil onto the rubber seal on the new filter, then manoeuvre it into position and screw it onto the engine until the seal just seats **(see illustrations)**. If a filter adaptor tool is available, tighten the filter to the torque setting specified at the beginning of the Chapter **(see illustration)**. Otherwise, tighten the filter as tight as possible by hand, or by the number of turns specified on the filter or its packaging. **Note:** *Do not use a strap or chain wrench to tighten the filter as you will damage it.*

4 Refill the engine to the proper level as described in Section 7, Steps 6 to 8.

5 Note the advice in Section 7 for disposing of used oil. Remember to drain all the old oil from the filter (you can punch a hole in the filter to ensure it drains fully) into the drain pan. Note that the old filter should be taken to the oil disposal facility rather than disposed of with the household rubbish.

22 Cooling system – draining, flushing and refilling

⚠️ **Warning: Allow the engine to cool completely before performing this maintenance operation. Also, don't allow antifreeze to come into contact with your skin or the painted surfaces of the motorcycle. Rinse off spills immediately with plenty of water. Antifreeze is highly toxic if ingested. Never leave antifreeze lying around in an open container or in puddles on the floor; children and pets are attracted by its sweet smell and may drink it. Check with local authorities (councils) about disposing of antifreeze. Many communities have collection centres where antifreeze can be disposed of safely. Antifreeze is also combustible, so don't store it near open flames.**

Draining

1 Secure the motorcycle upright using an auxiliary stand. Remove the lower fairing and the right-hand fairing side panel (see Chapter 8). Remove the pressure cap from the

21.3a Smear some clean oil onto the seal . . .

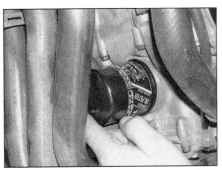

21.3b . . . then thread the filter onto the engine . . .

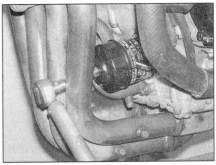

21.3c . . . and tighten it as described

1•22 Maintenance procedures

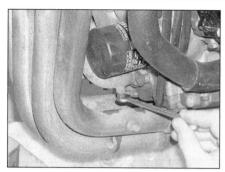

22.2a Unscrew the drain plug . . .

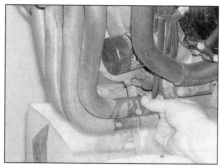

22.2b . . . and allow the coolant to drain

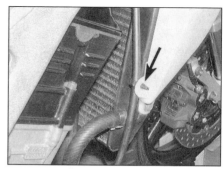

22.3a Release the clutch cable holder from the reservoir (arrowed)

22.3b Unscrew the mounting bolts (arrowed) . . .

22.3c . . . then remove the cap and drain the reservoir

22.12 Fill the system using the correct mixture as described

radiator filler neck by turning it anti-clockwise until it reaches a stop **(see illustration 16.7)**. If you hear a hissing sound (indicating there is still pressure in the system), wait until it stops. Now press down on the cap and continue turning the cap until it can be removed. Also remove the filler cap from the coolant reservoir.

2 Position a suitable container beneath the drain plug on the coolant pipe coming out of the front of the engine. Remove the drain plug and allow the coolant to completely drain from the system **(see illustrations)**. Retain the old sealing washer for use during flushing.

3 Position the container beneath the coolant reservoir. Release the clutch cable holder from the reservoir **(see illustration)**. Unscrew the reservoir mounting bolts, then remove the reservoir cap and tip the coolant out of the reservoir into the container, tipping it away from the reservoir breather hose so that none flows down it **(see illustrations)**. Remount the reservoir and fit the cable back into its holder. Fit the cap.

Flushing

4 Flush the system with clean tap water by inserting a garden hose in the radiator or filler neck. Allow the water to run through the system until it is clear and flows cleanly out of the drain hole. If the radiator is extremely corroded, remove it (see Chapter 3) and have it cleaned professionally.

5 Clean the drain hole, then install the drain plug using the old sealing washer.

6 Fill the cooling system with clean water mixed with a flushing compound. Make sure the flushing compound is compatible with aluminium components, and follow the manufacturer's instructions carefully. Install the pressure cap.

7 Start the engine and allow it to reach normal operating temperature. Let it run for about ten minutes.

8 Stop the engine. Let it cool for a while, then cover the pressure cap with a heavy rag and turn it anti-clockwise to the first stop, releasing any pressure that may be present in the system. Once the hissing stops, push down on the cap and remove it completely.

9 Drain the system once again.

10 Fill the system with clean water and repeat the procedure in Steps 8 to 10.

Refilling

11 Fit a new sealing washer onto the drain plug and tighten it to the torque setting specified at the beginning of the Chapter (the old washer can be reused if it is not damaged or deformed, though it is always better to use a new one) **(see illustration 22.2a)**.

12 Fill the system via the radiator filler neck with the proper coolant mixture (see this Chapter's Specifications) **(see illustration)**. *Note: Pour the coolant in slowly to minimise the amount of air entering the system.* When the system appears full, move the bike off its stand and shake it slightly to dissipate the coolant, then place the bike back on the auxiliary stand and top the system up.

13 When the system is full (all the way up to the top of the radiator filler neck), install the pressure cap **(see illustration 16.7)**. Now fill the coolant reservoir to the UPPER level mark (see *Daily (pre-ride) checks*).

14 Start the engine and allow it to idle for 2 to 3 minutes. Flick the throttle twistgrip part open 3 or 4 times, so that the engine speed rises to approximately 4000 – 5000 rpm, then stop the engine. Any air trapped in the system should have bled back to the top of the radiator via the small-bore air bleed hoses.

15 Let the engine cool, then remove the pressure cap as described in Step 1. Check that the coolant level is still up to the top of the radiator or filler neck. If it is low, add the specified mixture until it reaches the top. Refit the pressure cap.

16 Check the coolant level in the reservoir and top up if necessary.

17 Check the system for leaks.

18 Do not dispose of the old coolant by pouring it down the drain. Instead pour it into a heavy plastic container, cap it tightly and take it into an authorised disposal site or service station – see **Warning** at the beginning of this Section.

19 Install the fairing panels (see Chapter 8).

Maintenance procedures 1•23

Every 16,000 miles (24,000 km) or two years

23 Swingarm and suspension linkage bearings – re-greasing

1 Over a period of time the grease will harden and dirt will penetrate the bearings.
2 The rear suspension components are not equipped with grease nipples. Remove the swingarm and the suspension linkage as described in Chapter 6 for greasing of the bearings.

24 Steering head bearings – re-greasing

1 Over a period of time the grease will harden or may be washed out of the bearings by incorrect use of jet washes.
2 Disassemble the steering head for re-greasing of the bearings. Refer to Chapter 6 for details.

Every two years

25 Brake master cylinder and caliper seals – renewal

1 The seals will deteriorate over a period of time and lose their effectiveness, leading to sticky operation or fluid loss, or the ingress of air and dirt. Refer to Chapter 7 and disassemble the components for seal renewal every two years.

26 Brakes – fluid change

1 The brake fluid should be changed every two years or whenever a master cylinder or caliper overhaul is carried out. Refer to the brake bleeding section in Chapter 7, noting that all old fluid must be pumped from the fluid reservoir and hydraulic line before filling with new fluid.

> **HAYNES HINT**: Old brake fluid is invariably much darker in colour than new fluid, making it easy to see when all old fluid has been expelled from the system.

Every 20,000 miles (32,000 km)

27 In-line fuel filter – renewal

1 Refer to Section 6 for the filter renewal procedure.

Every 24,000 miles (40,000 km)

28 Valve clearances – check and adjustment

1 The engine must be completely cool for this maintenance procedure, so let the machine sit overnight before beginning.
2 Remove the lower fairing and the fairing side panels (see Chapter 8).
3 Remove the valve cover (see Chapter 2). Each cylinder is referred to by a number. They are numbered 1 to 4 from left to right, viewed as normally seated on the bike.
4 Make a chart or sketch of all valve positions so that a note of each clearance can be made against the relevant valve.
5 Unscrew the bolts securing the timing rotor/pick-up coil cover on the right-hand side of the engine and remove the cover, noting the wiring clamp and clutch cable holder (see illustration). Discard the gasket, as a new one must be used. Remove the dowels from either the crankcase or the cover if they are loose. Turn the engine using the timing rotor bolt, turning it in a clockwise direction only. Alternatively, place the motorcycle on an auxiliary stand so that the rear wheel is off the ground, select a high gear and rotate the rear wheel by hand in its normal direction of rotation.
6 Turn the engine until the 'T' mark faces back and aligns with the crankcase mating surfaces (see illustration), and the camshaft lobes for the No. 1 (left-hand) cylinder face away from each other, and the dot on each camshaft faces up and aligns with the arrow on the camshaft holder (see illustrations). If the cam lobes are facing towards each other and the dots are facing down, rotate the engine clockwise 360° (one full turn) so that the 'T' mark again aligns with the crankcase mating surfaces. The camshaft lobes will now

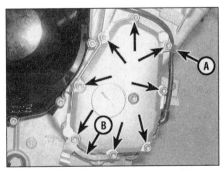

28.5 Unscrew the bolts (arrowed) and remove the cover, noting the wiring clamp (A) and cable holder (B)

28.6a Turn the engine until the 'T' mark aligns with the crankcase mating surfaces . . .

Maintenance procedures

28.6b ... and the camshaft lobes for No. 1 cylinder face away from each other as shown

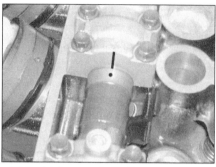

28.6c Intake camshaft alignment marks

28.6d Exhaust camshaft alignment marks

be facing away from each other and the dots will be facing up and aligned with the arrows, meaning that No. 1 cylinder is at TDC (Top Dead Centre) on the compression stroke.

7 With No. 1 cylinder at TDC on the compression stroke, check the clearances on the No. 1 cylinder intake and exhaust valves **(see illustration)**. Insert a feeler gauge of the same thickness as the correct valve clearance (see Specifications) between the camshaft lobe and follower of each valve and check that it is a firm sliding fit – you should feel a slight drag when the you pull the gauge out **(see illustration)**. If not, use the feeler gauges to obtain the exact clearance. Record the measured clearance on the chart.

8 Now turn the engine clockwise 180° (half a turn) so that the 'T' mark faces forward and

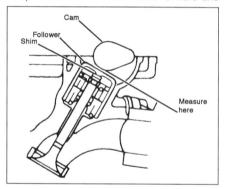

28.7a Make sure the cam lobes are not depressing the valves and are facing away from each other, then insert the feeler gauge between the cam and the follower

aligns with the crankcase mating surfaces and the camshaft lobes for the No. 2 cylinder are facing away from each other. The No. 2 cylinder is now at TDC on the compression stroke. Measure the clearances of the No. 2 cylinder valves using the method described in Step 7.

9 Now turn the engine clockwise 180° (half a turn) so that the 'T' mark faces back and aligns with the crankcase mating surfaces and the camshaft lobes for the No. 4 cylinder are facing away from each other. The No. 4 cylinder is now at TDC on the compression stroke. Measure the clearances of the No. 4 cylinder valves using the method described in Step 7.

10 Now turn the engine clockwise 180° (half a turn) so that the 'T' mark faces forward and aligns with the crankcase mating surfaces and the camshaft lobes for the No. 3 cylinder are facing away from each other. The No. 3 cylinder is now at TDC on the compression stroke. Measure the clearances of the No. 3

28.7b Measuring the clearance using a feeler gauge

cylinder valves using the method described in Step 7.

11 When all clearances have been measured and charted, identify whether the clearance on any valve falls outside that specified. If it does, the shim between the cam follower and the valve must be replaced with one of a thickness which will restore the correct clearance.

12 Shim replacement requires removal of the camshafts (see Chapter 2). There is no need to remove both camshafts if shims from only one side of the engine need replacing. Place rags over the spark plug holes and the cam chain tunnel to prevent a shim from dropping into the engine on removal.

13 With the camshaft removed, remove the cam follower of the valve in question, then retrieve the shim from inside the follower **(see illustrations)**. If it is not in the follower, pick it out of the top of the valve using either a magnet, a small screwdriver with a dab of grease on it (the shim will stick to the grease), or a screwdriver and a pair of pliers **(see illustration 27.16a)**. Do not allow the shim to fall into the engine.

14 A size mark should be stamped on the upper face of the shim – a shim marked 175 is 1.75 mm thick. If the mark is not visible, the shim thickness will have to be measured. It is recommended that the shim is measured, anyway, to check that it has not worn **(see illustration)**.

15 Using the appropriate shim selection chart, find where the measured valve clearance and existing shim thickness values intersect and read off the shim size required **(see illustrations)**. *Note: If the existing shim*

28.13a Lift out the follower ...

28.13b ... and remove the shim from inside the follower

28.14 Measure the shim using a micrometer

28.15a Shim selection chart – Intake camshaft

MEASURED CLEARANCE	INSTALLED PAD NUMBER																								
	120	125	130	135	140	145	150	155	160	165	170	175	180	185	190	195	200	205	210	215	220	225	230	235	240
0.00 ~ 0.02				120	125	130	135	140	145	150	155	160	165	170	175	180	185	190	195	200	205	210	215	220	225
0.03 ~ 0.07			120	125	130	135	140	145	150	155	160	165	170	175	180	185	190	195	200	205	210	215	220	225	230
0.08 ~ 0.10		120	125	130	135	140	145	150	155	160	165	170	175	180	185	190	195	200	205	210	215	220	225	230	235
0.11 ~ 0.20	RECOMMENDED CLEARANCE																							
0.21 ~ 0.22	125	130	135	140	145	150	155	160	165	170	175	180	185	190	195	200	205	210	215	220	225	230	235	240	
0.23 ~ 0.27	130	135	140	145	150	155	160	165	170	175	180	185	190	195	200	205	210	215	220	225	230	235	240		
0.28 ~ 0.32	135	140	145	150	155	160	165	170	175	180	185	190	195	200	205	210	215	220	225	230	235	240			
0.33 ~ 0.37	140	145	150	155	160	165	170	175	180	185	190	195	200	205	210	215	220	225	230	235	240				
0.38 ~ 0.42	145	150	155	160	165	170	175	180	185	190	195	200	205	210	215	220	225	230	235	240					
0.43 ~ 0.47	150	155	160	165	170	175	180	185	190	195	200	205	210	215	220	225	230	235	240						
0.48 ~ 0.52	155	160	165	170	175	180	185	190	195	200	205	210	215	220	225	230	235	240							
0.53 ~ 0.57	160	165	170	175	180	185	190	195	200	205	210	215	220	225	230	235	240								
0.58 ~ 0.62	165	170	175	180	185	190	195	200	205	210	215	220	225	230	235	240									
0.63 ~ 0.67	170	175	180	185	190	195	200	205	210	215	220	225	230	235	240										
0.68 ~ 0.72	175	180	185	190	195	200	205	210	215	220	225	230	235	240											
0.73 ~ 0.77	180	185	190	195	200	205	210	215	220	225	230	235	240												
0.78 ~ 0.82	185	190	195	200	205	210	215	220	225	230	235	240													
0.83 ~ 0.87	190	195	200	205	210	215	220	225	230	235	240														
0.88 ~ 0.92	195	200	205	210	215	220	225	230	235	240															
0.93 ~ 0.97	200	205	210	215	220	225	230	235	240																
0.98 ~ 1.02	205	210	215	220	225	230	235	240																	
1.03 ~ 1.07	210	215	220	225	230	235	240																		
1.08 ~ 1.12	215	220	225	230	235	240																			
1.13 ~ 1.17	220	225	230	235	240																				
1.18 ~ 1.22	225	230	235	240																					
1.23 ~ 1.27	230	235	240																						
1.28 ~ 1.32	235	240																							
1.33 ~ 1.37	240																								

28.15b Shim selection chart – Exhaust camshaft

Note revised clearance for 2000-on Europe models in Specifications section

MEASURED CLEARANCE	INSTALLED PAD NUMBER																								
	120	125	130	135	140	145	150	155	160	165	170	175	180	185	190	195	200	205	210	215	220	225	230	235	240
0.00 ~ 0.02				120	125	130	135	140	145	150	155	160	165	170	175	180	185	190	195	200	205	210	215		
0.03 ~ 0.07			120	125	130	135	140	145	150	155	160	165	170	175	180	185	190	195	200	205	210	215	220		
0.08 ~ 0.12		120	125	130	135	140	145	150	155	160	165	170	175	180	185	190	195	200	205	210	215	220	225		
0.13 ~ 0.17	120	125	130	135	140	145	150	155	160	165	170	175	180	185	190	195	200	205	210	215	220	225	230		
0.18 ~ 0.20	120	125	130	135	140	145	150	155	160	165	170	175	180	185	190	195	200	205	210	215	220	225	230	235	
0.21 ~ 0.30	RECOMMENDED CLEARANCE																							
0.31 ~ 0.32	125	130	135	140	145	150	155	160	165	170	175	180	185	190	195	200	205	210	215	220	225	230	235	240	
0.33 ~ 0.37	130	135	140	145	150	155	160	165	170	175	180	185	190	195	200	205	210	215	220	225	230	235	240		
0.38 ~ 0.42	135	140	145	150	155	160	165	170	175	180	185	190	195	200	205	210	215	220	225	230	235	240			
0.43 ~ 0.47	140	145	150	155	160	165	170	175	180	185	190	195	200	205	210	215	220	225	230	235	240				
0.48 ~ 0.52	145	150	155	160	165	170	175	180	185	190	195	200	205	210	215	220	225	230	235	240					
0.53 ~ 0.57	150	155	160	165	170	175	180	185	190	195	200	205	210	215	220	225	230	235	240						
0.58 ~ 0.62	155	160	165	170	175	180	185	190	195	200	205	210	215	220	225	230	235	240							
0.63 ~ 0.67	160	165	170	175	180	185	190	195	200	205	210	215	220	225	230	235	240								
0.68 ~ 0.72	165	170	175	180	185	190	195	200	205	210	215	220	225	230	235	240									
0.73 ~ 0.77	170	175	180	185	190	195	200	205	210	215	220	225	230	235	240										
0.78 ~ 0.82	175	180	185	190	195	200	205	210	215	220	225	230	235	240											
0.83 ~ 0.87	180	185	190	195	200	205	210	215	220	225	230	235	240												
0.88 ~ 0.92	185	190	195	200	205	210	215	220	225	230	235	240													
0.93 ~ 0.97	190	195	200	205	210	215	220	225	230	235	240														
0.98 ~ 1.02	195	200	205	210	215	220	225	230	235	240															
1.03 ~ 1.07	200	205	210	215	220	225	230	235	240																
1.08 ~ 1.12	205	210	215	220	225	230	235	240																	
1.13 ~ 1.17	210	215	220	225	230	235	240																		
1.18 ~ 1.22	215	220	225	230	235	240																			
1.23 ~ 1.27	220	225	230	235	240																				
1.28 ~ 1.32	225	230	235	240																					
1.33 ~ 1.37	230	235	240																						
1.38 ~ 1.42	235	240																							
1.43 ~ 1.47	240																								

Maintenance procedures

28.16a Fit the shim into the recess in the top of the valve . . .

28.16b . . . then install the follower

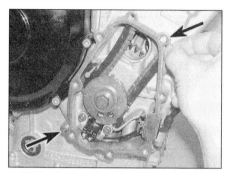

28.18a Locate the new gasket onto the dowels (arrowed)

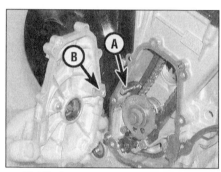

28.18b Locate the pivot pin (A) in the hole (B)

is marked with a number not ending in 0 or 5, round it up or down as appropriate to the nearest number ending in 0 or 5, so that the chart can be used. Shims are available in 0.05 mm increments from 1.20 mm to 2.40 mm. **Note:** *If the required replacement shim is greater than 2.40 mm (the largest available), the valve is probably not seating correctly due to a build-up of carbon deposits and should be checked and cleaned or resurfaced as required (see Chapter 2).*

16 Obtain the replacement shim, then lubricate it with molybdenum disulphide grease and fit it into its recess in the top of the valve, with the size marking on each shim facing up **(see illustration)**. Check that the shim is correctly seated, then lubricate the follower with molybdenum disulphide oil (a 50/50 mixture of molybdenum disulphide grease and engine oil) and install it onto the valve **(see illustration)**. Repeat the process for any other valves until the clearances are correct, then install the camshafts (see Chapter 2).

17 Rotate the crankshaft several turns to seat the new shim(s), then check the clearances again.

18 Install all disturbed components in a reverse of the removal sequence. Install the timing rotor/pick-up coil cover using a new gasket, and do not forget the dowels if removed **(see illustration)**. Locate the cam chain tensioner blade pivot pin in the hole in the cover **(see illustration)**. Tighten the cover bolts to the torque setting specified at the beginning of the Chapter, not forgetting the wiring clamp and the cable guide **(see illustration 27.5)**.

Every four years

29 Brake hoses – renewal

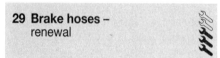

1 The hoses will deteriorate with age and should be replaced with new ones every four years regardless of their apparent condition.

2 Refer to Chapter 7 and disconnect the hoses from the master cylinders and calipers. Always replace the banjo union sealing washers with new ones.

Non-scheduled maintenance

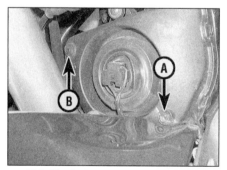

30.2 Vertical adjuster (A), horizontal adjuster (B)

30 Headlight aim – check and adjustment

Note: *An improperly adjusted headlight may cause problems for oncoming traffic or provide poor, unsafe illumination of the road ahead. Before adjusting the headlight aim, be sure to consult with local traffic laws and regulations – for UK models refer to MOT Test Checks in the Reference section.*

1 The headlight beam can be adjusted both horizontally and vertically. Before making any adjustment, check that the tyre pressures are correct and the suspension is adjusted as required. Make any adjustments to the headlight aim with the machine on level ground, with the fuel tank half full and with an assistant sitting on the seat. If the bike is usually ridden with a passenger on the back, have a second assistant to do this.

2 Vertical adjustment is made by turning the adjuster screw on the bottom outer corner of each headlight unit **(see illustration)**. Turn it anti-clockwise to raise the beam, and clockwise to lower it.

3 Horizontal adjustment is made by turning the adjuster knob on the top inner corner of each headlight unit **(see illustration 30.2)**. For the left-hand beam, turn it anti-clockwise to move the beam to the right, and clockwise to move it to the left. For the right-hand beam, turn it clockwise to move the beam to the right, and anti-clockwise to move it to the left.

Maintenance procedures

31 Cylinder compression – check

1 Among other things, poor engine performance may be caused by leaking valves, incorrect valve clearances, a leaking head gasket, or worn pistons, rings and/or cylinder walls. A cylinder compression check will help pinpoint these conditions and can also indicate the presence of excessive carbon deposits in the cylinder heads.
2 The only tools required are a compression gauge and a spark plug wrench. A compression gauge with a threaded end for the spark plug hole is preferable to the type which requires hand pressure to maintain a tight seal. Depending on the outcome of the initial test, a squirt-type oil can may also be needed.
3 Make sure the valve clearances are correctly set (see Section 28) and that the cylinder head nuts are tightened to the correct torque setting (see Chapter 2).
4 Refer to *Fault Finding Equipment* in the *Reference* section for details of the compression test. Refer to the specifications at the beginning of this Chapter for compression figures.

32 Engine oil pressure – check

1 None of the models covered in this manual are fitted with an oil pressure switch and warning light, only an oil level sensor and light (see Chapter 9 for further information). If a lubrication problem is suspected, first check the oil level (see *Daily (pre-ride) checks*).
2 If the oil level is correct, an oil pressure check must be carried out next. The check provides useful information about the condition of the engine's lubrication system.
3 To check the oil pressure, slacken the oil gallery bolt in the front of the left-hand side of the cylinder head – there is no need to remove it **(see illustration)**. Start the engine and allow it to idle.

4 After a short while oil should begin to seep out from the oil gallery plug. If no oil has appeared after one minute, stop the engine immediately. If the oil does not appear after one minute, either the pressure regulator is stuck open, the oil pump is faulty, the oil strainer or filter is blocked, or there is other engine damage.
5 Begin diagnosis by checking the oil filter, strainer and regulator, then the oil pump (see Chapter 2). If no problems are found with those items, it may well be that the bearing oil clearances are excessive and the engine needs to be overhauled.
6 If the oil appears very quickly and spurts out, the pressure may be too high, meaning either an oil passage is clogged, the regulator is stuck closed or the wrong grade of oil is being used.
7 Tighten the oil gallery bolt to the torque setting specified at the beginning of the Chapter.
8 Refer to Chapter 2 and rectify any problems before running the engine again.
9 If the oil pressure and oil level are both good, then the oil level sensor or its warning light may be faulty. Check them and the circuit and replace with a new one if necessary (see Chapter 9).

33 Fuel hoses – renewal

> **Warning: Petrol (gasoline) is extremely flammable, so take extra precautions when you work on any part of the fuel system. Don't smoke or allow open flames or bare light bulbs near the work area, and don't work in a garage where a natural gas-type appliance is present. If you spill any fuel on your skin, rinse it off immediately with soap and water. When you perform any kind of work on the fuel system, wear safety glasses and have a fire extinguisher suitable for a Class B type fire (flammable liquids) on hand.**

1 The fuel delivery hoses should be renewed

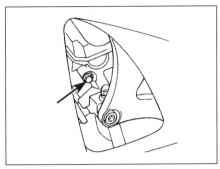

32.3 Oil gallery bolt (arrowed)

after a few years, regardless of their condition.
2 Remove the fuel tank (see Chapter 4). Disconnect the fuel hoses from the fuel tap, filter, fuel pump and from the carburettors, noting the routing of each hose and where it connects (see Chapter 4 if required). It is advisable to make a sketch of the various hoses before removing them, to ensure they are correctly installed.
3 Secure each new hose to its unions using new clamps. Run the engine and check that there are no leaks before taking the machine out on the road.

34 Front forks – oil change

1 Fork oil degrades over a period of time and loses its damping qualities. Refer to the following relevant sections of Chapter 6 for the procedure. Note that the forks do not need to be completely disassembled.
2 Remove the forks from the yokes as described in Chapter 6, Section 6. Follow the procedure in Section 7 of Chapter 6 (steps 1, 3 to 8) to remove the fork top bolt, spacer, spring seat, spring and damper adjuster rod, then invert the fork leg over a suitable container and pump the fork and damper cartridge to expel as much oil as possible.
3 Refill the fork with new oil as described in steps 23 to 28 of the same section. Refit the forks as described in Section 6 of Chapter 6.

Notes

Chapter 2
Engine, clutch and transmission

Contents

Alternator – check, removal and installation ....see Chapter 9	Oil and filter – change ....see Chapter 1
Cam chain, tensioner, tensioner blade and guides – removal, inspection and installation 9	Oil cooler – removal and installation 7
Camshafts and followers – removal, inspection and installation 10	Oil level – check ....see Daily (pre-ride) checks
Clutch – check ....see Chapter 1	Oil pressure – check ....see Chapter 1
Clutch – removal, inspection and installation 14	Oil pump – removal, inspection and installation 19
Clutch cable – removal and installation 15	Oil sump, oil strainer and pressure relief valve – removal, inspection and installation 18
Connecting rods – removal, inspection and installation 24	Operations possible with the engine in the frame 2
Crankcase halves – separation and reassembly 20	Operations requiring engine removal 3
Crankcase halves and cylinder bores – inspection and servicing ... 21	Pick-up coil assembly – removal and installation ......see Chapter 5
Crankshaft and main bearings – removal, inspection and installation 23	Piston rings – inspection and installation 26
Cylinder head – removal and installation 11	Pistons – removal, inspection and installation 25
Cylinder head and valves – disassembly, inspection and reassembly 13	Recommended running-in procedure 31
Engine – cylinder compression check ....see Chapter 1	Regulator/rectifier – check, removal and installation ....see Chapter 9
Engine – removal and installation 5	Selector drum and forks – removal, inspection and installation 29
Engine disassembly and reassembly – general information 6	Spark plug gap – check and adjustment ....see Chapter 1
Gearchange mechanism – removal, inspection and installation 17	Starter clutch and idle/reduction gear – check, removal, inspection and installation 16
General information 1	Starter motor – removal and installation ....see Chapter 9
Idle speed – check and adjustment ....see Chapter 1	Transmission shafts – disassembly, inspection and reassembly ... 28
Initial start-up after overhaul 30	Transmission shafts – removal and installation 27
Main and connecting rod bearings – general information 22	Valve clearances – check and adjustment ....see Chapter 1
Major engine repair – general information 4	Valve cover – removal and installation 8
Neutral switch – check, removal and installation ....see Chapter 9	Valves/valve seats/valve guides – servicing 12

Degrees of difficulty

| Easy, suitable for novice with little experience | Fairly easy, suitable for beginner with some experience | Fairly difficult, suitable for competent DIY mechanic | Difficult, suitable for experienced DIY mechanic | Very difficult, suitable for expert DIY or professional |

Specifications

General

Type	Four-stroke in-line four
Capacity	998 cc
Bore	74.0 mm
Stroke	58.0 mm
Compression ratio	11.8 to 1
Cylinder numbering	1 to 4 from left to right
Cooling system	Liquid cooled
Clutch	Wet multi-plate
Transmission	Six-speed constant mesh
Final drive	Chain

Camshafts

Intake lobe height
  Standard .... 32.50 to 32.60 mm
  Service limit (min) .... 32.40 mm
Exhaust lobe height
  Standard .... 32.95 to 33.05 mm
  Service limit (min) .... 32.85 mm
Journal diameter
  1998 and 1999 models .... 24.437 to 24.450 mm
  2000-on models .... 24.459 to 24.472 mm
Holder diameter .... 24.500 to 24.521 mm
Journal oil clearance
  1998 and 1999 models .... 0.050 to 0.084 mm
  2000-on models .... 0.028 to 0.062 mm
Runout (max) .... 0.03 mm

Cylinder head

Warpage (max) .... 0.10 mm

Valves, guides and springs

Valve clearances .... see Chapter 1
Intake valve
  Stem diameter
    Standard .... 3.975 to 3.900 mm
    Service limit (min) .... 3.945 mm
  Guide bore diameter
    Standard .... 4.000 to 4.012 mm
    Service limit (max) .... 4.050 mm
  Stem-to-guide clearance
    Standard .... 0.010 to 0.037 mm
    Service limit (max) .... 0.08 mm
  Head diameter .... 22.9 to 23.1 mm
  Face width .... 1.76 to 2.90 mm
  Seat width .... 0.9 to 1.1 mm
  Margin thickness .... 0.5 to 0.9 mm
  Valve lift .... 7.45 to 7.65 mm
Exhaust valve
  Stem diameter
    1998 and 1999 models
      Standard .... 4.460 to 4.475 mm
      Service limit (min) .... 4.430 mm
    2000-on models
      Standard .... 4.465 to 4.480 mm
      Service limit (min) .... 4.430 mm
  Guide bore diameter
    Standard .... 4.500 to 4.512 mm
    Service limit (max) .... 4.550 mm
  Stem-to-guide clearance
    1998 and 1999 models
      Standard .... 0.025 to 0.052 mm
      Service limit (max) .... 0.10 mm
    2000-on models
      Standard .... 0.020 to 0.047 mm
      Service limit (max) .... 0.10 mm
  Head diameter .... 24.4 to 24.6 mm
  Face width .... 1.76 to 2.90 mm
  Seat width .... 0.9 to 1.1 mm
  Margin thickness .... 0.5 to 0.9 mm
  Valve lift .... 7.75 to 7.95 mm
Valve stem runout (max) .... 0.01 mm
Valve springs free length
  Intake .... 38.90 mm
  Exhaust .... 40.67 mm
Valve spring bend (max)
  Intake .... 1.7 mm
  Exhaust .... 1.8 mm

Clutch
Friction plates
  Quantity
    1998 models 7
    1999-on models 8
  Thickness
    Standard .... 2.9 to 3.1 mm
    Service limit (min) .... 2.8 mm
Plain plates
  Quantity
    1998 models 6
    1999-on models 7
  Thickness .... 1.9 to 2.1 mm
  Warpage (max) .... 0.1 mm

Lubrication system
Oil pressure .... see Chapter 1
Relief valve opening pressure .... 69.7 to 81.1 psi (4.8 to 5.6 bars)
Oil pump
  Inner rotor tip-to-outer rotor clearance .... 0.09 to 0.15 mm
  Outer rotor-to-body clearance .... 0.03 to 0.08 mm

Cylinder bores
Bore .... 74.000 to 74.010 mm
Ovality (out-of-round) (max) .... 0.05 mm
Taper (max) .... 0.05 mm
Cylinder compression .... see Chapter 1

Crankshaft and bearings
Main bearing oil clearance .... 0.029 to 0.053 mm
Runout (max) .... 0.03 mm

Connecting rods
Big-end side clearance .... 0.160 to 0.262 mm
Big-end oil clearance .... 0.032 to 0.056 mm

Pistons
Piston diameter (measured 5.0 mm up from skirt, at 90° to
  piston pin axis) .... 73.955 to 73.970 mm
Piston-to-bore clearance
  Standard .... 0.030 to 0.055 mm
  Service limit (max) .... 0.12 mm
Piston pin diameter .... 16.991 to 17.000 mm
Piston pin bore diameter in piston .... 17.002 to 17.013 mm
Piston pin-to-piston pin bore clearance
  Standard .... 0.002 to 0.022 mm
  Service limit .... 0.072 mm

Piston rings
Top ring
  Type .... Barrel
  Ring width .... 2.75 mm
  Ring thickness .... 0.9 mm
  Ring end gap (installed) .... 0.19 to 0.31 mm
  Piston ring-to-groove clearance .... 0.030 to 0.065 mm
2nd ring
  Type .... Taper
  Ring width .... 2.8 mm
  Ring thickness .... 0.8 mm
  Ring end gap (installed) .... 0.30 to 0.45 mm
  Piston ring-to-groove clearance .... 0.020 to 0.055 mm
Oil ring
  Ring width .... 2.6 mm
  Ring thickness .... 1.5 mm
  Side-rail end gap (installed) .... 0.10 to 0.35 mm

Transmission

Gear ratios (no. of teeth)
  1998 and 1999 models
    Primary reduction .... 1.581 to 1 (68/43T)
    Final reduction .... 2.688 to 1 (43/16T)
    1st gear .... 2.600 to 1 (39/15T)
    2nd gear .... 1.842 to 1 (35/19T)
    3rd gear .... 1.500 to 1 (30/20T)
    4th gear .... 1.333 to 1 (28/21T)
    5th gear .... 1.200 to 1 (30/25T)
    6th gear .... 1.115 to 1 (29/26T)
  2000-on models
    Primary reduction .... 1.581 to 1 (68/43T)
    Final reduction .... 2.688 to 1 (43/16T)
    1st gear .... 2.500 to 1 (35/14T)
    2nd gear .... 1.842 to 1 (35/19T)
    3rd gear .... 1.500 to 1 (30/20T)
    4th gear .... 1.333 to 1 (28/21T)
    5th gear .... 1.200 to 1 (30/25T)
    6th gear .... 1.115 to 1 (29/26T)
Shaft runout (max) .... 0.08 mm

Selector drum and forks

Selector fork shaft runout (max) .... 0.1 mm

Torque wrench settings

Engine mounting bolts
  1998 and 1999 models
    Upper rear mounting bolt nut .... 55 Nm
    Lower rear mounting bolt nut .... 55 Nm
    Left-hand front mounting bolts .... 40 Nm
    Right-hand front mounting bolt .... 55 Nm
    Mounting bolt lug pinchbolts .... 24 Nm
  2000-on models
    Upper rear mounting bolt nut .... 55 Nm
    Lower rear mounting bolt nut .... 55 Nm
    Left-hand front mounting bolts .... 40 Nm
    Right-hand front mounting bolt .... 40 Nm
    Mounting bolt lug pinchbolts .... 24 Nm
Coolant pipe bolts .... 10 Nm
Oil cooler bolt .... 35 Nm
Valve cover bolts .... 12 Nm
Cam chain tensioner mounting bolts .... 10 Nm
Cam chain tensioner cap bolt .... 10 Nm
Camshaft holder bolts .... 10 Nm
Camshaft sprocket bolts .... 24 Nm
Timing rotor/pick-up coil cover bolts .... 12 Nm
Cylinder head nuts
  1998 and 1999 models .... 50 Nm
  2000-on models
    Domed nuts .... 65 Nm
    All other nuts .... 50 Nm
Cylinder head bolts .... 12 Nm
Oil pump drive chain guard .... 10 Nm
Clutch nut .... 70 Nm
Clutch diaphragm spring bolts .... 8 Nm
Clutch cover bolts .... 12 Nm
Starter clutch bolts .... 12 Nm
Gearchange shaft centralising spring locating pin .... 22 Nm
Oil strainer bolts .... 10 Nm
Oil sump bolts .... 10 Nm
Oil pump housing bolts .... 10 Nm
Oil/water pump assembly driven sprocket bolt .... 15 Nm
Oil/water pump assembly mounting bolts and sprocket cover bolts ... 10 Nm
Oil pipe bolts .... 10 Nm
Coolant outlet pipe bolt .... 10 Nm

Engine, clutch and transmission 2•5

Torque wrench settings (continued)

Crankcase 9 mm bolts	
1998 and 1999 models	32 Nm
2000-on models	15 Nm + 45 to 50°
Crankcase 8 mm bolts	24 Nm
Crankcase 6 mm bolts (Nos. 16 and 24)	14 Nm
Crankcase 6 mm bolts (all other bolts)	12 Nm
Front sprocket cover support plate bolt	10 Nm
Oil baffle plate bolts	10 Nm
Connecting rod cap nuts (if refitting the original connecting rods)	
Initial setting	15 Nm
Final setting	36 Nm
Connecting rod cap nuts (if fitting new connecting rods)	
Initial setting	20 Nm
Final setting	angle-tighten through 120°
Transmission input shaft bearing housing Torx screws	12 Nm
Selector drum retaining plate bolts	10 Nm

1 General information

The engine/transmission unit is a liquid-cooled in-line four, with five valves per cylinder. The valves are operated by double overhead camshafts which are chain driven off the right-hand end of the crankshaft. The engine/transmission assembly is constructed from aluminium alloy. The crankcase is divided horizontally.

The crankcase incorporates a wet sump, pressure-fed lubrication system which uses a chain-driven, dual-rotor oil pump, an oil filter, a relief valve and an oil level sensor. The pump is driven by a chain and sprockets, with the drive sprocket locating into the starter driven sprocket on the back of the clutch housing. The oil is cooled by a cooler which is fed off the engine cooling system.

The alternator is on the left-hand end of the crankshaft, and the starter clutch is on the back of the clutch housing.

Power from the crankshaft is routed to the transmission via the clutch. The clutch is of the wet, multi-plate type and is gear-driven off the crankshaft. The transmission is a six-speed constant-mesh unit. Final drive to the rear wheel is by chain and sprockets.

2 Operations possible with the engine in the frame

The components and assemblies listed below can be removed without having to remove the engine/transmission assembly from the frame. If however, a number of areas require attention at the same time, removal of the engine is recommended.

Valve cover
Camshafts
Cam chain
Clutch and starter clutch
Gearchange mechanism
Alternator
Starter motor
Pick-up coil assembly
Oil filter and cooler
Oil sump, oil strainer and oil pressure relief valve
Oil pump and water pump

3 Operations requiring engine removal

It is necessary to remove the engine/transmission assembly from the frame to gain access to the following components.

Cylinder head
Crankshaft and bearings
Connecting rods and bearings
Cylinder bores, pistons and piston rings
Transmission shafts
Selector drum and forks

4 Major engine repair – general information

1 It is not always easy to determine when, or if, an engine should be completely overhauled, as a number of factors must be considered.

2 High mileage is not necessarily an indication that an overhaul is needed, while low mileage, on the other hand, does not preclude the need for an overhaul. Frequency of servicing is probably the single most important consideration. An engine that has regular and frequent oil and filter changes, as well as other required maintenance, will most likely give many miles of reliable service. Conversely, a neglected engine, or one which has not been run-in properly, may require an overhaul very early in its life.

3 Exhaust smoke and excessive oil consumption are both indications that piston rings and/or valve guides are in need of attention, although make sure that the fault is not due to oil leakage.

4 If the engine is making obvious knocking or rumbling noises, the connecting rods and/or main bearings are probably at fault.

5 Loss of power, rough running, excessive valve train noise and high fuel consumption rates may also point to the need for an overhaul, especially if they are all present at the same time. If a complete tune-up does not remedy the situation, major mechanical work is the only solution.

6 An engine overhaul generally involves restoring the internal parts to the specifications of a new engine. During a major overhaul, the piston rings and main and connecting rod bearings are usually replaced with new ones and the cylinder walls honed. Generally, the valve seats are re-ground, since they are usually in less than perfect condition at this point. The end result should be a like new engine that will give as many trouble-free miles as the original.

7 Before beginning the engine overhaul, read through the related procedures to familiarise yourself with the scope and requirements of the job. Overhauling an engine is not all that difficult, but it is time-consuming. Plan on the motorcycle being tied up for a minimum of two weeks. Check on the availability of parts and make sure that any necessary special tools, equipment and supplies are obtained in advance.

8 Most work can be done with typical workshop hand tools, although a number of precision measuring tools are required for inspecting parts to determine if they must be renewed. Often a dealer will handle the inspection of parts and offer advice concerning reconditioning and replacement. As a general rule, time is the primary cost of an overhaul so it does not pay to install worn or substandard parts.

9 As a final note, to ensure maximum life and minimum trouble from a rebuilt engine, everything must be assembled with care in a spotlessly clean environment.

2•6 Engine, clutch and transmission

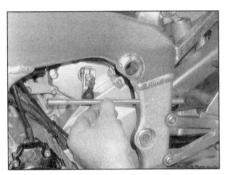

5.6a Unscrew the bolt and detach the earth lead . . .

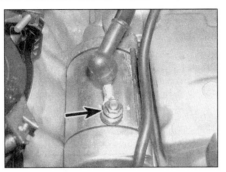

5.6b . . . then unscrew the nut (arrowed) and detach the power lead

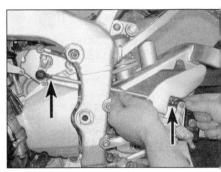

5.8a Slacken the locknuts (arrowed), then thread the rod off the lever and arm . . .

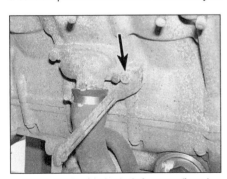

5.8b . . . and withdraw it from the frame

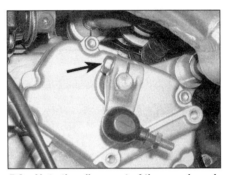

5.8c Note the alignment of the punchmark with the slit, then unscrew the bolt (arrowed) and slide the arm off the shaft

5 Engine – removal and installation

Caution: *The engine is very heavy. Engine removal and installation should be carried out with the aid of at least one assistant. Personal injury or damage could occur if the engine falls or is dropped. An hydraulic or mechanical floor jack should be used to support and lower or raise the engine, if possible.*

Removal

1 Support the motorcycle securely in an upright position using an auxiliary stand. Work can be made easier by raising the machine to a suitable working height on an hydraulic ramp or a suitable platform. Make sure the motorcycle is secure and will not topple over (see Section 1 of *Tools and Workshop Tips* in the *Reference* section). When disconnecting any wiring, cables and hoses, it is advisable to mark or tag them as a reminder to where they connect.

2 If the engine is dirty, particularly around its mountings, wash it thoroughly before starting any major dismantling work. This will make work much easier and rule out the possibility of caked-on lumps of dirt falling into some vital component.

3 Remove the seat, seat cowling, lower fairing, fairing side panels and fairing (see Chapter 8). **Note:** *The fairing can stay on the bike, though it is wise to remove it as it could get in the way and could be damaged – if it's not there, it can't do either!*

4 Remove the fuel tank (see Chapter 4).

5 Drain the engine oil and the coolant (see Chapter 1).

6 Disconnect the negative (–ve) lead from the battery, then disconnect the positive (+ve) lead (see Chapter 9). If required, remove the starter motor (see Chapter 9). If the starter motor is left in place, remove the rear mounting bolt and detach the earth lead, then remove the terminal nut and detach the power lead **(see illustrations)**. Secure the leads clear of the engine.

7 Remove the carburettors (see Chapter 4). Plug the intake manifolds with clean rag. Also remove the fuel pump and filter assembly (see Chapter 4).

8 Slacken the gearchange lever linkage rod locknuts, then unscrew the rod and separate it from the lever and the arm (the rod is reverse-threaded on one end, so will unscrew from both lever and arm simultaneously when turned in the one direction) **(see illustration)**. Note how far the rod is threaded into the lever and arm, as this determines the height of the lever relative to the footrest. Withdraw the rod from the frame **(see illustration)**. If required unscrew the linkage arm pinch bolt and slide the arm off the shaft, noting how the punchmark on the shaft aligns with the slit in the arm **(see illustration)**. If no mark is visible, make your own before removing the arm, so that it can be correctly aligned with the shaft on installation.

9 Remove the front sprocket (see Chapter 6).

10 Make a note or sketch of all the cooling system hoses between the radiator and the engine. Remove the radiator (see Chapter 3) but detach the hoses from their unions on the engine components rather than from the radiator itself, then remove the radiator along with the hoses, noting their routing, and releasing them from any ties. To prevent the possibility of damaging the front mudguard when removing the engine, unscrew the bolt securing the radiator mounting bracket to the front of the engine and remove it, noting how it fits **(see illustration)**. If required, also remove the thermostat housing along with the outlet pipes from the valve cover (see Chapter 3). If required, remove the oil cooler (see Section 7). If required, unscrew the bolt securing the water pump inlet pipe to the left-hand side of the engine and pull it out **(see illustrations)**. Also if required, slacken the clamps securing the coolant hoses to the inlet union and the oil

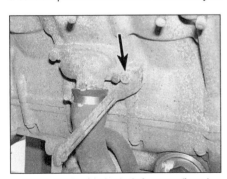

5.10a Unscrew the bolt (arrowed) and remove the bracket

5.10b Unscrew the bolt (arrowed) . . .

Engine, clutch and transmission 2•7

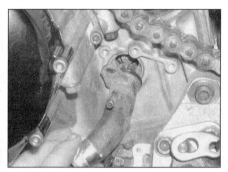

5.10c ... and draw the pipe out of the engine

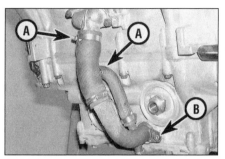

5.10d Slacken the clamp screws (A) and detach the hoses, then unscrew the bolt (B) ...

5.10e ... and draw the pipe out of the engine

cooler (if not removed) on the front of the engine and detach the hoses, then unscrew the bolt securing the coolant pipe to the engine and pull it out **(see illustrations)**. Discard the pipe O-rings, as new ones must be used.

11 Remove the exhaust system (see Chapter 4).

12 Trace the alternator wiring from the alternator and disconnect it at the connector **(see illustration)**. Release the wiring from any clips or ties, noting its routing, and coil it so that it does not impede engine removal.

13 Trace the oil level sensor wiring from the bottom of the engine and disconnect it at the connector **(see illustration)**. Release the wiring from any clips or ties, noting its routing, and coil it so that it does not impede engine removal.

14 Trace the ignition pick-up coil wiring from the coil on the right-hand side of the engine

and disconnect it at the connector **(see illustration)**. Release the wiring from any clips or ties, noting its routing, and coil it so that it does not impede engine removal.

15 Trace the neutral switch wiring from the back of the engine and disconnect it at the connector **(see illustration)**. Release the wiring from any clips or ties, noting its routing, and coil it so that it does not impede engine removal.

16 Trace the speed sensor wiring from the back of the engine and disconnect it at the connector **(see illustration)**. Release the wiring from any clips or ties, noting its routing, and feed it down to the switch.

17 Detach the clutch cable from the release mechanism arm (see Section 15).

18 If required, detach the crankcase breather hose from the crankcase and remove it.

19 At this point, position an hydraulic or mechanical jack under the engine with a block

of wood between the jack head and crankcase. Make sure the jack is centrally positioned so the engine will not topple in any direction when the last mounting bolt is removed. Take the weight of the engine on the jack. It is also advisable to place a block of wood between the rear wheel and the ground, or under the swingarm, in case the bike tilts back onto the rear wheel when the engine is removed. Check around the engine and frame to make sure that all wiring, cables and hoses that need to be disconnected have been disconnected, and that any that remain connected to the engine are not retained by any clips, guides or brackets connected to the frame. Check that any protruding mounting brackets will not get in the way, and remove them if necessary.

20 Slacken the pinchbolt on the mounting lug for the right-hand front mounting bolt **(see illustration)**. Unscrew and remove the right-hand front mounting bolt and washer, noting

5.12 Disconnect the alternator wiring connector ...

5.13 ... the oil level sensor wiring connector ...

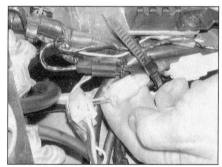

5.14 ... the pick-up coil wiring connector ...

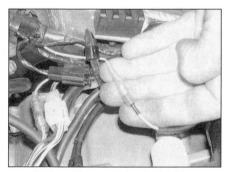

5.15 ... the neutral switch wiring connector ...

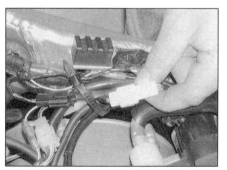

5.16 ... and the speed sensor wiring connector

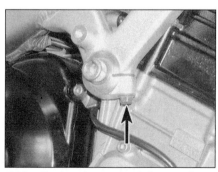

5.20a Slacken the pinchbolt (arrowed) ...

2•8 Engine, clutch and transmission

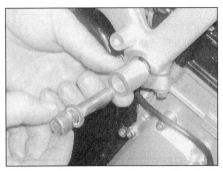

5.20b ... then unscrew the mounting bolt and remove the spacer

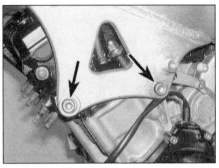

5.20c Unscrew the left-hand front mounting bolts (arrowed)

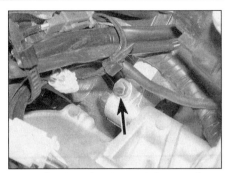

5.20d Slacken the pinchbolt (arrowed) ...

the spacer between the frame and the engine – remove the spacer now if it is loose, or retrieve it later **(see illustration)**. Unscrew and remove the left-hand front mounting bolts and washers **(see illustration)**. Slacken the pinchbolt on the mounting lug for the upper rear mounting bolt collar **(see illustration)**. Unscrew the nuts on the upper and lower rear mounting bolts, but do not yet withdraw the bolts **(see illustration)**. Make sure the engine is properly supported on the jack, and have an assistant support it as well, then withdraw the upper and lower rear mounting bolts **(see illustration)**. Carefully lower the engine a little, then bring it forward slightly, then lower it more and manoeuvre it out from the right-hand side **(see illustration)**. Either remove the collar from the upper rear mounting bolt lug for safekeeping, noting which way round it fits, or tighten the pinchbolt so that it is held securely and cannot fall out **(see illustration)**.

Installation

21 Installation is the reverse of removal, noting the following points:

a) Make sure no wires, cables or hoses become trapped between the engine and the frame when installing the engine.
b) Many of the engine mounting bolts are of different size and length. Make sure the correct bolt is installed in its correct location, with its washer, if fitted. Lubricate the threads of the upper and lower rear mounting bolts with lithium-based grease. Install all of the bolts and nuts finger-tight only until they are all located, then tighten them in the order below to their torque settings as specified at the beginning of the Chapter.
c) Before the engine is mounted, install the collar for the upper rear mounting bolt lug if it was removed, making sure the shouldered end faces the inside, or slacken the pinchbolt to release it if it wasn't removed **(see illustration 5.20h)**. With the aid of an assistant, place the engine unit on top of the jack and block of wood and carefully raise the engine into position in the frame, making sure the mounting bolt holes align **(see illustration 5.20g)**. Also make sure no wires, cables or hoses become trapped between the engine and the frame. Locate all the mounting bolts, not forgetting the spacer and washer with the right-hand front mounting bolt **(see illustration 5.20b)**, and the washers with the left-hand front mounting bolts **(see illustration)**, and tighten them finger-tight.
d) First tighten the upper and lower rear mounting bolt nuts to the torque setting specified at the beginning of the Chapter **(see illustration 5.20e)**. Now tighten the left-hand front mounting bolts to the specified torque **(see illustration 5.20c)**. Next tighten the right-hand front mounting bolt to the specified torque. Finally tighten the pinchbolts for the right-hand front and upper rear mounting bolt lugs to the specified torque **(see illustrations 5.20 a and d)**.
e) Use new gaskets on the exhaust pipe connections, and new O-rings smeared with grease on the coolant pipes **(see illustration)**. Tighten the pipe bolts to the specified torque setting.

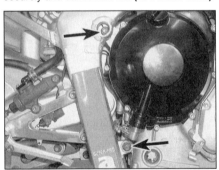

5.20e ... then unscrew the rear mounting bolt nuts (arrowed) ...

5.20f ... and withdraw the bolts (arrowed)

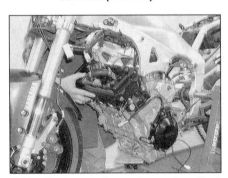

5.20g Manoeuvre the engine as described and remove it

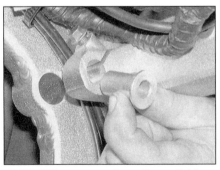

5.20h Either remove the collar or tighten the pinchbolt to secure it

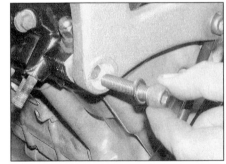

5.21a Do not forget the washers with the left-hand front bolts

Engine, clutch and transmission

f) Make sure all wires, cables and hoses are correctly routed and connected, and secured by any clips or ties.
g) Refill the engine with oil and coolant (see Chapter 1).
h) Adjust the throttle and clutch cable freeplay and engine idle speed (see Chapter 1).
i) Adjust the drive chain slack (see Chapter 1).
j) Start the engine and check for any oil or coolant leaks before installing the body panels.

6 Engine disassembly and reassembly – general information

Disassembly

1 Before disassembling the engine, the external surfaces of the unit should be thoroughly cleaned and degreased. This will prevent contamination of the engine internals, and will also make working a lot easier and cleaner. A high flash-point solvent, such as paraffin (kerosene) can be used, or better still, a proprietary engine degreaser such as Gunk. Use old paintbrushes and toothbrushes to work the solvent into the various recesses of the engine casings. Take care to exclude solvent or water from the electrical components and from the intake and exhaust ports.

Warning: *The use of petrol (gasoline) as a cleaning agent should be avoided because of the risk of fire.*

2 When clean and dry, arrange the unit on the workbench, leaving a suitable clear area for working. Gather a selection of small containers and plastic bags so that parts can be grouped together in an easily identifiable manner. Some paper and a pen should be on hand so that notes can be made and labels attached where necessary. A supply of clean rag is also required.

3 Before commencing work, read through the appropriate section so that some idea of the necessary procedure can be gained. When removing components it should be noted that

5.21b Fit new O rings smeared with grease onto the coolant pipes

great force is seldom required, unless specified. In many cases, a component's reluctance to be removed is indicative of an incorrect approach or removal method – if in any doubt, re-check with the text.

4 An engine support stand can be made from short lengths of 2 x 4 inch wood bolted together into a rectangle to help support the engine if required **(see illustration)** – though the engine will sit nicely on the flat bottom of the sump, and there are two pegs at the front to keep it stable. The perimeter of the mount should be just big enough to accommodate the sump within it, so that the engine rests on its crankcase.

5 When disassembling the engine, keep 'mated' parts together (including gears, pistons, connecting rods, valves, etc. that have been in contact with each other during engine operation). These 'mated' parts must be reused or renewed as an assembly.

6 A complete engine/transmission disassembly should be done in the following general order with reference to the appropriate Sections (or Chapters, where indicated).

Remove the valve cover
Remove the camshafts
Remove the cylinder head
Remove the clutch and starter clutch
Remove the pick-up coil assembly (see Chapter 5)
Remove the alternator (see Chapter 9)
Remove the starter motor (see Chapter 9)
Remove the gearchange mechanism
Remove the oil sump
Remove the oil pump/water pump

6.4 An engine support made from pieces of 2 x 4 inch wood

Separate the crankcase halves
Remove the crankshaft
Remove the connecting rods and pistons
Remove the transmission output shaft
Remove the selector drum and forks
Remove the transmission input shaft

Reassembly

7 Reassembly is accomplished by reversing the general disassembly sequence.

7 Oil cooler – removal and installation

Note: *The oil cooler can be removed with the engine in the frame. If the engine has been removed, ignore the steps which do not apply.*

Removal

1 The cooler is located on the front of the engine. Drain the engine oil and the cooling system (see Chapter 1).
2 Remove the exhaust system (see Chapter 4).
3 Slacken the clamp securing each hose to the cooler and detach the hoses **(see illustration)**.
4 Bend down the tabs on the cooler bolt lockwasher **(see illustration)**. Unscrew the bolt using a socket and remove the lockwasher **(see illustration)**. Remove the cooler, noting how the tab locates between the lugs on the crankcase. Discard the cooler and bolt O-rings, as new ones must be used.

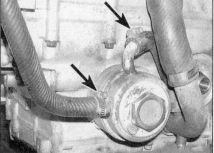

7.3 Slacken the clamp screws (arrowed) and detach the hoses

7.4a Bend down the lockwasher tab(s) . . .

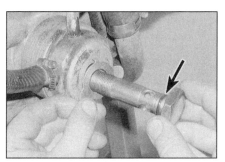

7.4b . . . then unscrew the bolt and remove the washer, noting the O-ring (arrowed) on the bolt

2•10 Engine, clutch and transmission

7.6a Fit a new O-ring into the groove in the cooler

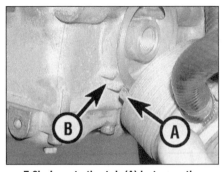

7.6b Locate the tab (A) between the lugs (B)

7.6c Tighten the bolt to the specified torque

7.6d Bend up the tabs to secure the bolt

8.2a Unscrew the cover bolts (arrowed) . . .

5 Check the cooler body for cracks and dents and any evidence of coolant leakage and replace it with a new one, if necessary. Also check the hoses for splits, cracks, hardening and deterioration and fit new ones, if required.

Installation

6 Installation is the reverse of removal, noting the following:
 a) Clean the mating surfaces of the crankcase and the cooler with a rag and some thinners.
 b) Use a new O-ring on the cooler body and smear it with clean engine oil. Make sure it seats in its groove (see illustration).
 c) Locate the tab on the cooler between the lugs on the crankcase (see illustration).
 d) Use a new lockwasher if the tabs on it are weakened or deformed – Yamaha recommend using a new one regardless. Use a new O-ring on the cooler bolt (see illustration 7.4b). Tighten the bolt to the torque setting specified at the beginning of the Chapter (see illustration).
 e) Bend up the lockwasher tabs to secure the bolt (see illustration).
 f) Make sure the coolant hoses are pressed fully onto their unions and are secured by the clamps (see illustration 7.3).
 g) Fill the engine with oil (see Chapter 1).
 h) Refill the cooling system (see Chapter 1).

8 Valve cover – removal and installation

Note: *The valve cover can be removed with the engine in the frame. If the engine has been removed, ignore the steps which do not apply.*

Removal

1 Remove the carburettors (see Chapter 4), and the radiator, the thermostat housing and the outlet pipes from the valve cover (see Chapter 3).
2 Unscrew the bolts securing the valve cover and remove it (see illustrations). If the cover is stuck, do not try to lever it off with a screwdriver. Tap it gently around the sides with a rubber hammer or block of wood to dislodge it.

Installation

3 Examine the valve cover gasket for signs of damage or deterioration and fit a new one, if necessary. Similarly check the rubber grommets on the cover bolts for cracks, hardening and deterioration (see illustration).

8.2b . . . and lift the cover off the engine

8.3 Check the grommets on the bolts for cracks and deterioration

Engine, clutch and transmission 2•11

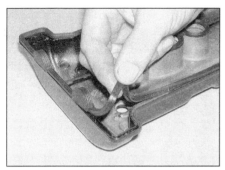

8.5 Make sure the gasket locates correctly in the cover

8.6 Apply the sealant to the cutouts in the head

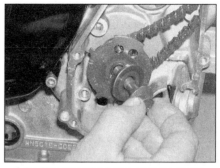

9.2a Unscrew the bolt and remove the washer . . .

4 Clean the mating surfaces of the cylinder head and the valve cover with lacquer thinner, acetone or brake system cleaner.

5 Fit the gasket into the valve cover, making sure it locates correctly into the groove **(see illustration)**. Use a few dabs of grease to keep the gasket in place while the cover is fitted.

6 Apply a suitable sealant to the cut-outs in the cylinder head where the gasket half-circles fit **(see illustration)**. Position the valve cover on the cylinder head, making sure the gasket stays in place **(see illustration 8.2b)**. Install the cover bolts and tighten them to the torque setting specified at the beginning of the Chapter.

7 Install the remaining components in the reverse order of removal.

9 Cam chain, tensioner, tensioner blade and guides – removal, inspection and installation

Note: *The cam chain, tensioner and guides can be removed with the engine in the frame.*

Cam chain

Removal

1 Remove the camshafts (see Section 10).
2 Unscrew the bolt securing the timing rotor and remove the rotor, noting how it fits **(see illustrations)**. To prevent the crankshaft turning, either select a gear and apply the rear brake (if the engine is in the frame), or remove the alternator cover (see Chapter 9) and use a rotor holding strap **(see illustration)**.
3 Draw the cam chain off its sprocket and out of the engine **(see illustration)**. If required, draw the sprocket off the crankshaft, noting which way round it fits, and remove the timing rotor/sprocket locating pin **(see illustrations)**.

Inspection

4 Check the chain for binding, kinks and any obvious damage and replace it with a new one, if necessary. Check the camshaft and crankshaft sprocket teeth for wear and renew the cam chain, camshaft sprockets and crankshaft sprocket as a set, if necessary.

Installation

5 Installation of the chain is the reverse of removal. Make sure the marked side of the timing rotor faces out. Tighten the rotor bolt to the torque setting specified at the beginning of the Chapter, counter-holding the crankshaft as on removal **(see illustration)**.

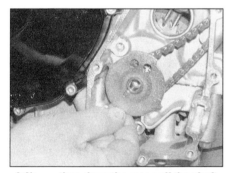

9.2b . . . then draw the rotor off the shaft

9.2c Counter-holding the crankshaft using a strap on the alternator rotor

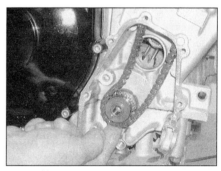

9.3a Draw the chain off the sprocket and out of the engine

9.3b Draw the sprocket off the shaft . . .

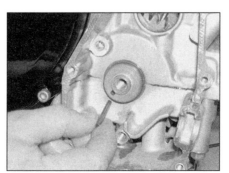

9.3c . . . and remove the locating pin

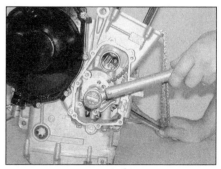

9.5 Tighten the rotor bolt to the specified torque

2•12 Engine, clutch and transmission

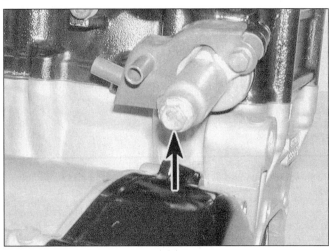

9.7 Unscrew the bolt (arrowed) and remove the washer

9.8a Slacken the bolts (arrowed) . . .

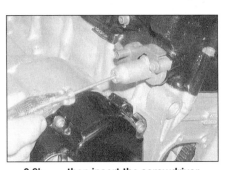

9.8b . . . then insert the screwdriver, retract the plunger and remove the tensioner as described

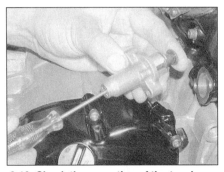

9.10 Check the operation of the tensioner as described

Cam chain tensioner

Removal

6 To access the tensioner, remove the carburettors (see Chapter 4).

7 Unscrew the tensioner cap bolt and remove the sealing washer **(see illustration)**.

8 Slacken the tensioner mounting bolts slightly **(see illustration)**. Insert a small flat-bladed screwdriver in the end of the tensioner so that it engages the slotted plunger **(see illustration)**. Turn the screwdriver clockwise until the plunger is fully retracted and hold it in this position while unscrewing the tensioner mounting bolts. Note the carburettor vent hose holder secured by the bolts. Withdraw the tensioner from the engine, then release the screwdriver – the plunger will spring back out once the screwdriver is removed, but can be easily reset on installation.

9 Discard the gasket, as a new one must be used on installation. Do not dismantle the tensioner.

Inspection

10 Check that the plunger moves smoothly when drawn into the tensioner (by turning the screwdriver) and springs back out freely when released **(see illustration)**.

11 If the tensioner or any of its components are worn or damaged, or if the plunger does not run smoothly and freely or is seized in the body, the tensioner must be replaced with a new one – individual internal components are not available.

Installation

12 Ensure the tensioner and cylinder block surfaces are clean and dry and fit a new gasket onto the tensioner.

13 Insert a small flat-bladed screwdriver in the end of the tensioner so that it engages the slotted plunger, then turn it clockwise until the plunger is fully retracted **(see illustration 9.10)**. Hold the screwdriver so the plunger remains retracted while the tensioner and its bolts are installed and while the bolts are tightened to the torque setting specified at the beginning of the Chapter – make sure the tensioner is installed with the 'UP' mark facing up, and do not forget to secure the carburettor vent hose holder with the bolts **(see illustrations)**. Relax your grip on the screwdriver and check that it turns as the tensioner extends itself, then remove the screwdriver. Install the tensioner cap bolt with a new sealing washer and tighten it to the specified torque **(see illustration)**.

14 Install the carburettors (see Chapter 4).

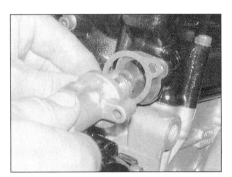

9.13a Install the tensioner using a new gasket . . .

9.13b . . . and do not forget the vent hose holder

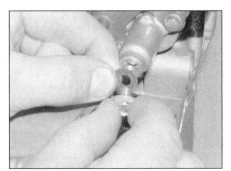

9.13c Install the cap bolt using a new washer

Engine, clutch and transmission 2•13

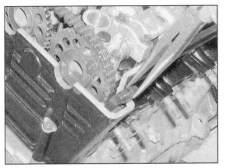

9.16 Lift the front guide out of the engine using pliers

9.17a Withdraw the pivot pin . . .

9.17b . . . and lift the blade out of the top of the engine

Tensioner blade and guides

Removal

15 Remove the valve cover (see Section 8).
16 To remove the cam chain front guide, lift it out of the front of the cam chain tunnel, noting which way round it fits and how it locates **(see illustration)**.
17 To remove the cam chain tensioner blade, first remove the cam chain tensioner (see above) and the intake camshaft (see Section 10). Withdraw the tensioner blade pivot pin, then draw the blade out of the top of the engine, noting which way round it fits **(see illustrations)**.

Inspection

18 Check the sliding surfaces of the blades for excessive wear, deep grooves, cracking and other obvious damage, and renew them if necessary.

Installation

19 Slide the front guide blade into the front of the cam chain tunnel **(see illustration 9.16)**, making sure it locates correctly onto its seat and its lugs locate in their cut-outs **(see illustration)**.
20 Apply some clean engine oil to the tensioner blade pivot pin, then install the blade and insert the pin **(see illustrations 9.17b and a)**.

21 Install the valve cover (see Section 8).

10 Camshafts and followers – removal, inspection and installation

Note: *The camshafts can be removed with the engine in the frame. Place rags over the spark plug holes and the cam chain tunnel to prevent components from dropping into the engine on removal.*

Removal

1 Remove the lower fairing and the fairing side panels (see Chapter 8). Remove the valve cover (see Section 8).
2 Unscrew the bolts securing the timing rotor/pick-up coil cover on the right-hand side of the engine and remove the cover, noting the wiring clamp and cable guide **(see illustration)**. Discard the gasket, as a new one must be used. Remove the dowels from either the crankcase or the cover, if they are loose. The engine can be turned using the timing rotor bolt, turning it in a clockwise direction only. Alternatively, place the motorcycle on an auxiliary stand so that the rear wheel is off the ground, select a high gear and rotate the rear wheel by hand in its normal direction of rotation.

9.19 Make sure the guide blade locates correctly

3 Turn the engine until the 'T' mark faces back and aligns with the crankcase mating surfaces **(see illustration)**, and the camshaft lobes for the No. 1 (left-hand) cylinder face away from each other, and the dot on each camshaft faces up and aligns with the arrow on the camshaft holder **(see illustrations)**. If the cam lobes are facing towards each other and the dots are facing down, rotate the engine clockwise 360° (one full turn) so that the 'T' mark again aligns with the crankcase mating surfaces. The camshaft lobes will now be facing away from each other and the dots will be facing up and aligned with the arrows, meaning the No. 1 cylinder is at TDC (Top Dead Centre) on the compression stroke.

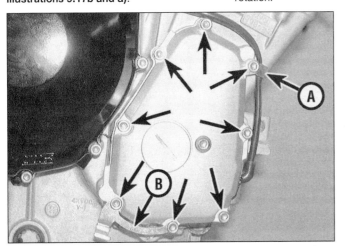

10.2 Unscrew the bolts (arrowed) and remove the cover, noting the wiring clamp (A) and cable guide (B)

10.3a Turn the engine until the 'T' mark aligns with the crankcase mating surfaces . . .

10.3b ... and the camshaft lobes for No. 1 cylinder face away from each other as shown

10.3c Intake camshaft alignment marks

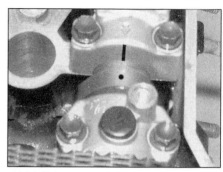

10.3d Exhaust camshaft alignment marks

10.7 Camshaft holder identification

10.8a Unscrew the bolts and lift off the holders ...

4 Before disturbing the camshafts, make a note of the timing markings described above and how they align. If you are in any doubt as to the alignment of the markings, or if they are not visible, make your own alignment marks between all components, and also between a tooth on each sprocket (including the crankshaft sprocket) and its corresponding link on the chain, before disturbing them. These markings ensure that the valve timing can be correctly set up on assembly without difficulty. As it is easy to be a tooth out on installation, marking between a tooth on each sprocket and its link in the chain is especially useful.

10.8b ... then remove the camshafts

5 Remove the cam chain tensioner (see Section 9).
6 Remove the cam chain guide blade (see Section 9).
7 Before removing the camshaft holders, make a note of which fits where, as they must be installed in the same place. The large holders for each camshaft are marked 'L' and 'R' (for Left and Right), and the intake ones have eight bolts each while the exhaust ones only have four each. The small holders on the right-hand end are marked 'I' and 'E' for Intake and Exhaust respectively. If the marks are not visible, make your own before disturbing them – it is essential that they are returned to their original locations on installation (see illustration).
8 Unscrew the camshaft holder bolts for the camshaft being worked on, slackening them evenly and a little at a time in a criss-cross pattern, starting from the outside and working towards the centre. Slacken the bolts above any lobes that are pressing onto a valve last in the sequence so that the pressure from the open valves cannot cause the camshaft to bend.

Caution: A camshaft could break if the holder bolts are not slackened as described and the pressure from a depressed valve causes the shaft to bend. Also, if the holder does not come squarely away from the head, the holder is likely to break. If this happens the cylinder head must be renewed; the holders are matched to the head and cannot be renewed separately.

Remove the bolts, then lift off the camshaft holders, noting how they fit (see illustration). Retrieve the dowels from either the holder or the cylinder head, if they are loose. Remove the camshafts, rotating them towards the centre of the engine as you do so (see illustration). Keep all 'mated' parts together. While the camshafts are out, don't allow the cam chain to go slack and do not rotate the crankshaft – the chain may drop down and bind between the crankshaft and case, which could damage these components. Wire the chain to another component, or secure it using a rod of some sort to prevent it from dropping.

9 If the followers and shims are being removed from the cylinder head, obtain a container which is divided into twenty compartments, and label each compartment with the location of its corresponding valve in the cylinder head and whether it belongs with an intake or an exhaust valve. If a container is not available, use labelled plastic bags (egg cartons also work very well!). Remove the cam

Engine, clutch and transmission 2•15

10.9a Lift out the follower . . .

10.9b . . . and remove the shim from inside

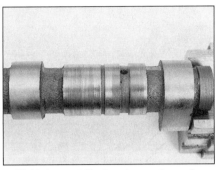

10.10 Inspect the bearing surfaces for scratches or wear

follower of the valve in question, then retrieve the shim from the inside of the follower **(see illustrations)**. If it is not in the follower, pick it out of the top of the valve using either a magnet, a small screwdriver with a dab of grease on it (the shim will stick to the grease), or a screwdriver and a pair of pliers **(see illustration 10.21a)**. Do not allow the shim to fall into the engine.

Inspection

10 Inspect the bearing surfaces of the cylinder head and holders and the corresponding journals on the camshaft **(see illustration)**. Look for score marks, deep scratches and evidence of spalling (a pitted appearance). If damage is noted or wear is excessive, replace the cylinder head and holders as a set with new ones – individual holders are not available.

11 Check the camshaft lobes for heat discoloration (blue appearance), score marks, chipped areas, flat spots and spalling **(see illustration)**. Measure the height of each lobe with a micrometer **(see illustration)** and compare the results to the minimum lobe height listed in this Chapter's Specifications. If damage is noted or wear is excessive, the camshaft must be replaced with a new one. Also, be sure to check the condition of the followers.

12 Check the amount of camshaft runout by supporting each end of the camshaft on V-blocks, and measuring any runout using a dial gauge. If the runout exceeds the specified limit the camshaft must be replaced with a new one.

 Refer to Tools and Workshop Tips (Section 3) in the Reference section for details of how to read a micrometer and dial gauge.

13 Next, check the camshaft journal oil clearances. Check each camshaft in turn rather than at the same time – the exhaust camshaft is identifiable by its two lobes per cylinder, and the intake camshaft by its three lobes per cylinder. Clean the camshaft, the bearing surfaces in the cylinder head and camshaft holders with a clean lint-free cloth, then lay the camshaft in place in the head.

14 Cut some strips of Plastigauge and lay one piece on each journal, parallel with the camshaft centreline **(see illustration)**. Make sure the camshaft holder dowels are installed. Lay the holders in their correct place in the case (see Step 7). Lubricate the threads of the holder bolts with engine oil, then install them in the holders. Tighten the bolts evenly and a little at a time in a criss-cross pattern to the torque setting specified at the beginning of the Chapter, working from the centre of the camshaft outwards, (i.e. starting with the bolts that are above valves that will be opened when the camshaft is tightened down). Whilst tightening the bolts, make sure the holders are being pulled squarely down and are not binding on the dowels. Whilst doing this, don't let the camshaft rotate.

15 Now unscrew the bolts evenly and a little at a time in a criss-cross pattern, starting from the outside and working towards the centre, and carefully lift off the camshaft holders.

16 To determine the oil clearance, compare the crushed Plastigauge (at its widest point) on each journal to the scale printed on the Plastigauge container **(see illustration)**. Compare the results to this Chapter's Specifications. If the oil clearance is greater than specified, measure the diameter of the camshaft journal with a micrometer **(see illustration)**. If the journal diameter is less than the specified limit, replace the camshaft with a new one and recheck the clearance. If the clearance is still too great, or if the

10.11a Check the lobes of the camshaft for wear – here is an example of damage requiring camshaft repair or renewal

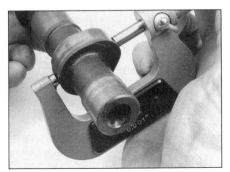

10.11b Measure the height of the camshaft lobes with a micrometer

10.14 Lay a strip of Plastigauge across each bearing journal, parallel with the camshaft centreline

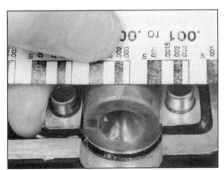

10.16a Compare the width of the crushed Plastigauge to the scale printed on the Plastigauge container

10.16b Measure the cam bearing journals with a micrometer

10.21a Fit the shim into the recess in the top of the valve . . .

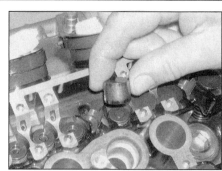

10.21b . . . then install the follower

camshaft journal is within its limit, replace the cylinder head and holders as a set with new ones – individual holders are not available.

17 Except in cases of oil starvation, the cam

 Before renewing the camshafts, cylinder head or holders because of damage, check with local machine shops specialising in motorcycle engine work. In the case of the camshafts, it may be possible for cam lobes to be welded, reground and hardened, at a cost far lower than that of a new camshaft. If the bearing surfaces in the head or holders are damaged, it may be possible for them to be bored out to accept bearing inserts. Due to the cost of new components it is recommended that all options are explored!

chain wears very little. If the chain has stretched excessively, which makes it difficult to maintain proper tension, or if it is stiff or the links are binding or kinking, replace it with a new one. Refer to Section 9 for replacement.

18 Check the sprockets for wear, cracks and other damage. If the sprockets are worn, the cam chain is also worn, and so is the sprocket on the crankshaft. If severe wear is apparent, the entire engine should be disassembled for inspection.

19 Inspect the cam chain guide blade and tensioner blade (see Section 9).

20 Inspect the outer surfaces of the cam followers for evidence of scoring or other damage. If a follower is in poor condition, it is probable that the bore in which it works is also damaged. Check for clearance between the followers and their bores. Whilst no specifications are given, if slack is excessive, renew the followers. If the bores are seriously out-of-round or tapered, then replace the cylinder head and holders, and the followers, with new ones.

Installation

21 If removed, lubricate each shim and its follower with molybdenum disulphide oil (a 50/50 mixture of molybdenum disulphide grease and engine oil) and fit each shim into its recess in the top of the valve, with the size marking on each shim facing up **(see illustration)**. Make sure each shim is correctly seated in the top of the valve assembly, then install each follower, making sure it fits squarely in its bore **(see illustration)**. *Note: It is most important that the shims and followers are returned to their original valves, otherwise the valve clearances will be inaccurate.*

22 Make sure the bearing surfaces on the camshafts and in the cylinder head are clean, then apply molybdenum disulphide oil (a 50/50 mixture of molybdenum disulphide grease and engine oil) to each of them. Also apply it to the camshaft lobes.

23 Check that the 'T' mark on the rotor still aligns with the crankcase mating surfaces (see Step 3) **(see illustration 10.3a)**.

24 Fit the exhaust camshaft, identifiable by its two lobes per cylinder **(see illustration)**, onto the front of the head, making sure the No. 1 (left-hand) cylinder lobes are facing forwards and the punchmark on the camshaft is facing up **(see illustrations 10.3b and d)**. Fit the cam chain around the sprocket, aligning the marks between sprocket tooth and link if made (in which case, first make sure that the mark between the crankshaft sprocket and chain align). When fitting the chain, pull up on the front run to remove all slack from it.

25 Now fit the intake camshaft, identifiable by its three lobes per cylinder **(see illustration 10.8b)**, onto the back of the head, making sure the No. 1 cylinder lobes are facing back and the punchmark on the camshaft end is facing up **(see illustrations 10.3b and c)**. Fit the cam chain around the intake sprocket, aligning the marks between sprocket tooth and link, if made. When fitting the chain, pull it tight to make sure there is no slack between the two camshaft sprockets – any slack in the chain must lie in the rear run, so that it is taken up by the tensioner.

26 Fit the camshaft holder dowels into the holder or cylinder head. Make sure the bearing surfaces in the holders are clean, then apply molybdenum disulphide oil (a 50/50 mixture of molybdenum disulphide grease and engine oil) to them. Lay the holders in their correct place in the case (see Step 7) **(see illustration)**. Lubricate the threads of the holder bolts with engine oil, then install them in the holders **(see illustration)**. Tighten the bolts evenly and a little at a time in a criss-cross pattern to the torque setting specified at

10.24 Install the camshaft as described

10.26a Fit the holders . . .

10.26b . . . then install the bolts and tighten them as described

Engine, clutch and transmission 2•17

10.27 Use a wooden dowel in the tensioner orifice to take up the slack in the chain

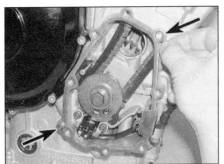

10.31a Locate the new gasket onto the dowels (arrowed)

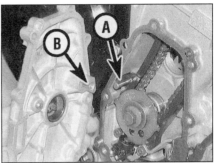

10.31b Locate the pivot pin (A) in the hole (B)

the beginning of the Chapter, working from the centre of the camshaft outwards, (i.e. starting with the bolts that are above valves that will be opened when the camshaft is tightened down). Whilst tightening the bolts, make sure the holders are being pulled squarely down and are not binding on the dowels.

Caution: *The camshaft is likely to break if it is tightened down onto the closed valves before the open valves. The holders are likely to break if they are not tightened down evenly and squarely.*

27 Using a piece of wooden dowel, press on the back of the cam chain tensioner blade via the tensioner bore in the crankcase to ensure that any slack in the cam chain between the crankshaft and the intake camshaft, and between the two camshafts, is taken up and transferred to the rear run of the chain (where it will later be taken up by the tensioner) **(see illustration)**. At this point, check that all the timing marks are still in **exact** alignment as described in Steps 3 and 4. Note that it is easy to be slightly out (one tooth on the sprocket) without the marks appearing drastically out of alignment. If it is necessary to turn the engine, keep the wooden dowel pressed onto the tensioner blade as the chain is likely to jump around on the sprocket. If the marks are out, verify which sprocket is misaligned, then, if there is enough slack in the chain, turn the camshaft and slip the chain round on the sprocket as you do. If there is not enough slack, unscrew the sprocket bolts and slide it off the camshaft, then disengage it from the chain. Move the camshaft round as required, then fit the sprocket back into the chain and onto the camshaft, and check the marks again. With everything correctly aligned, tighten the sprocket bolts to the torque setting specified at the beginning of the Chapter.

Caution: *If the marks are not aligned exactly as described, the valve timing will be incorrect and the valves may strike the pistons, causing extensive damage to the engine.*

28 Install the cam chain guide blade (see Section 9).
29 Install the cam chain tensioner (see Section 9). Turn the engine anti-clockwise through two full turns and check again that all the timing marks still align (see Steps 3 and 4).
30 Check the valve clearances and adjust them if necessary (see Chapter 1).
31 Install the timing rotor/pick-up coil cover using a new gasket, and do not forget the dowels if removed **(see illustration)**. Locate the cam chain tensioner blade pivot pin in the hole in the cover **(see illustration)**. Tighten the cover bolts to the torque setting specified at the beginning of the Chapter, not forgetting the wiring clamp and the cable guide **(see illustration 10.2)**.
32 Install the valve cover (see Section 8), and the fairing panels (see Chapter 8).

11 Cylinder head – removal and installation

Caution: *The engine must be completely cool before beginning this procedure or the cylinder head may become warped.*
Note: *To remove the cylinder head the engine must be removed from the frame.*

Removal

1 Remove the engine from the frame (see Section 5).
2 Remove the valve cover (see Section 8) and the camshafts and followers (see Section 10). On 2000-on models, slacken the clamps securing the air induction system (AIS) pipes to the unions on the front of the cylinder head and draw them off. Remove the sealing collars and replace them with new ones, if they are deformed or damaged.
3 The cylinder head is secured by two bolts and ten nuts **(see illustration)**. First unscrew and remove the bolts on the right-hand end of the head. Now slacken the nuts evenly and a little at a time in a criss-cross pattern, starting from the outside and working inwards (see *Tools and Workshop Tips* (Section 4) in the *Reference* section), until they are all slack, then remove the nuts and their washers. Use either a magnet, a screwdriver, or a piece of wire hooked over at the end to lift them out where necessary. Note which nut fits where, as there are different types **(see illustration)**.

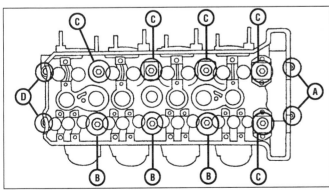

11.3a Cylinder head bolts (A), sleeve nuts (B), shouldered nuts (C) and domed nuts (D)

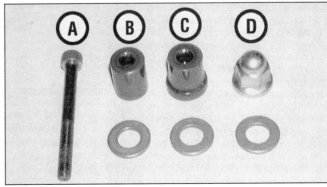

11.3b Cylinder head bolts (A), sleeve nuts (B), shouldered nuts (C) and domed nuts (D)

2•18 Engine, clutch and transmission

11.4 Carefully lift the head up off the block

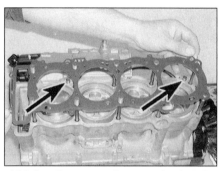

11.8 Lay the new gasket over the dowels (arrowed) and onto the head

12.1 Check all around the edge of each valve for solvent leakage

4 Pull the cylinder head up off the crankcase (see illustration). If it is stuck, tap around the joint faces of the head with a soft-faced mallet to free it. Do not attempt to free the head by inserting a screwdriver between it and the crankcase – you will damage the sealing surfaces. Remove the old cylinder head gasket and discard it, as a new one must be used.

5 If they are loose, remove the dowels from the cylinder block (see illustration 11.8). If they appear to be missing they are probably stuck in the underside of the cylinder head.

6 Check the cylinder head gasket and the mating surfaces on the cylinder head and block for signs of leakage, which could indicate warpage. Refer to Section 13 and check the flatness of the cylinder head.

Installation

7 Clean all traces of old gasket material from the cylinder head and block. If a scraper is used, take care not to scratch or gouge the soft aluminium. Be careful not to let any of the gasket material fall into the crankcase, the cylinder bores or the oil passages. Lubricate the cylinder bores with new engine oil.

 Refer to Tools and Workshop Tips (Section 7) for details of gasket removal methods.

8 Ensure both cylinder head and crankcase mating surfaces are clean. If removed, fit the dowels into the crankcase, then lay the new head gasket in place, making sure it locates correctly over the dowels, and all the holes are correctly aligned (see illustration). Never re-use the old gasket.

9 Carefully fit the cylinder head onto the crankcase, making sure it locates correctly onto the dowels (see illustration 11.4).

10 Lubricate the threads and seating surfaces of the cylinder head nuts with clean engine oil. Install the nuts with their washers, making sure they are correctly located, and tighten them finger-tight (see illustrations 11.3b and a). Now tighten the nuts evenly and in two stages, working in a criss-cross pattern starting from the centre and moving outwards (see *Tools and Workshop Tips* (Section 4) in the *Reference* section), to the torque setting specified at the beginning of the Chapter.

11 Now install the bolts and tighten them to the specified torque setting.

12 Install the remaining components in a reverse of their removal sequence, referring to the relevant Sections or Chapters (see Steps 1 and 2).

12 Valves/valve seats/ valve guides – servicing

1 Because of the complex nature of this job and the special tools and equipment required, most owners leave servicing of the valves, valve seats and valve guides to a professional. However, you can make an initial assessment of whether the valves are seating, and therefore sealing, correctly by pouring a small amount of solvent into each of the valve ports. If the solvent leaks past any valve into the combustion chamber area the valve is not seating and sealing correctly (see illustration).

2 You can also remove the valves from the cylinder head, clean the components, check them for wear to assess the extent of the work needed, and, unless a valve service is required, grind in the valves (see Section 13). The head can then be reassembled.

3 The dealer service department will remove the valves and springs, renew the valves and guides, recut the valve seats, check and renew the valve springs, spring retainers and collets (as necessary), replace the valve seals with new ones and reassemble the valve components.

4 After the valve service has been performed, the head will be in like-new condition. When the head is returned, be sure to clean it again very thoroughly before installation on the engine, to remove any metal particles or abrasive grit that may still be present from the valve service operations. Use compressed air, if available, to blow out all the holes and passages.

13 Cylinder head and valves – disassembly, inspection and reassembly

1 As mentioned in the previous section, valve overhaul should be left to a Yamaha dealer. However, disassembly, cleaning and inspection of the valves and related components can be done (if the necessary special tools are available) by the home mechanic. This way, no expense is incurred if the inspection reveals that overhaul is not required at this time.

2 To disassemble the valve components without the risk of damaging them, a valve spring compressor is absolutely essential. Make sure it is suitable for motorcycle work.

Disassembly

3 Before proceeding, arrange to label and store the valves along with their related components in such a way that they can be returned to their original locations without getting mixed up (see illustration). A good way to do this is to use the same container as

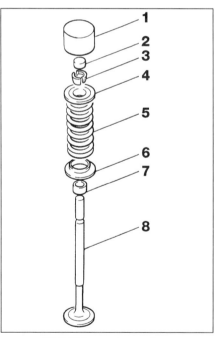

13.3 Valve components – exhaust valve shown

1 Follower
2 Shim
3 Collets
4 Spring retainer
5 Valve spring
6 Spring seat – (intake spring seat differs)
7 Valve stem oil seal
8 Valve

Engine, clutch and transmission 2•19

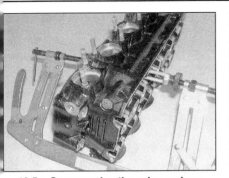

13.5a Compressing the valve springs using a valve spring compressor

13.5b Make sure the compressor is a good fit both on the top . . .

13.5c . . . and the bottom of the valve assembly

the shims and followers are stored in (see Section 10), or to obtain a separate container which is divided into twenty compartments, and to label each compartment with the identity of the valve which will be stored in it (i.e. number of cylinder, intake or exhaust side, inner, centre or outer valve). Alternatively, labelled plastic bags will do just as well.

4 Clean all traces of old gasket material from the cylinder head. If a scraper is used, take care not to scratch or gouge the soft aluminium.

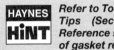 **Refer to Tools and Workshop Tips (Section 7) in the Reference section for details of gasket removal methods.**

13.5d Remove the collets with needle-nose pliers, tweezers, a magnet or a screwdriver with a dab of grease on it

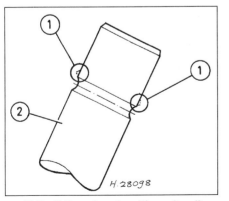

13.5e If the valve stem (2) won't pull through the guide, deburr the area above the collet groove (1)

5 Compress the valve spring on the first valve with a spring compressor, making sure it is correctly located onto each end of the valve assembly **(see illustrations)**. On the underside of the head, make sure the plate on the compressor only contacts the valve and not the soft aluminium of the head – if the plate is too big for the valve, use a spacer between them. Do not compress the springs any more than is absolutely necessary. Remove the collets, using either needle-nose pliers, tweezers, a magnet or a screwdriver with a dab of grease on it **(see illustration)**. Carefully release the valve spring compressor and remove the spring retainer, noting which way up it fits, the spring from the top of the head, and the valve from the underside of the head **(see illustrations 13.26c, b and a)**. If the valve binds in the guide (won't pull through), push it back into the head and deburr the area around the collet groove with a very fine file or whetstone **(see illustration)**.

6 Pull the valve stem seals off the top of the valve guides with pliers and discard them (the old seals should never be reused) **(see illustration)**. Remove the spring seat, noting which way up it fits – using a magnet is the easiest way to lift the seat off the head **(see illustration)**.

7 Repeat the procedure for the remaining valves. Remember to keep the parts for each valve together and in order so they can be reinstalled in the same location.

8 Next, clean the cylinder head with solvent and dry it thoroughly. Compressed air will speed the drying process and ensure that all holes and recessed areas are clean.

13.6a Pull the seal off the top of the guide . . .

9 Clean all the valve springs, collets, retainers and spring seats with solvent and dry them thoroughly. Do the parts from one valve at a time so that no mixing of parts between valves occurs.

10 Scrape off any deposits that may have formed on the valve, then use a motorised wire brush to remove deposits from the valve heads and stems. Again, make sure the valves do not get mixed up.

Inspection

11 Inspect the head very carefully for cracks and other damage. If cracks are found, a new head will be required. Check the cam bearing surfaces for wear and evidence of seizure. Check the camshafts for wear as well (see Section 10).

12 Using a precision straight-edge and a feeler gauge set to the warpage limit listed in the specifications at the beginning of the Chapter, check the head gasket mating surface for warpage. Refer to *Tools and Workshop Tips* (Section 3) in the *Reference* section for details of how to use the straight-edge. If the head is warped beyond the limit, but not excessively so, it should be possible to resurface it using a 400 to 600 grade wet and dry paper on a large surface plate, moving the cylinder head in a figure of eight rotation, and turning it frequently to ensure an even coverage. Take the head to a specialist repair shop if you lack the necessary equipment.

13 Examine the valve seats in the combustion chamber. If they are pitted, cracked or burned, the head will require work

13.6b . . . then remove the spring seat

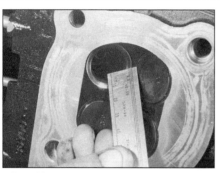

13.13 Measure the valve seat width with a ruler (or for greater precision use a vernier caliper)

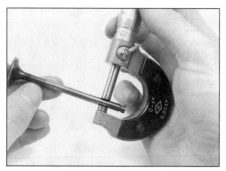

13.14a Measure the valve stem diameter with a micrometer

13.14b Insert a small hole gauge into the valve guide and expand it so there's a slight drag when it's pulled out

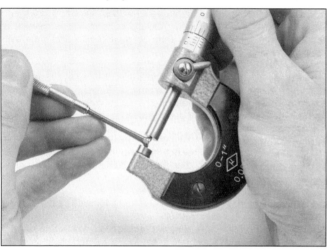

13.14c Measure the small hole gauge with a micrometer

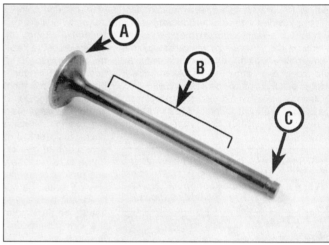

13.15a Check the valve face (A), stem (B) and collet groove (C) for signs of wear and damage

beyond the scope of the home mechanic. Measure the valve seat width and compare it to this Chapter's Specifications **(see illustration)**. If it exceeds the service limit, or if it varies around its circumference, valve overhaul is required. If available, use Prussian blue to determine the extent of valve seat wear. Uniformly coat the seat with the Prussian blue, then install the valve and rotate it back and forth using a lapping tool. Remove the valve and check whether the ring of blue on the valve is uniform and continuous around the valve, and of the correct width as specified.

14 Measure the valve stem diameter **(see illustration)**. Clean the valve guides to remove any carbon build-up, then measure the inside diameters of the guides (at both ends and the centre of the guide) with a small hole gauge and micrometer **(see illustrations)**. The guides are measured at the ends and at the centre to determine if they are worn in a bell-mouth pattern (more wear at the ends). Subtract the stem diameter from the valve guide diameter to obtain the valve stem-to-guide clearance. If the stem-to-guide clearance is greater than listed in this Chapter's Specifications, renew whichever components are worn beyond their specification limits. If the valve guide is within specifications, but is worn unevenly, it should be renewed.

15 Carefully inspect each valve face, stem and collet groove area for cracks, pits and burned spots **(see illustration)**. Measure the valve margin thickness and compare it to the specifications **(see illustrations)**. If it is thinner than specified, replace the valve with a new one. The margin is the portion of the valve head which is below the valve seat.

16 Rotate the valve and check for any obvious indication that it is bent. Using V-blocks and a

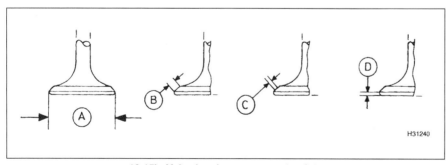

13.15b Valve head measurement points

A Head diameter   B Face width   C Seat width   D Margin thickness

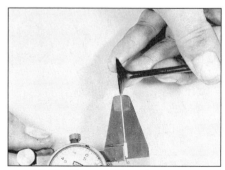

13.15c Measure the valve margin thickness as shown

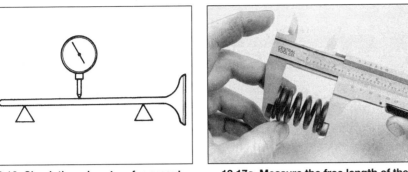

13.16 Check the valve stem for runout using V-blocks and a dial gauge

13.17a Measure the free length of the valve springs

13.17b Check the valve springs for squareness

dial gauge if available, measure the valve stem runout and compare the results to the specifications **(see illustration)**. If the measurement exceeds the service limit specified, the valve must be replaced with a new one. Check the end of the stem for pitting and excessive wear. The presence of any of the above conditions indicates the need for valve servicing. The stem end can be ground down, provided that the amount of stem above the collet groove after grinding is sufficient.

17 Check the end of each valve spring for wear and pitting. Measure the spring free length and compare it to that listed in the specifications **(see illustration)**. If any spring is shorter than specified it has sagged and must be replaced with a new one. Also place the spring upright on a flat surface and check it for bend by placing a ruler against it **(see illustration)**. If the bend in any spring exceeds the specified limit, it must be replaced with a new one.

18 Check the spring retainers and collets for obvious wear and cracks. Any questionable parts should not be reused, as extensive damage will occur in the event of failure during engine operation.

19 If the inspection indicates that no overhaul work is required, the valve components can be reinstalled in the head. If any new valves are being installed, they can be identified as follows: the exhaust valves are marked '4XV', the outer intake valves are marked '4XV:', and the centre intake valve is marked '4XV.' **(see illustration)**.

Reassembly

20 Unless a valve service has been performed, before installing the valves in the head they should be ground in (lapped) to ensure a positive seal between the valves and seats. This procedure requires coarse and fine valve grinding compound and a valve grinding tool. If a grinding tool is not available, a piece of rubber or plastic hose can be slipped over the valve stem (after the valve has been installed in the guide) and used to turn the valve.

21 Apply a small amount of coarse grinding compound to the valve face and some molybdenum disulphide oil (a 50/50 mixture of molybdenum disulphide grease and engine oil) to the valve stem, then slip the valve into the guide **(see illustration)**. **Note:** *Make sure each valve is installed in its correct guide and be careful not to get any grinding compound on the valve stem.*

22 Attach the grinding tool (or hose) to the valve and rotate the tool between the palms of your hands. Use a back-and-forth motion (as though rubbing your hands together) rather than a circular motion (i.e. so that the valve rotates alternately clockwise and anti-clockwise, rather than in one direction only) **(see illustration)**. Lift the valve off the seat and turn it at regular intervals to distribute the grinding compound properly. Continue the grinding procedure until the valve face and seat contact area is of uniform and correct width and unbroken around the entire circumference of the valve face and seat **(see illustrations)**. If required, use Prussian Blue as described in Step 13 to check the valve seat.

23 Carefully remove the valve from the guide and wipe off all traces of grinding compound.

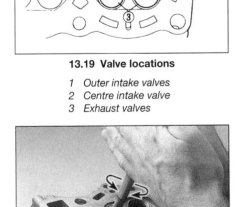

13.19 Valve locations

1 Outer intake valves
2 Centre intake valve
3 Exhaust valves

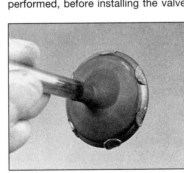

13.21 Apply the lapping compound very sparingly, in small dabs, to the valve face only

13.22a Rotate the valve grinding tool back and forth between the palms of your hands

13.22b The valve face and seat should show a uniform unbroken ring . . .

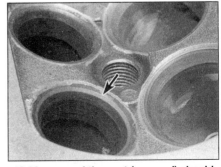

13.22c . . . and the seat (arrowed) should be the specified width all the way round

2•22 Engine, clutch and transmission

13.25a Fit the spring seat . . .

13.25b . . . followed by the valve stem seal

13.25c Press the seal into position using a suitable deep socket

Use solvent to clean the valve and wipe the seat area thoroughly with a solvent soaked cloth.

24 Repeat the procedure with fine valve grinding compound, then repeat the entire procedure for the remaining valves.

25 Working on one valve at a time, lay the spring seat in place in the cylinder head **(see illustration)**. Fit the intake valve seats with the dished side up so the spring sits in it. Fit the exhaust valve seats with the shouldered side up so the spring locates around it. Fit a new valve stem seal onto the guide **(see illustration)**. Use an appropriate size deep socket to push the seal over the end of the valve guide until it is felt to clip into place **(see illustration)**. Don't twist or cock the seal, or it will not seal properly against the valve stem. Also, don't remove it again or it will be damaged. Note that there are different size seals for the intake and exhaust valves – make sure you have the correct size for the valve being installed (the intake valve seals are slightly smaller).

26 Coat the valve stem with molybdenum disulphide oil (a 50/50 mixture of molybdenum disulphide grease and engine oil), then install it into its guide, rotating it slowly to avoid damaging the seal **(see illustration)**. Check that the valve moves up and down freely in the guide. Next, install the spring, with its closer-wound coils facing down into the cylinder head, followed by the spring retainer, with its shouldered side facing down so that it fits into the top of the spring **(see illustrations)**.

27 Apply a small amount of grease to the inside of the collets – this will help to help hold them in place as the pressure is released from the spring **(see illustration)**. Compress the spring with the valve spring compressor and install the collets **(see illustration 13.5d)**.

When compressing the spring, do so only as far as is necessary to slip the collets into place. Make certain that the collets are securely locked in the retaining groove **(see illustration)**.

28 Support the cylinder head on blocks so the valves can't contact the workbench top, then very gently tap the top of the valve stem to help seat the collets in the groove **(see illustration)**.

 HAYNES HiNT *Check for proper sealing of the valves by pouring a small amount of solvent into each of the valve ports. If the solvent leaks past any valve into the combustion chamber area the valve grinding operation on that valve should be repeated.*

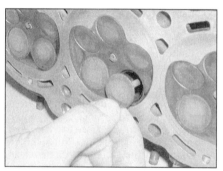

13.26a Lubricate the stem and slide the valve into its correct location

13.26b Fit the valve spring with its closer wound coils facing down . . .

13.26c . . . then fit the spring retainer

13.27a A small dab of grease will help to keep the collets in place on the valve while the spring is released

13.27b Compress the springs and install the collets, making sure they locate in the groove

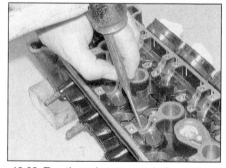

13.28 Tap the valve stem to help seat the collets in the groove

Engine, clutch and transmission

29 Repeat the procedure for the remaining valves. Remember to keep the parts for each valve together and separate from the other valves so they can be reinstalled in the same location.

14 Clutch – removal, inspection and installation

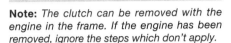

Note: *The clutch can be removed with the engine in the frame. If the engine has been removed, ignore the steps which don't apply.*

Removal

1 Remove the lower fairing and the right-hand fairing side panel (see Chapter 8). Drain the engine oil (see Chapter 1).
2 Detach the clutch cable from the release mechanism arm (see Section 15).
3 Working evenly in a criss-cross pattern, unscrew the clutch cover bolts **(see illustration)**. Remove the cover, being prepared to catch any residual oil.
4 Remove the gasket and discard it. Note the positions of the two locating dowels and remove them for safe-keeping if they are loose – they could be in either the cover or the crankcase.
5 Working in a criss-cross pattern, gradually slacken the clutch diaphragm spring retainer bolts until the spring pressure is released **(see illustrations)**. Counter-hold the clutch housing to prevent it turning. Remove the bolts, retainer, diaphragm spring and spring seat, then remove the clutch pressure plate **(see illustration 14.17)**. Remove the pull-rod from the pressure plate, noting the bearing **(see illustration 14.27a)**.
6 Grasp the complete set of clutch plates and remove them as a pack. Unless the plates are being replaced with new ones, keep them in their original order. On 1999-on models, do not yet remove the inner plain and friction plates that are part of the anti-judder assembly.

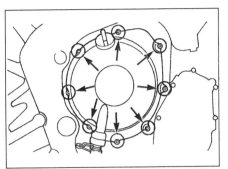

14.3 Unscrew the bolts (arrowed) and remove the cover

14.5a Unscrew the spring retainer bolts (arrowed) as described

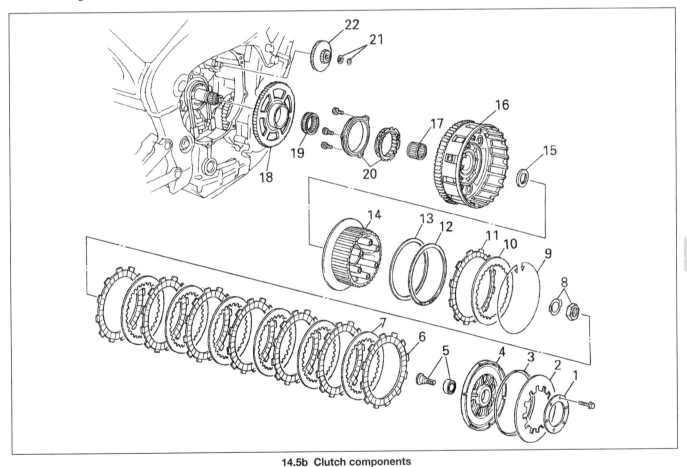

14.5b Clutch components

1 Spring retainer	6 Friction plates	10 Plain plate	15 Thrust washer	20 Starter clutch assembly
2 Diaphragm spring	7 Plain plates	11 Friction plate	16 Clutch housing	21 Circlip and washer
3 Spring seat	8 Clutch nut and lockwasher	12 Anti-judder spring	17 Needle bearing	22 Idle/reduction gear
4 Pressure plate		13 Spring seat	18 Starter driven gear	
5 Pullrod and bearing	9 Wire retainer	14 Clutch centre	19 Needle bearing	

2•24 Engine, clutch and transmission

14.7a Bend back the lockwasher tabs

7 Bend back the tabs on the clutch nut lockwasher **(see illustration)**. To remove the clutch nut, the transmission input shaft must be locked. This can be done in several ways. If the engine is in the frame, engage 1st gear and have an assistant hold the rear brake on hard with the rear tyre in firm contact with the ground. Alternatively, the Yamaha service tool – Pt. No. 90890-04086 (European models) or YM-91042 (USA models) – or a similar commercially available or home-made tool **(see Tool tip)**, can be used to stop the clutch centre from turning whilst the nut is slackened. Unscrew the nut and remove the lockwasher from the input shaft, noting how it fits **(see illustration)**. Discard the lockwasher, as a new one must be used on installation.

8 Slide the clutch centre and the thrust washer off the shaft **(see illustrations 14.25a and 14.24)**.

9 Slide the clutch housing/starter clutch assembly off the shaft, noting how it engages with the primary drive gear, the starter driven gear and the oil pump drive sprocket **(see illustration)**. If the needle bearing did not come away with the housing, slide it off the shaft **(see illustration 14.23a)**. Otherwise, remove it from the housing.

10 If required (e.g. if the crankcase is being separated), refer to Section 19 and remove the oil pump drive sprocket and chain from the input shaft.

11 If required, separate the starter driven gear and starter clutch from the back of the clutch housing (see Section 16).

Inspection

12 After an extended period of service the clutch friction plates will wear and promote clutch slip. Measure the thickness of each friction plate using a vernier caliper **(see illustration)**. If any plate has worn to or beyond the service limit given in the Specifications at the beginning of the Chapter, the friction plates must be renewed as a set. Also, if any of the plates smell burnt or are glazed, they must be renewed as a set.

13 The plain plates should not show any signs of excess heating (bluing). Check for warpage using a flat surface and feeler gauges **(see illustration)**. If any plate exceeds the maximum permissible amount of warpage, or shows signs of bluing, all plain plates must be renewed as a set.

14 Check the clutch diaphragm spring and spring seat for wear, damage or deformation and replace them with new ones, if necessary **(see illustration)**.

15 Inspect the clutch assembly for burrs and indentations on the edges of the protruding tangs of the friction plates and/or slots in the edge of the housing with which they engage. Similarly check for wear between the inner tongues of the plain plates and the slots in the clutch centre. Wear of this nature will cause clutch drag and slow disengagement during gear changes, as the plates will snag when the pressure plate is lifted. With care, a small amount of wear can be corrected by dressing

An alternative to the Yamaha tool can be fabricated from some steel strap, bent at the ends and bolted together in the middle

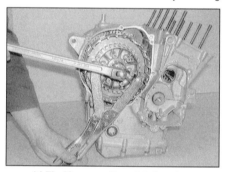

14.7b Unscrew the clutch nut as described

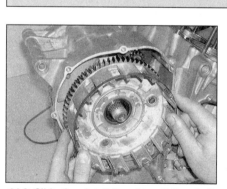

14.9 Slide the clutch housing off the shaft

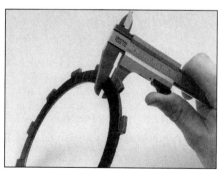

14.12 Measuring clutch friction plate thickness

14.13 Check the plain plates for warpage

Engine, clutch and transmission 2•25

14.17 Check the pressure plate and its bearing (arrowed) as described

14.18 Release the wire retainer (arrowed) to remove the anti-judder assembly

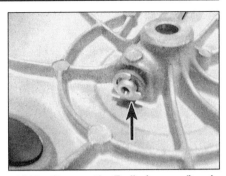

14.19a Remove the E-clip (arrowed) and draw the shaft out of the cover

14.19b Remove the E-clip and draw the arm off the shaft, noting how the spring fits

14.19c Locate the return spring ends as shown

14.23a Slide the bearing onto the shaft . . .

with a fine file, but if this is excessive the worn components should be renewed.

16 Inspect the needle roller bearing in conjunction with the internal bearing surface of the clutch housing. If there are any signs of wear, pitting or other damage the affected parts must be renewed.

17 Check the pressure plate and its bearing for signs of wear or damage and roughness **(see illustration)**. Check the pull-rod for signs of roughness, wear or damage. Replace any parts, as necessary, with new ones.

18 On all 1999-on models, if the clutch has been juddering, remove the wire retainer from the clutch centre and slide off the inner plain plate and friction plate, the anti-judder spring and spring seat **(see illustration)**. Check the plates as described above. Replace the spring and seat with new ones. Install the spring with the OUTSIDE mark facing out. Do not mix up the inner friction plate with the rest – it is identifiable by its larger internal diameter. Install the friction plate over the spring and spring seat, then fit the plain plate. Secure the assembly with the retainer ring, making sure it locates properly in its groove.

19 Check the clutch release mechanism in the clutch cover for smooth operation. Check the pinion and pull-rod teeth for signs of damage. If necessary, prise off the E-clip and remove the washer securing the actuating shaft, and withdraw the shaft from the cover **(see illustration)**. Check the condition of the oil seal and replace it with a new one, if necessary. Check the two bearings for roughness, wear or damage. If they need to be replaced with new ones, first remove the oil seal, then heat the cover in very hot water to ease removal and drift the bearings out. If required, prise off the E-clip and remove the washer securing the arm on the top of the shaft, then release the return spring ends, noting how they fit, and draw the arm and spring off **(see illustration)**. Clean all components and lubricate the seal and bearings with grease. Installation is the reverse of removal. When installing the arm on the shaft, make sure the 'UP' mark on the arm faces up the shaft, so that it will be facing up when installed in the cover, and do not forget to fit the return spring, locating it as shown **(see illustration)**.

Installation

20 Remove all traces of old gasket from the crankcase and clutch cover surfaces.

21 If removed, install the oil pump drive sprocket and chain (see Section 19).

22 If separated, fit the starter clutch and starter driven gear onto the back of the clutch housing (see Section 16).

23 Lubricate the needle roller bearing with clean engine oil and slide it onto the shaft **(see illustration)**. Install the housing, making sure the primary drive and driven gears, and the starter idle/reduction and driven gears engage correctly, and that the tabs on the oil pump drive sprocket locate into the slots in the back of the starter driven gear **(see illustration)**. The idle/reduction gear can be turned by hand to ease installation – remove the starter motor and turn the gear via the orifice if required, or use a screwdriver to turn the gear. To check that the housing has engaged correctly, turn it by hand – the oil pump drive chain should turn. If it doesn't, draw the housing away slightly and turn it, then press it in and continue to turn it until the housing is felt to engage with the sprocket and the chain starts to turn.

24 Lubricate the thrust washer with clean engine oil and fit it onto the shaft **(see illustration)**.

14.23b . . . then install the housing as described

14.24 Fit the thrust washer . . .

14.25a ... the clutch centre ...

14.25b ... the lockwasher ...

14.25c ... and the clutch nut ...

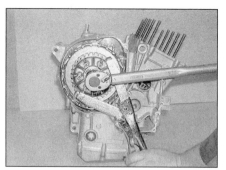

14.25d ... then tighten the nut to the specified torque ...

14.25e ... and secure it by bending up the lockwasher tabs

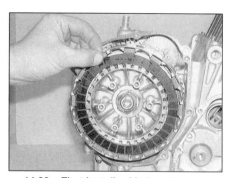

14.26a First install a friction plate ...

25 Slide the clutch centre onto the shaft splines, then fit the new lockwasher **(see illustrations)**. Install the clutch nut with the shouldered side on the inside, then tighten it to the torque setting specified at the beginning of the Chapter using the method employed on removal to lock the input shaft (see Step 7), **(see illustration)**. Note: *Check that the clutch centre rotates freely after tightening.* Bend up the tabs of the lockwasher to secure the nut **(see illustration)**.

26 Coat each clutch plate with engine oil prior to installation. Build up the plates as follows: first fit a friction plate, then a plain plate, then alternate friction and plain plates until all are installed **(see illustrations)**.

27 Lubricate the bearing in the pressure plate with some clean oil. Fit the pull-rod into the back of the pressure plate **(see illustration)**. Fit the pressure plate onto the clutch, making sure the castellations in the rim locate into the slots in the clutch centre (when it is properly located, there will be no freeplay between the plates) **(see illustration)**.

28 Fit the spring seat, diaphragm spring and spring retainer **(see illustrations)**. When installing the spring, locate the end of each tang against the flat on each bolt lug **(see illustration)**. Install the bolts and tighten them evenly and a little at a time in a criss-cross

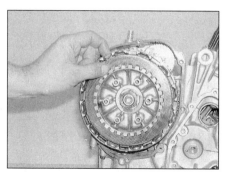

14.26b ... then a plain plate, and so on

14.27a Fit the pullrod into the back of the pressure plate ...

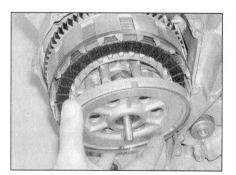

14.27b ... then install the plate, making sure it locates correctly

14.28a Fit the diaphragm spring seat ...

14.28b ... followed by the spring ...

Engine, clutch and transmission 2•27

14.28c ... locating the tangs against the flats (arrowed)

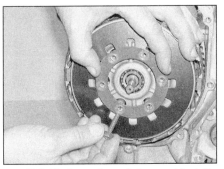
14.28d Fit the spring retainer and install the bolts ...

14.28e ... and tighten them to the specified torque

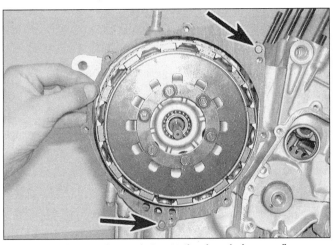
14.29a Locate the gasket onto the dowels (arrowed) ...

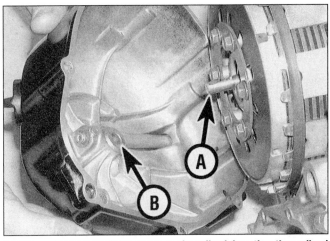
14.29b ... then install the cover as described, locating the pullrod end (A) in the orifice (B)

sequence to the specified torque setting **(see illustration)**. Counter-hold the clutch housing to prevent it turning when tightening the spring bolts. Set the pull-rod so that the teeth point towards the rear and are angled up slightly.

29 Insert the dowels in the crankcase and fit the new gasket onto them **(see illustration)**. When installing the clutch cover, getting the release shaft to align correctly with the pullrod can be tricky. Set the release arm so that it is pointing out, then as the cover is installed and the teeth engage, the arm should move in. Install the cover, making sure the end of the pullrod locates in the orifice in the cover, and tighten its bolts evenly in a criss-cross sequence to the specified torque setting **(see illustration)**.

30 To check that the release mechanism is correctly aligned, push the release arm forward until all the freeplay in the release mechanism has been taken up. At this point the punchmark on the arm should align with the triangle mark on the cover **(see illustration)**. If the marks do not align, remove the E-clip and the arm, noting how the spring fits **(see illustrations 14.19b and c)**, and move the arm around on the splines of the shaft until they do. Make sure the spring is correctly set and install the E-clip.

31 Fit the clutch cable into the release arm (see Section 15).

32 Refill the engine with oil (see Chapter 1).
33 Install the fairing side panel and the lower fairing (see Chapter 8).

15 Clutch cable – removal and installation

Removal

1 Remove the lower fairing (see Chapter 8).
2 Slacken the locknut securing the cable in the bracket on the engine and slip the cable out of the bracket **(see illustrations)**.

14.30 The release mechanism is correctly set if the punchmark aligns with the triangle when the arm is pushed in

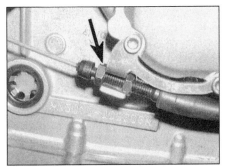
15.2a Slacken the locknut (arrowed) ...

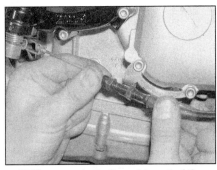
15.2b ... and slip the cable out of the bracket

2•28 Engine, clutch and transmission

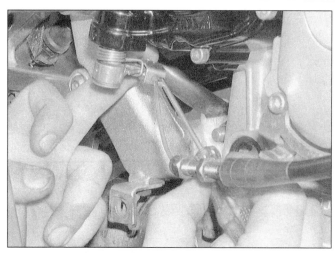

15.3a Bend back the tab on the retainer...

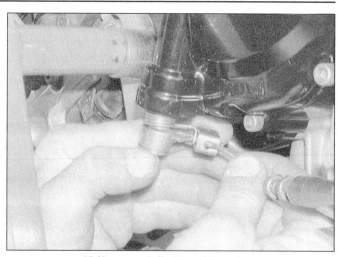

15.3b ... and slip the cable end out

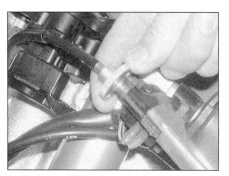

15.4 Set the adjuster as described...

15.5a ... then draw the cable out of the adjuster...

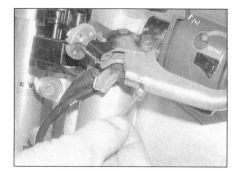

15.5b ... and free it from the lever

3 Bend out the tab in the cable retainer on the end of the release mechanism arm, then slip the cable end out of the retainer, noting how it fits **(see illustrations)**.
4 Screw the adjuster at the handlebar end of the cable fully in to the lever bracket, then turn it out to align the slot in the adjuster with that in the lever bracket **(see illustration)**.
5 Pull the outer cable end from the socket in the adjuster and release the inner cable from the lever **(see illustrations)**. Remove the cable from the machine, noting its routing and any guides or clips.

> **HAYNES HINT** Before removing the cable from the bike, tape the lower end of the new cable to the upper end of the old cable. Slowly pull the lower end of the old cable out, guiding the new cable down into position. Using this method will ensure the cable is routed correctly.

Installation

6 Installation is the reverse of removal. Apply grease to the cable ends. Make sure the cable is correctly routed. Turn the adjuster on the lever bracket so that the slots are not aligned **(see illustration 15.4)**. Bend in the retainer on the release mechanism arm to secure the cable end **(see illustration 15.3a)**. Check the clutch release mechanism for smooth operation and any signs of wear or damage. Remove it for cleaning and re-greasing if required (see Section 14).
7 Adjust the clutch lever freeplay (see Chapter 1). Install the lower fairing (see Chapter 8).

16 Starter clutch and idle/reduction gear – check, removal, inspection and installation

Check

1 The operation of the starter clutch can be checked in situ. Remove the starter motor (see Chapter 9). Insert your finger into the starter motor orifice and turn the idle/reduction gear by hand – it should turn freely clockwise (as you look at it from the orifice), and lock when turned anti-clockwise. If not, remove it for inspection.

Removal

2 Remove the clutch – the starter clutch is mounted on the back of the clutch housing (see Section 14). If the starter driven gear does not come away with the starter clutch, slide it off the shaft.
3 If required, remove the circlip and washer securing the idle/reduction gear on its shaft and slide it off, noting which way round it fits (place some rag around the bottom of the clutch casing to prevent the circlip or washer falling into the sump) **(see illustration)**.

Inspection

4 If separated, fit the starter driven gear into the starter clutch, rotating it anti-clockwise as you do to spread the sprags and allow it to enter **(see illustration 16.10b)**. With the clutch housing face down on a workbench, check that the starter driven gear rotates freely in an anti-clockwise direction and locks against the sprag assembly in a clockwise direction **(see illustration)**.

16.3 Remove the circlip (arrowed) and slide off the washer and gear

Engine, clutch and transmission 2•29

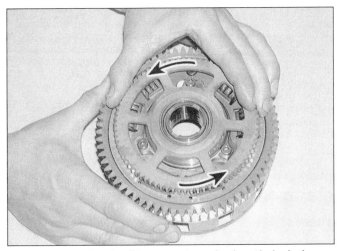

16.4 Check that the driven gear turns freely anti-clockwise

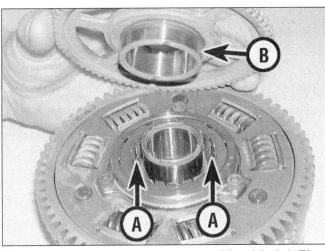

16.6a Check the condition of the sprags (A) and the hub (B)

5 If it doesn't, withdraw the starter driven gear from the starter clutch, rotating it anti-clockwise as you do. Withdraw the needle bearing from the centre of the clutch **(see illustration 16.10a)**.

6 Check the condition of the sprags inside the starter clutch and the corresponding surface on the driven gear hub **(see illustration)**. If they are damaged, marked or flattened at any point, the sprag assembly should be replaced with a new one. To remove it, hold the clutch housing and unscrew the retainer bolts, then remove the retainer and lift out the sprag assembly, noting how they fit **(see illustrations)**. Install the new starter clutch in a reverse sequence.

Apply clean engine oil to the sprags. Apply a suitable non-permanent thread locking compound to the bolts and tighten them to the torque setting specified at the beginning of the Chapter **(see illustration)**.

7 Check the needle roller bearing and the bearing surfaces in the starter driven gear hub and on the crankshaft **(see illustration 16.10a)**. If the bearing surface shows signs of excessive wear or the bearing itself is worn or damaged, they should be replaced with new ones.

8 Check the teeth of the starter idle/reduction gear and the corresponding teeth of the starter driven gear and starter motor drive shaft. Replace the gears and/or starter motor armature with new ones if worn or chipped teeth are discovered on related gears. Also check the idle/reduction gear shaft for damage, and check that the gear is not a sloppy fit on the shaft. Replace the gear and/or shaft with a new one if necessary – the shaft is secured from the inside of the crankcase by a bolt.

Installation

9 Lubricate the idle/reduction gear shaft with clean engine oil. Slide the gear onto the shaft, making sure the smaller pinion faces outwards, and the teeth of the larger pinion mesh correctly with the teeth of the starter motor shaft **(see illustration)**. Fit the washer and secure the gear with the circlip, making sure it locates in the groove **(see illustrations)**.

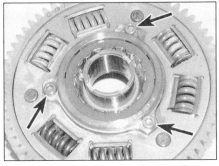

16.6b Unscrew the bolts (arrowed) . . .

16.6c . . . and remove the retainer and sprag assembly

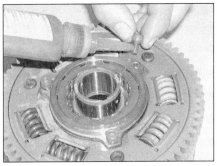

16.6d Apply a threadlock to the retainer bolts

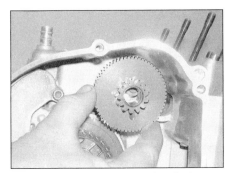

16.9a Slide the gear onto the shaft . . .

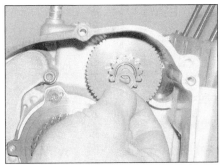

16.9b . . . then fit the washer . . .

16.9c . . . and secure them with the circlip

16.10a Fit the bearing into the clutch . . .

16.10b . . . then install the driven gear

10 Lubricate the needle roller bearing with clean engine oil and fit it into the starter clutch **(see illustration)**. Lubricate the outside of the starter driven gear hub with clean engine oil, then fit it into the starter clutch, rotating it anti-clockwise as you do so to spread the sprags and allow the hub to enter **(see illustration)**.

11 Install the clutch (see Section 14).

17 Gearchange mechanism – removal, inspection and installation

Note: *The gearchange mechanism can be removed with the engine in the frame. If the engine has been removed, ignore the steps which don't apply.*

Removal

1 Make sure the transmission is in neutral. Remove the lower fairing and the left-hand fairing side panel (see Chapter 8).

2 Slacken the gearchange lever linkage rod locknuts, then unscrew the rod and separate it from the lever and the arm (the rod is reverse-threaded on one end, so will unscrew from both lever and arm simultaneously when turned in the one direction) **(see illustration)**. Note how far the rod is threaded into the lever and arm, as this determines the height of the lever relative to the footrest. Withdraw the rod from the frame **(see illustration)**. Unscrew the gearchange lever linkage arm pinch bolt and slide the arm off the shaft, noting how the punchmark on the shaft aligns with the slit in the arm **(see illustration)**. If no mark is visible, make your own before removing the arm so that it can be correctly aligned with the shaft on installation.

3 Unscrew the bolts securing the front sprocket cover and remove it **(see illustration)**.

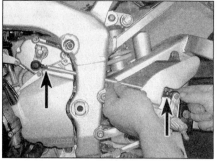

17.2a Slacken the locknuts (arrowed), then thread the rod off the lever and arm . . .

17.2b . . . and withdraw it from the frame

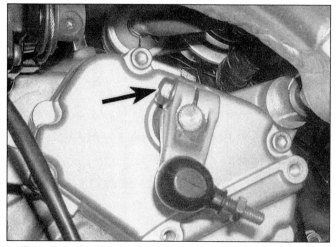

17.2c Note the alignment of the punchmark with the slit, then unscrew the bolt (arrowed) and slide the arm off the shaft

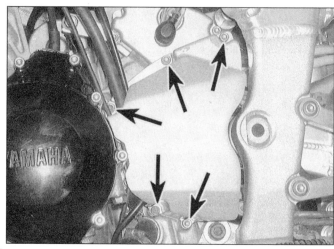

17.3 Unscrew the bolts (arrowed) and remove the cover. Note how the wiring and coolant hose are routed

Engine, clutch and transmission 2•31

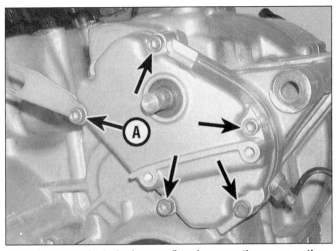

17.4 Unscrew the bolts (arrowed) and remove the cover, noting the idle speed adjuster holder (A)

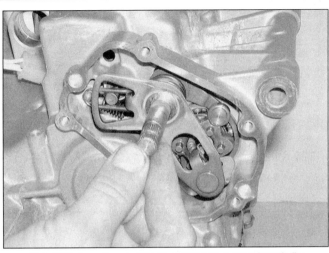

17.5 Withdraw the shaft/arm assembly, noting how it fits

4 Unscrew the bolts securing the gearchange mechanism cover and remove it, noting how one of the bolts secures the idle speed adjuster holder **(see illustration)**. Discard the gasket, as a new one must be used. Remove the dowels from either the cover or the crankcase if they are loose.

5 Note how the gearchange shaft centralising spring ends fit on each side of the locating pin in the crankcase, and how the pawls on the selector arm locate onto the pins on the end of the selector drum **(see illustration 17.11)**. Grasp the end of the shaft and withdraw the shaft/arm assembly **(see illustration)**.

6 Note how the stopper arm spring ends locate and how the roller on the arm locates in the neutral detent on the selector drum **(see illustration 17.10b)**. Unhook the spring from its anchor, then lift the stopper arm off the drum and remove it **(see illustrations)**. Retrieve the washer from the rim of the shaft bore in the crankcase or from the end of the collar **(see illustration)**.

Inspection

7 Check the selector arm for cracks, distortion and wear of its pawls, and check for any corresponding wear on the selector pins on the selector drum **(see illustration)**. Also check the stopper arm roller and the detents in the selector drum for any wear or damage, and make sure the roller turns freely **(see illustration)**. Replace any components that are worn or damaged with new ones. On 1998 and 1999 models, separate the stopper arm from the collar by removing the circlip and washer and sliding the arm off. To refit, slide the arm onto the collar, making sure it is the correct way round, then fit the washer and secure them with the circlip.

8 Inspect the shaft centralising spring and the stopper arm return spring for fatigue, wear or damage. If any faults are found, renew the components. Slide the centralising spring off the shaft, noting the spacer. Also check that the centralising spring locating pin in the crankcase is securely tightened. If it is loose, remove it and apply a non-permanent thread locking compound to its threads, then tighten it to the torque setting specified at the beginning of the Chapter.

9 Check the gearchange shaft/arm assembly for distortion and damage to the splines. If the assembly is bent you can attempt to straighten it, but if the splines are damaged it must be renewed. Also check the condition of the shaft oil seal and bearing in the cover. If the oil seal is damaged, deteriorated or shows

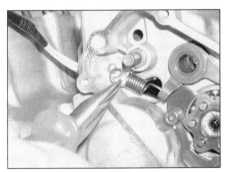

17.6a Unhook the spring . . .

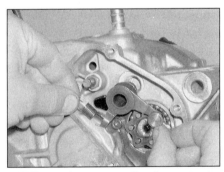

17.6b . . . then remove the stopper arm, noting how it fits . . .

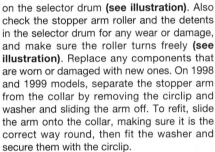

17.6c . . . and retrieve the washer

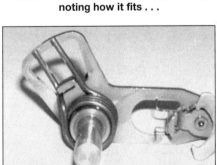

17.7a Check the gearchange shaft/selector arm assembly . . .

17.7b . . . and the stopper arm assembly as described

2•32 Engine, clutch and transmission

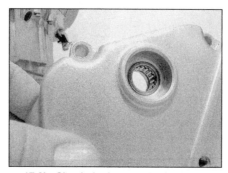

17.9a Lever the old seal out

17.9b Check the bearing in the cover

17.9c Press or drive the new seal into place

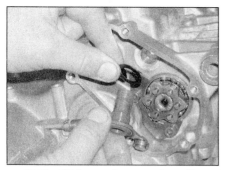

17.10a Fit the washer over the collar

17.10b Installed stopper arm assembly

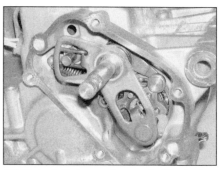

17.11 Installed gearchange shaft/selector arm assembly

signs of leakage it must be replaced with a new one. Lever out the old seal with a flat-bladed screwdriver **(see illustration)**. If the bearing is damaged or does not run smoothly and freely, it must be replaced with a new one – refer to Section 5 of *Tools and Workshop Tips* in the *Reference* section for information on bearing check, removal and installation methods **(see illustration)**. Drive the new seal squarely into place, with its lip facing inward, using a seal driver or suitable socket **(see illustration)**.

Installation

10 Lubricate the inside and outside of the stopper arm collar with clean engine oil. Fit the washer onto the end of the collar **(see illustration)**. Install the stopper arm assembly, locating the arm onto the neutral detent on the selector drum **(see illustration 17.6b)**. Fit the return spring **(see illustration 17.6a)**, making sure the spring ends are positioned correctly **(see illustration)**.
11 Make sure the centralising spring ends are correctly positioned on the selector arm **(see illustration 17.7a)**. Slide the shaft/arm assembly into place **(see illustration 17.5)**, locating the selector arm pawls onto the pins on the selector drum and the centralising spring ends onto each side of the locating pin **(see illustration)**.
12 If removed, fit the gearchange mechanism cover dowels into the crankcase. Install the cover using a new gasket, making sure they locate correctly onto the dowels, and tighten the cover bolts securely **(see illustrations)**. Do not forget to secure the idle speed adjuster holder with the front bolt **(see illustration 17.4)**.
13 Install the front sprocket cover, making sure the wiring and hose are correctly routed, and tighten the bolts securely **(see illustration)**. Yamaha specify to use a suitable non-permanent thread locking compound on the front bolt **(see illustration)**.
14 Slide the gearchange lever linkage arm onto the shaft, aligning the punchmark on the shaft

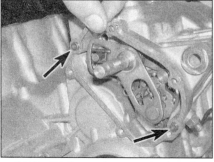

17.12a Fit the gasket over the dowels (arrowed) . . .

17.12b . . . then install the cover

17.13a Install the cover . . .

17.13b . . . and apply a threadlock to the front bolt

Engine, clutch and transmission 2•33

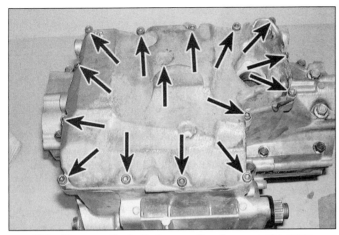
18.3a Unscrew the bolts (arrowed) and remove the sump

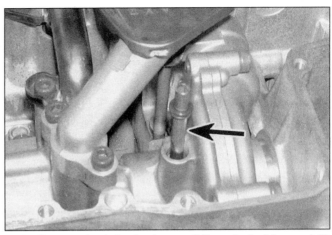
18.3b Remove the drain pipe (arrowed) if required – it is a push fit

18.4 Unscrew the bolts (arrowed) and remove the strainer

18.5 Remove the pressure relief valve – it is a push fit

with the slit in the arm (see illustration 17.2c). Tighten the pinchbolt securely. Slide the linkage rod into the frame and thread it onto the lever and arm (the rod is reverse-threaded on one end, so will screw onto both lever and arm simultaneously when turned in the one direction) (see illustration 17.2b). Set it as noted on removal to give the same gearchange lever height, then tighten the locknuts securely (see illustration 17.2a). To adjust the lever height, slacken the locknuts and thread the rod in or out as required, then retighten the nuts.

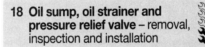
18 Oil sump, oil strainer and pressure relief valve – removal, inspection and installation

Note: *The oil sump, strainer and pressure relief valve can be removed with the engine in the frame. If the engine has been removed, ignore the steps which don't apply.*

Removal

1 Remove the exhaust system (see Chapter 4).
2 Drain the engine oil (see Chapter 1). Either remove the oil level sensor if required (see Chapter 9), or trace the wire from the switch and disconnect it at the single bullet connector – you may have to remove the front sprocket cover to access it (see Section 17, Steps 2 and 3).
3 Unscrew the sump bolts, slackening them evenly in a criss-cross sequence to prevent distortion, and remove the sump (see illustration). Note the position of the oil level sensor wiring clamp. Discard the gasket, as a new one must be used. Note the positions of the dowels and remove them if they are loose. Note how the water pump drain pipe fits between the pump and sump and remove it if required (see illustration).
4 Unscrew the oil strainer bolts and remove the strainer (see illustration).
5 Pull the pressure relief valve out of the crankcase (see illustration). Discard the O-ring, as a new one must be used.

Inspection

6 Remove all traces of gasket from the sump and crankcase mating surfaces, and clean the inside of the sump with solvent.
7 Clean the oil strainer in solvent and remove any debris caught in the strainer mesh. Inspect the strainer for any signs of wear or damage and replace it with a new one, if necessary – separate it from the housing by levering it off.
8 Push the relief valve plunger into the valve body and check that it moves smoothly and freely against the spring pressure (see illustration). If not, remove the circlip, noting that it is under spring pressure, and remove the spring seat, spring and plunger (see illustrations). Clean all the components in solvent and check them for scoring, wear or

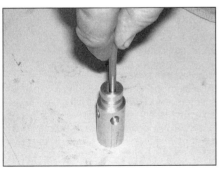
18.8a Check that the plunger is not seized in the valve body

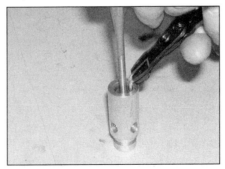
18.8b Press down on the spring seat to remove the circlip

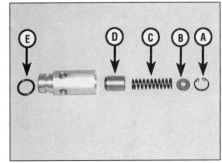
18.8c Circlip (A), spring seat (B), spring (C), plunger (D), O-ring (E)

18.9 Install the relief valve using a new O-ring

18.10 The arrow on the strainer points to the front of the engine

18.11a Fit new O-rings onto the drain pipe if necessary

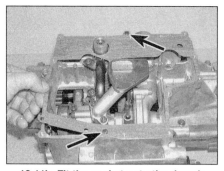

18.11b Fit the gasket onto the dowels (arrowed)

18.12a Install the sump . . .

18.12b . . . making sure the drain pipe locates correctly

damage. If any is found, replace the relief valve with a new one – individual components are not available. Otherwise, coat the inside of the valve body and the plunger with clean engine oil, then insert the plunger, spring and spring seat and secure them with the circlip. Check the action of the valve plunger again – if it is still suspect, replace the valve with a new one.

Installation

9 Fit a new O-ring onto the relief valve and smear it with grease (see illustration). Push the valve into its socket in the crankcase (see illustration 18.5).
10 Install the oil strainer, making sure the arrow points to the front of the engine (see illustration). Apply a suitable non-permanent thread locking compound to the bolts and tighten them to the specified torque setting.
11 Check the condition of the water pump drain pipe O-rings and replace them with new

ones if they are in any way damaged, deformed or deteriorated. Fit the pipe into the sump or water pump as required (see illustration). If removed, fit the sump dowels into the crankcase. Lay a new gasket onto the sump (if the engine is in the frame) or onto the crankcase (if the engine has been removed and is positioned upside down on the work surface) (see illustration). Make sure the holes in the gasket align correctly with the bolt holes.
12 Position the sump on the crankcase, making sure the water pump drain pipe locates correctly (see illustrations). Apply a suitable non-permanent thread locking compound to the centre and front sump bolts. Install the bolts, not forgetting the oil level sensor wiring clamp, and tighten them evenly and a little at a time in a criss-cross pattern to the specified torque setting (see illustration).
13 Either install the oil level sensor if removed (see Chapter 9), or connect the wire

at the connector. Install the sprocket cover if removed (see Section 17, Steps 13 and 14).
14 Fill the engine with the correct type and quantity of oil (see Chapter 1).
15 Install the exhaust system (see Chapter 4). Start the engine and check for leaks around the sump.

19 Oil pump – removal, inspection and installation

Note: *The oil pump and water pump come as an assembly and are removed as such. They can be removed with the engine in the frame. If the engine has been removed, ignore the steps which don't apply.*

Removal

1 Drain the engine oil and the coolant (see Chapter 1).
2 Remove the clutch (see Section 14), the sump and the oil strainer (see Section 18). If the water pump drain pipe does not come away with the sump, remove it from the pump (see illustration 18.11a).
3 Slacken the clamps securing the coolant hoses to the water pump outlet pipe on the front of the engine and detach the hoses. Unscrew the bolt securing the pipe to the front of the engine and pull it out (see illustrations). Discard the O-rings, as new ones must be used.
4 Unscrew the two bolts securing the large bore oil pipe and the single bolt securing the

18.12c Apply a threadlock to the front and centre bolts

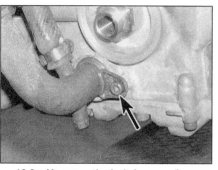

19.3a Unscrew the bolt (arrowed) . . .

Engine, clutch and transmission 2•35

19.3b ... and draw the pipe out of the engine

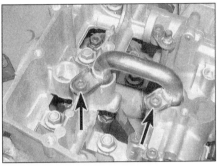

19.4a Unscrew the bolts (arrowed) and remove the large bore pipe ...

19.4b ... and the small bore pipe

19.5a Unscrew the bolts (arrowed) and remove the cover

19.5b Unscrew the bolts (arrowed) and remove the guide

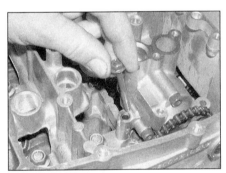

19.6a Unscrew the bolt ...

19.6b ... and withdraw the long dowel

small bore pipe **(see illustrations)**. Remove the pipes, noting how they fit.

5 Unscrew the bolts securing the oil pump driven sprocket cover and remove the cover, noting how it fits **(see illustration)**. Unscrew the bolts securing the chain guide and remove it **(see illustration)**.

6 Unscrew the remaining bolt securing the pump assembly, then draw out the long dowel using pliers **(see illustrations)**. Draw the pump out of the coolant inlet union in the crankcase, then tilt it to get some slack in the chain and disengage the chain first from the drive sprocket, then from the driven sprocket on the pump, and remove it from the engine, being prepared to catch any residual oil and/or coolant **(see illustrations)**.

19.6c Draw the pump out of the case and tilt it ...

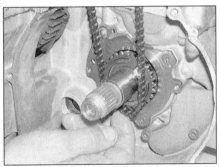

19.6d ... then disengage the chain from the drive sprocket to create slack

19.6e Slip the chain off the driven sprocket and remove the pump

2•36 Engine, clutch and transmission

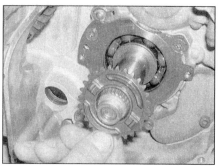

19.7a Slide the drive sprocket off the shaft ...

19.7b ... and remove the thrust washer

7 Remove the chain and slide the drive sprocket off the shaft, noting which way round it fits **(see illustration)**. Also slide the thrust washer off the shaft **(see illustration)**.

Inspection

8 If required, the pump can be disassembled for cleaning and inspection **(see illustration)**. Unscrew the driven sprocket bolt and remove the sprocket – use a screwdriver inserted through one of the holes in the sprocket and lodge it against the pump body to prevent the sprocket turning **(see illustration)**. Retrieve the washer from the end of the shaft or the back of the sprocket **(see illustration)**.

9 Unscrew the three bolts securing the oil pump housing **(see illustration)**. Remove the housing and its locating pins, then remove the outer and inner rotors – they may come away with the housing or remain on the shaft **(see illustrations)**. Note which way round the rotors fit, as it is best to install them the same

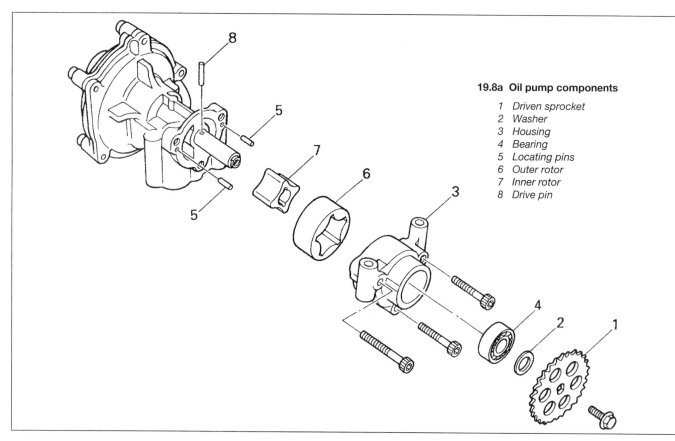

19.8a Oil pump components

1. Driven sprocket
2. Washer
3. Housing
4. Bearing
5. Locating pins
6. Outer rotor
7. Inner rotor
8. Drive pin

19.8b Unscrew the bolt and remove the sprocket ...

19.8c ... and the washer

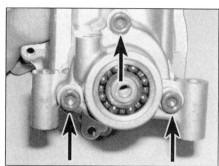

19.9a Unscrew the bolts (arrowed) ...

Engine, clutch and transmission 2•37

19.9b ... and remove the housing ...

19.9c ... and its locating pins

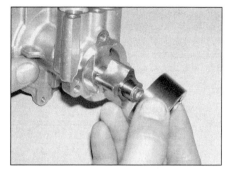
19.9d Remove the outer and inner rotors ...

19.9e ... and the drive pin

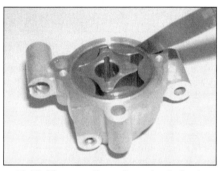
19.12 Measure the outer rotor to body clearance as shown

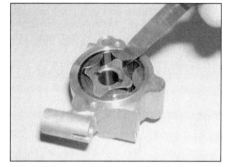
19.13 Measure the inner rotor tip to outer rotor clearance as shown

way round. Withdraw the drive pin from the shaft, noting how it locates in the slots in the inner rotor (see illustration).
10 Clean all components in solvent. Check that the oil pipes are clear by blowing them through with compressed air.
11 Inspect the pump housing and rotors for scoring and wear. If any damage, scoring or uneven or excessive wear is evident, replace them with new ones.
12 Fit the outer rotor into the pump housing, then fit the inner rotor into the outer rotor. Measure the clearance between the outer rotor and the pump body with a feeler gauge and compare it to the maximum clearance listed in the specifications at the beginning of the Chapter (see illustration). If the clearance measured is greater than the maximum listed, replace the housing and rotors with new ones.
13 Measure the clearance between the inner rotor tip and the outer rotor with a feeler gauge and compare it to the maximum clearance listed in the specifications at the beginning of the Chapter (see illustration). If the clearance measured is greater than the maximum listed, replace the rotors with new ones.
14 Check the pump drive and driven sprockets and the chain for wear or damage, and replace them with new ones if necessary. Also check the bearing in the pump housing – refer to Section 5 of Tools and Workshop Tips in the Reference section for information on bearing checks and renewal.
15 If the pump is good, make sure all the components are clean, then lubricate them with new engine oil. Fit the drive pin through the hole in the shaft – draw the shaft out a bit against the mechanical seal spring in the water pump to access the hole (or press the water pump impeller in if the cover has been removed) (see illustration 19.9e). Slide the inner rotor onto the shaft, with the slots in the rotor facing down so that they locate over the drive pin (see illustration). Fit the outer rotor onto the inner rotor (see illustration). Fit the housing locating pins into the body, then fit the pump housing over the rotors (see

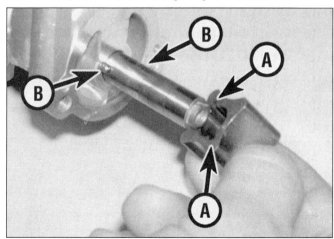
19.15a Locate the slots in the inner rotor (A) over the drive pin ends (B) ...

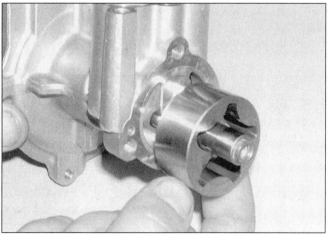
19.15b ... then fit the outer rotor over the inner rotor

2•38 Engine, clutch and transmission

19.15c Install and tighten the housing bolts to the specified torque

19.16a Locate the sprocket over the flats on the shaft end . . .

19.16b . . . and tighten the bolt to the specified torque

illustrations 19.9c and b). Install the bolts and tighten them to the torque setting specified at the beginning of the Chapter **(see illustration)**.

16 Fit the washer onto the end of the shaft **(see illustration 19.8c)**, then locate the driven sprocket, aligning the flats, and with the marked side facing in **(see illustration)**. Apply a suitable non-permanent thread locking compound to the sprocket bolt and tighten it to the specified torque setting, using a screwdriver as on removal to prevent it turning **(see illustration)**.

17 Rotate the pump shaft by hand and check that the rotors turn smoothly and freely. If not, recheck the pump, then refer to Chapter 3 and disassemble the water pump for inspection.

Installation

18 If removed, slide the inner thrust washer onto the shaft, followed by the oil pump drive sprocket, making sure its drive tabs face out **(see illustrations 19.7b and a)**.

19 Before installing the pump, prime it by filling it with clean engine oil **(see illustration)**. Check the condition of the large bore oil pipe and coolant outlet pipe O-rings and replace them with new ones if they are in any way damaged, deformed or deteriorated. Similarly check the O-ring in the water pump inlet orifice **(see illustration)**. It is a good idea to use new O-rings whatever the condition of the old ones. Smear all the O-rings with lithium-based grease before installing them.

20 Locate the drive chain on the input shaft, but not on the drive sprocket – this gives it enough slack to engage with the driven sprocket on the pump **(see illustration)**. Install the pump, tilting it to engage the drive chain on the driven sprocket, then slip the chain onto the drive sprocket **(see illustrations 19.6e and d)**. Slide the pump into the crankcase, making sure the O-ring does not come out of its groove **(see illustration 19.6c)**. Fit the dowel into the pump and crankcase **(see illustration)**. Apply a suitable non-permanent thread locking compound to the mounting bolt and sprocket cover bolts. Install the mounting bolt **(see illustration 19.6a)**, then locate the cover and tighten all the bolts to the torque setting specified at the beginning of the Chapter **(see illustration)**. Install the chain guide with the 'UP' marking facing up and out and tighten its bolts to the torque setting specified at the beginning of the Chapter **(see illustration 19.5b)**.

21 Install the small bore oil pipe, using a thread lock on its bolt and tightening it to the specified torque **(see illustrations)**. Fit the O-rings onto

19.19a Prime the pump with new oil

19.19b It is best to fit a new O-ring

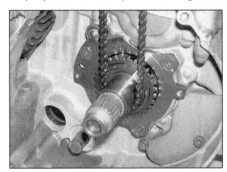

19.20a Locate the chain as shown when installing the pump

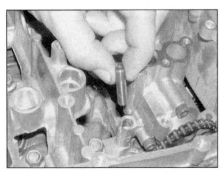

19.20b Fit the long dowel into the pump and crankcase

19.20c Fit the sprocket cover

19.21a Install the pipe . . .

Engine, clutch and transmission 2•39

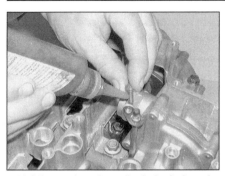

19.21b ... then threadlock the bolt and tighten it to the specified torque

19.21c Fit the O-rings ...

19.21d ... then install the pipe

the large bore oil pipe and fit it into its sockets **(see illustrations)**. Apply a suitable non-permanent thread locking compound to the pipe bolts and tighten them to the torque setting specified at the beginning of the Chapter.

22 Fit the O-rings onto the coolant outlet pipe and fit it into its socket **(see illustration 19.3b)**. Apply a suitable non-permanent thread locking compound to the pipe bolt and tighten it to the torque setting specified at the beginning of the Chapter **(see illustration 19.3a)**. Attach the coolant hoses and secure them with the clamps.

23 Install the clutch (see Section 14), the oil strainer and the sump (see Section 18).

20.4 Unscrew the bolt and remove the support plate

24 Fill the engine with the specified quantity and type of new engine oil and coolant (see Chapter 1).

20 Crankcase halves – separation and reassembly

Note: *To separate the crankcase halves, the engine must be removed from the frame.*

Separation

1 To access the connecting rods, pistons and rings, crankshaft, bearings, transmission shafts, and selector drum and forks, the crankcase must be split into two parts.

2 To enable the crankcases to be separated, the engine must be removed from the frame (see Section 5). Before the crankcases can be separated the following components must be removed:
  a) Camshafts (Section 10).
  b) Cylinder head (Section 11).
  c) Oil sump (Section 18).
  d) Clutch (Section 14).
  e) Oil/water pump assembly and drive chain (Section 19).
  f) Alternator rotor and stator (Chapter 9).

3 If the crankcases are being separated as part of a complete engine overhaul, remove the following components:
  a) Oil cooler (Section 7).
  b) Gearchange mechanism (Section 17).
  c) Starter motor (Chapter 9).
  d) Oil pressure relief valve (Section 18).
  e) Neutral switch, speed sensor and oil level sensor (Chapter 9).
  f) Pick-up coil (Chapter 5).
  g) Air induction system components – 2000-on models (see Chapter 4).

4 Unscrew the remaining bolt securing the front sprocket cover support plate to the left-hand side of the engine and remove the plate **(see illustration)**.

5 Turn the engine upside down so that it rests on the back of the upper crankcase and the cylinder head studs. The crankcases are joined by ten 9 mm bolts, two 8 mm bolts and sixteen 6 mm bolts. Make a cardboard template of the crankcase and punch a hole for each bolt location. This will ensure all bolts are installed correctly on reassembly – this is important, as many of them are of slightly differing length.

6 Unscrew the crankcase bolts evenly and a little at a time in a **reverse** of the numerical sequence shown and as marked on the crankcase (the number of each bolt is cast into the crankcase), until they are finger-tight, then remove them **(see illustrations)**. Note

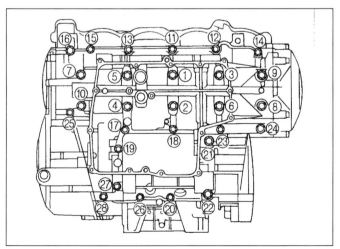

20.6a Crankcase bolt location and TIGHTENING sequence – slacken them in REVERSE order

20.6b Each bolt number is cast into the crankcase

2•40 Engine, clutch and transmission

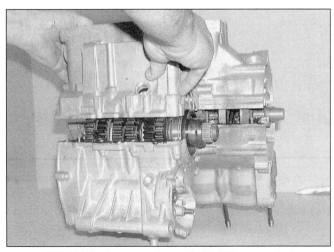

20.7 Lift the lower crankcase half off the upper half

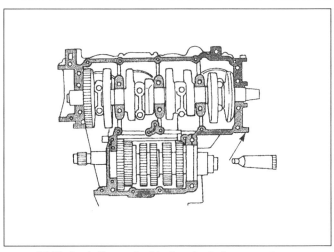

20.13 Apply the sealant to the shaded area as shown

the washers fitted with some of the bolts. **Note:** *As each bolt is removed, store it in its relative position, with its washer where applicable, in the cardboard template of the crankcase halves.*

7 Carefully lift the lower crankcase half off the upper half, using a soft-faced hammer to tap around the joint to initially separate the halves, if necessary **(see illustration)**. **Note:** *If the halves do not separate easily, make sure all fasteners have been removed. Do not try and separate the halves by levering against the crankcase mating surfaces as they are easily scored and will leak oil. Tap around the joint faces with a soft-faced mallet.*

8 Remove the three locating dowels from the crankcase (they could be in either half) **(see illustration 20.14)**.

9 Refer to Sections 21 to 29 for the removal and installation of the components housed within the crankcases.

Reassembly

10 Remove all traces of sealant from the crankcase mating surfaces.

11 Ensure that all components and their bearings are in place in the upper and lower crankcase halves (see Section 21). If the transmission shafts have not been removed, check the condition of the oil seal on the left-hand end of the output shaft and replace it with a new one if it is damaged, deformed or deteriorated **(see illustration 27.7b)**. Check that the selector drum is in the neutral position (see Section 29).

12 Generously lubricate the crankshaft, transmission shafts and selector drum and forks, particularly around the bearings, with clean engine oil, then use a rag soaked in high flash-point solvent to wipe over the mating surfaces of both crankcase halves to remove all traces of oil.

13 Apply a small amount of suitable sealant (such as Yamaha Bond 1215) to the mating surface of one crankcase half as shown **(see illustration)**.

Caution: *Do not apply an excessive amount of sealant as it will ooze out when the case halves are assembled and may obstruct oil passages. Do not apply the sealant on or too close (within 2 to 3 mm) to any of the bearing inserts or surfaces.*

14 If removed, fit the three locating dowels into the crankcase **(see illustration)**.

15 Check again that all components are in position, particularly that the bearing shells are still correctly located in the lower crankcase half. If the timing rotor is still on the crankshaft, make sure the cam chain is installed with the crankshaft, otherwise the rotor will have to be removed. Carefully fit the lower crankcase half onto the upper crankcase half, making sure the dowels locate correctly **(see illustration 20.7)**.

16 Check that the lower crankcase half is correctly seated. **Note:** *The crankcase halves should fit together without being forced. If the casings are not correctly seated, remove the lower crankcase half and investigate the problem. Do not attempt to pull them together using the crankcase bolts as the casing will crack and be ruined.* Make sure the output shaft oil seal is correctly seated.

17 On 1998 and 1999 models, clean the threads of the 9 mm lower crankcase bolts (Nos. 1 to 10) and apply new engine oil to their threads **(see illustration)**. Insert them with their washers in their original locations shown and as marked on the crankcase (the number of each bolt is cast into the crankcase) **(see illustrations 20.6a and b)**. Secure the bolts finger-tight at first, then tighten them evenly and a little at a time in the correct numerical sequence to the torque setting specified at the beginning of the Chapter **(see illustration)**.

18 On 2000-on models, clean the threads of the 9 mm lower crankcase bolts (Nos. 1 to 10) and apply new engine oil to their threads **(see illustration 20.17a)**. Insert them with their

20.14 Fit the dowels (arrowed) into the crankcase

20.17a Oil the bolt threads and do not forget the washers . . .

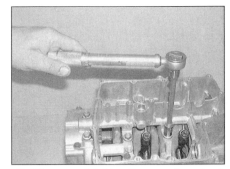

20.17b . . . and tighten them as described

washers in their original locations shown and as marked on the crankcase (the number of each bolt is cast into the crankcase) **(see illustrations 20.6a and b)**. Secure the bolts finger-tight at first, then tighten them evenly and a little at a time in the correct numerical sequence to the torque setting specified at the beginning of the Chapter **(see illustration 20.17b)**. Now tighten them by a further 45 to 50° (1/8 of a turn) using a degree disc (see Section 4 of *Tools and Workshop Tips* in the *Reference* section).

19 On all models, clean the threads of the 8 and 6 mm crankcase bolts and apply new engine oil to their threads. Insert them in their original locations shown and as marked on the crankcase (the number of each bolt is cast into the crankcase) **(see illustrations 20.6a and b)**. Secure all bolts finger-tight at first, then tighten them evenly and a little at a time in the correct numerical sequence to the torque settings specified at the beginning of the Chapter. Be sure to distinguish correctly between the 8 and 6 mm bolts, and between the Nos. 16 and 24 6 mm bolts and the rest, as they have different settings (see Torque Wrench Settings, in Specifications).

20 With all crankcase fasteners tightened, check that the crankshaft and transmission shafts rotate smoothly and easily. Check that the transmission shafts rotate freely and independently in neutral, then rotate the selector drum by hand and select each gear in turn whilst rotating the input shaft. Check that all gears can be selected and that the shafts rotate freely in every gear. If there are any signs of undue stiffness, tight or rough spots, or of any other problem, the fault must be rectified before proceeding further.

21 Apply a suitable non-permanent thread locking compound to the threads of the front sprocket cover support plate bolt, then fit the plate and tighten the bolt to the specified torque setting **(see illustration 20.4)**.

22 Install all other removed assemblies in the reverse of the sequences given in Steps 2 and 3, according to your procedure.

21 Crankcase halves and cylinder bores – inspection and servicing

Crankcase halves

1 After the crankcases have been separated, remove the crankshaft and bearings, connecting rods and pistons, transmission shafts, selector drum and forks, and any other components or assemblies not already removed, referring to the relevant Sections of this and other Chapters (see Steps 2 and 3 in Section 20).

2 Withdraw the oil feed pipe from the upper crankcase – it is a push-fit **(see illustration)**. Check the condition of the pipe O-rings and replace them with new ones if they are in any way damaged, deformed or deteriorated. If

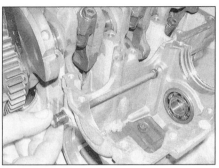

21.2 Withdraw the oil pipe from the crankcase

required, unscrew the bolts securing the two oil baffle plates in the upper crankcase half and remove them, noting how they fit. If required, unscrew the oil cooler bolt from the front of the lower crankcase half.

3 Clean the crankcases thoroughly with new solvent and dry them with compressed air. Blow out all oil passages and pipes with compressed air.

4 Remove all traces of old gasket sealant from the mating surfaces. Minor damage to the surfaces can be cleaned up with a fine sharpening stone or grindstone.

Caution: Be very careful not to nick or gouge the crankcase mating surfaces, or oil leaks will result. Check both crankcase halves very carefully for cracks and other damage.

5 Small cracks or holes in aluminium castings can be repaired with an epoxy resin adhesive as a temporary measure. Permanent repairs can only be effected by argon-arc welding, and only a specialist in this process is in a position to advise on the economy or practical aspect of such a repair. If any damage is found that can't be repaired, renew the crankcase halves as a set.

6 Damaged threads can be economically reclaimed by using a diamond section wire insert, of the Heli-Coil type, which is easily fitted after drilling and re-tapping the affected thread.

7 Sheared studs or screws can be removed with stud or screw extractors, and they usually succeed in dislodging the most stubborn of them.

> **HAYNES HiNT** *Refer to Tools and Workshop Tips (Section 2) in the Reference section for details of installing a thread insert and using screw extractors.*

8 Install the oil baffle plates, then apply a suitable non-permanent thread locking compound to the threads of the bolts and tighten them to the torque setting specified at the beginning of the Chapter. Apply a suitable sealant (such as Yamaha Bond 1215) to the groove in the edge of the seating side of the large baffle plate. Install the oil feed pipe using new O-rings if necessary and push it fully in, locating the tab on the outer end in the cutout in the crankcase **(see illustration)**.

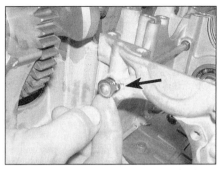

21.8 Locate the tab in the cutout in the crankcase (arrowed)

9 Install the crankshaft, pistons and bearings, cam chain, transmission shafts and selector drum and forks, before reassembling the crankcase halves.

Cylinder bores

Caution: Do not attempt to separate the liners from the cylinder block.

10 Check the cylinder walls carefully for scratches and score marks.

11 Using telescoping gauges and a micrometer (see Section 3 of *Tools and Workshop Tips* in the *Reference* section), check the dimensions of each cylinder to assess the amount of wear, taper and ovality. Measure near the top (but below the level of the top piston ring at TDC), centre and bottom (but above the level of the oil ring at BDC) of the bore, both parallel to and across the crankshaft axis **(see illustration)**. Compare the results to the specifications at the beginning of the Chapter.

12 If the precision measuring tools are not available, take the crankcase to a Yamaha dealer or specialist motorcycle repair shop for assessment and advice.

13 If the cylinders are worn beyond the service limit, or badly scratched, scuffed or scored, renew the crankcases as a set. Yamaha do not supply an oversize piston and ring set. If new crankcase are fitted, new piston rings must be used, and it is a good idea to fit new pistons as well.

14 If the cylinders are in good condition and the piston-to-bore clearance is within specifications (see Section 25), the cylinders

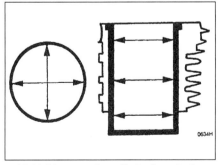

21.11 Measure the cylinder bore in the directions shown with a telescoping gauge, then measure the gauge with a micrometer

should be honed (de-glazed). To perform this operation you will need the proper size flexible hone with fine stones (see Specialist Tools in *Tools and Workshop Tips* in the *Reference* section), or a bottle-brush type hone, plenty of light oil or honing oil, some clean rags and an electric drill motor.

15 Hold the block sideways (so that the bores are horizontal rather than vertical) in a vice with soft jaws or cushioned with wooden blocks. Mount the hone in the drill motor, compress the stones and insert the hone into the cylinder. Thoroughly lubricate the cylinder, then turn on the drill and move the hone up and down in the cylinder at a pace which produces a fine cross-hatch pattern on the cylinder wall with the lines intersecting at an angle of approximately 60°. Be sure to use plenty of lubricant and do not take off any more material than is necessary to produce the desired effect. Do not withdraw the hone from the cylinder while it is still turning. Switch off the drill and continue to move it up and down in the cylinder until it has stopped turning, then compress the stones and withdraw the hone. Wipe the oil from the cylinder and repeat the procedure on the other cylinder. Remember, do not take too much material from the cylinder wall.

16 Wash the cylinders thoroughly with warm soapy water to remove all traces of the abrasive grit produced during the honing operation. After rinsing, dry the cylinders thoroughly and apply a thin coat of light, rust-preventative oil to all machined surfaces.

17 If you do not have the equipment or desire to perform the honing operation, take the crankcase to a Yamaha dealer or specialist motorcycle repair shop.

22 Main and connecting rod bearings – general information

1 Even though main and connecting rod bearings are generally replaced with new ones during the engine overhaul, the old bearings should be retained for close examination as they may reveal valuable information about the condition of the engine.

2 Bearing failure occurs mainly because of lack of lubrication, the presence of dirt or other foreign particles, overloading the engine and/or corrosion. Regardless of the cause of bearing failure, it must be corrected before the engine is reassembled to prevent it from happening again.

3 When examining the bearings, match them with their corresponding journal on the crankshaft to help identify the cause of any problem.

4 Dirt and other foreign particles get into the engine in a variety of ways. It may be left in the engine during assembly or it may pass through filters or breathers. It may get into the oil and from there into the bearings. Metal chips from machining operations and normal

23.3 Carefully lift the crankshaft out of the crankcase

engine wear are often present. Abrasives are sometimes left in engine components after reconditioning operations, especially when parts are not thoroughly cleaned using the proper cleaning methods. Whatever the source, these foreign objects often end up imbedded in the soft bearing material and are easily recognised. Large particles will not imbed in the bearing and will score or gouge the bearing and journal. The best prevention for this cause of bearing failure is to clean all parts thoroughly and keep everything spotlessly clean during engine reassembly. Frequent and regular oil and filter changes are also recommended.

5 Lack of lubrication or lubrication breakdown has a number of interrelated causes. Excessive heat (which thins the oil), overloading (which squeezes the oil from the bearing face) and oil leakage or throw off (from excessive bearing clearances, worn oil pump or high engine speeds) all contribute to lubrication breakdown. Blocked oil passages will also starve a bearing and destroy it. When lack of lubrication is the cause of bearing failure, the bearing material is wiped or extruded from the steel backing of the bearing. Temperatures may increase to the point where the steel backing and the journal turn blue from overheating.

 Refer to Tools and Workshop Tips (Section 5) in the Reference section for bearing fault finding.

23.4 To remove a main bearing shell, push it sideways and lift it out

6 Riding habits can have a definite effect on bearing life. Full throttle low speed operation, or labouring the engine, puts very high loads on bearings, which tend to squeeze out the oil film. These loads cause the bearings to flex, which produces fine cracks in the bearing face (fatigue failure). Eventually the bearing material will loosen in pieces and tear away from the steel backing. Short trip riding leads to corrosion of bearings, as insufficient engine heat is produced to drive off the condensed water and corrosive gases produced. These products collect in the engine oil, forming acid and sludge. As the oil is carried to the engine bearings, the acid attacks and corrodes the bearing material.

7 Incorrect bearing installation during engine assembly will lead to bearing failure as well. Tight fitting bearings which leave insufficient bearing oil clearances result in oil starvation. Dirt or foreign particles trapped behind a bearing insert result in high spots on the bearing which lead to failure.

8 To avoid bearing problems, clean all parts thoroughly before reassembly, double check all bearing clearance measurements and lubricate the new bearings with clean engine oil during installation.

23 Crankshaft and main bearings – removal, inspection and installation

Note: *To remove the crankshaft the engine must be removed from the frame and the crankcase halves separated.*

Removal

1 Remove the engine from the frame (see Section 5) and separate the crankcase halves (see Section 20).

2 Refer to Section 24, Steps 2, 3 and 4, then unscrew the connecting rod cap nuts and separate the caps from the crankpin. Keep the caps in order and the correct way round, as they must be installed in their original location. **Note:** *If no work is to be carried out on the piston/connecting rod assemblies there is no need to remove them from the bores, but push them up to the top of the bores so that the bottom ends are clear of the crankshaft.*

3 Lift the crankshaft out of the upper crankcase half, taking care not to dislodge the main bearing shells, then remove the cam chain **(see illustration)**.

4 If required, remove the bearing shells from the crankcase halves by pushing their centres to the side, then lifting them out **(see illustration)**. Keep the shells in order.

Inspection

5 Clean the crankshaft with solvent, using a rifle-cleaning brush to scrub out the oil passages. If available, blow the crank dry with compressed air, and also blow through the oil passages. Check the primary drive gear for wear or damage. If any of the teeth are

Engine, clutch and transmission 2•43

23.13 Place a strip of Plastigauge on each bearing journal

23.19 Measure the crushed Plastigauge using the scale on the pack – be sue to use the correct one as both metric and imperial are included

excessively worn, chipped or broken, the crankshaft must be replaced with a new one. Check the primary driven gear on the back of the clutch housing for corresponding wear or damage. Also check the cam chain sprocket, the sprockets on the camshafts and the cam chain itself and replace them with new ones, if necessary.

6 Refer to Section 22 and examine the main bearing shells. If they are scored, badly scuffed or appear to have been seized, new bearings must be installed. Always renew the main bearings as a set. If they are badly damaged, check the corresponding crankshaft journals. Evidence of extreme heat, such as discoloration, indicates that lubrication failure has occurred. Be sure to thoroughly check the oil pump and pressure relief valve as well as all oil holes and passages before reassembling the engine.

7 Give the crankshaft journals a close visual examination, paying particular attention where damaged bearings have been discovered. If the journals are scored or pitted in any way, a new crankshaft will be required. Note that undersizes are not available, precluding the option of re-grinding the crankshaft.

8 Place the crankshaft on V-blocks and check the runout at the main bearing journals using a dial gauge. Compare the reading to the maximum specified at the beginning of the Chapter. If the runout exceeds the limit, the crankshaft must be renewed.

Oil clearance check

9 Whether new bearing shells are being fitted or the original ones are being re-used, the main bearing oil clearance should be checked before the engine is reassembled. Main bearing oil clearance is measured with a product known as Plastigauge.

10 If not already done, remove the bearing shells from the crankcase halves (see Step 4). Clean the backs of the shells and the bearing housings in both crankcase halves, and the main bearing journals on the crankshaft.

11 Press the bearing shells into their cut-outs, ensuring that the tab on each shell engages in the notch in the crankcase **(see illustration 23.26)**. Make sure the bearings are fitted in the correct locations and take care not to touch any shell's bearing surface with your fingers.

12 Ensure the shells and crankshaft are clean and dry. Lay the crankshaft in position in the upper crankcase **(see illustration 23.3)**.

13 Cut several lengths of the appropriate size Plastigauge (they should be slightly shorter than the width of the crankshaft journals). Place a strand of Plastigauge on each journal, making sure it will be clear of the oil holes in the shells when the lower crankcase is installed **(see illustration)**. Make sure the crankshaft is not rotated.

14 If removed, fit the dowels into the crankcase **(see illustration 20.14)**. Carefully fit the lower crankcase half onto the upper half, making sure the dowels locate correctly and the Plastigauge is not disturbed **(see illustration 20.7)**. Check that the lower crankcase half is correctly seated. **Note:** *Do not tighten the crankcase bolts if the casing is not correctly seated.*

15 On 1998 and 1999 models, clean the threads of the 9 mm lower crankcase bolts (Nos. 1 to 10) and apply new engine oil to their threads **(see illustration 20.17a)**. Insert them with their washers in their original locations shown and as marked on the crankcase (the number of each bolt is cast into the crankcase) **(see illustrations 20.6a and b)**. Secure the bolts finger-tight at first, then tighten them evenly and a little at a time in the correct numerical sequence to the torque setting specified at the beginning of the Chapter **(see illustration 20.17b)**.

16 On 2000-on models, clean the threads of the 9 mm lower crankcase bolts (Nos. 1 to 10) and apply new engine oil to their threads **(see illustration 20.17a)**. Insert them with their washers in their original locations shown and as marked on the crankcase (the number of each bolt is cast into the crankcase) **(see illustrations 20.6a and b)**. Secure the bolts finger-tight at first, then tighten them evenly and a little at a time in the correct numerical sequence to the torque setting specified at the beginning of the Chapter **(see illustration 20.17b)**. Now tighten them by a further 45 to 50° (1/8 of a turn) using a degree disc (see Section 4 of *Tools and Workshop Tips* in the *Reference* section).

17 On all models, clean the threads of the 8 and 6 mm crankcase bolts and apply new engine oil to their threads. Insert them in their original locations shown and as marked on the crankcase (the number of each bolt is cast into the crankcase) **(see illustrations 20.6a and b)**. Secure all bolts finger-tight at first, then tighten them evenly and a little at a time in the correct numerical sequence to the torque settings specified at the beginning of the Chapter. Be sure to distinguish correctly between the 8 and 6 mm bolts, and between the Nos. 16 and 24 6 mm bolts and the rest, as they have different settings (see *Torque Wrench Settings* in Specifications).

18 Now unscrew the crankcase bolts evenly and a little at a time in a reverse of the numerical sequence shown and as marked on the crankcase (the number of each bolt is cast into the crankcase), until they are finger-tight, then remove them **(see illustrations 20.6a and b)**. Note the washers fitted with some of the bolts. **Note:** *As each bolt is removed, store it in its relative position, with its washer where applicable, in the cardboard template of the crankcase halves.* Carefully lift off the lower crankcase half, making sure the Plastigauge is not disturbed.

19 Compare the width of the crushed Plastigauge on each crankshaft journal to the scale printed on the Plastigauge envelope to obtain the main bearing oil clearance **(see illustration)**. Compare the reading to the specifications at the beginning of the Chapter.

20 On completion carefully scrape away all traces of the Plastigauge material from the

2•44 Engine, clutch and transmission

23.23a Main bearing journal numbers (A), big-end bearing journal numbers (B)

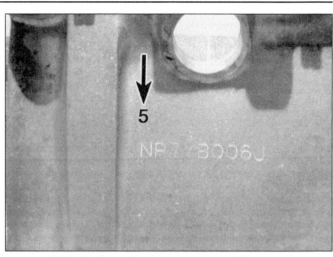

23.23b Main bearing housing number(s) (arrowed)

crankshaft journal and bearing shells; use a fingernail or other object which is unlikely to score them.

21 If the oil clearance falls into the specified range, no bearing shell renewal is required (provided they are in good condition). If the clearance is beyond the service limit, refer to the marks on the case and the marks on the crankshaft and select new bearing shells (see Steps 23 and 24). Install the new shells and check the oil clearance once again (the new shells may bring bearing clearance within the specified range). Always renew all of the shells at the same time.

22 If the clearance is still greater than the service limit listed in this Chapter's Specifications (even with replacement shells), the crankshaft journal is worn and the crankshaft should be renewed.

Main bearing shell selection

23 Replacement bearing shells for the main bearings are supplied on a selective fit basis. Code numbers stamped on various components are used to identify the correct replacement bearings. The crankshaft journal size numbers are stamped on the outside of the crankshaft web on the left-hand end **(see illustration)**. The left-hand block of five numbers are for the main bearing journals (the right-hand block of four numbers are for the big-end bearing journals). The first number of the five is for the left-hand (No. 1) journal, and so on. The main bearing housing numbers are stamped into the back of the lower crankcase half **(see illustration)**. The first number of the five is for the left-hand (No. 1) journal, and so on. Note that if there is only one number stamped into the crankcase, it means that all the journals are the same number.

24 A range of bearing shells is available. To select the correct bearing for a particular journal, subtract the main bearing journal number (stamped on the crank web) from the main bearing housing number (stamped on the crankcase), and then subtract 2 from the result (i.e. crankcase number minus crankweb number minus 2). Compare the bearing number calculated with the table below to find the colour coding of the replacement bearing required.

Number	Colour
-1	pink/violet
0	pink/white
1	pink/blue
2	pink/black
3	pink/brown

Installation

25 If not already done, remove the bearing shells from the crankcase halves (see Step 4). Clean the backs of the shells and the bearing cut-outs in both crankcase halves, and the main bearing journals on the crankshaft. If new shells are being fitted, ensure that all traces of the protective grease are cleaned off using paraffin (kerosene). Wipe the shells and crankcase halves dry with a lint-free cloth. Make sure all the oil passages and holes are clear, and blow them through with compressed air if it is available.

26 Press the bearing shells into their locations. Make sure the tab on each shell engages in the notch in the casing **(see illustration)**. Make sure the bearings are fitted in the correct locations and take care not to touch any shell's bearing surface with your fingers. Lubricate each shell with clean engine oil.

27 If the timing rotor is still on the crankshaft, fit the cam chain onto its sprocket. If the rotor has been removed, the cam chain can be installed later. Lower the crankshaft into position in the upper crankcase, feeding the cam chain into its tunnel if applicable, and making sure all bearings remain in place **(see illustration 23.3)**.

28 Refer to Section 24, Steps 30 to 32, and fit the connecting rods on to the crankshaft.

29 Check that the crankshaft rotates smoothly and freely. If there are any signs of roughness or tightness, detach the rods and recheck the assembly, and if that is good, the oil main and big-end clearances (see above and Section 24). Sometimes tapping the bottom of the connecting rod cap will relieve tightness.

30 Reassemble the crankcase halves (see Section 20).

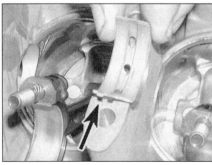

23.26 Press each shell into place, locating the tab in the notch (arrowed)

24 Connecting rods – removal, inspection and installation

Note: *To remove the connecting rods the engine must be removed from the frame and the crankcases separated.*

Removal

1 Remove the engine from the frame (see Section 5) and separate the crankcase halves (see Section 20).

2 Before separating the rods from the crankshaft, measure the side clearance on each rod with a feeler gauge **(see illustration)**. If the clearance on any rod is greater than the service limit listed in this Chapter's Specifications, replace that rod with a new one.

Engine, clutch and transmission 2•45

24.2 Measure the connecting rod side clearance using a feeler gauge

24.3 The arrow on the piston points to the front of the engine and the 'Y' mark on the connecting rod faces the left-hand side

24.4a Unscrew the nuts (arrowed) . . .

3 Using paint or a felt marker pen, mark the relevant cylinder identity across the join between each connecting rod and cap. Note the 'Y' mark on each connecting rod which must face to the left-hand side of the engine, and the arrow on the piston which points to the front of the engine – these ensure that the cap and rod are fitted the correct way around on reassembly **(see illustration)**. Note that the number and letter already across the rod and cap indicate rod size and weight grade respectively, not cylinder number.

4 Unscrew the connecting rod cap nuts and separate the cap and rod from the crankpin **(see illustrations)**. If the cap appears stuck, tap it on one end with a hammer while pulling it. Lift the crankshaft out of the upper crankcase half, taking care not to dislodge the main bearing shells **(see illustration 23.3)**.

5 Turn the crankcase on its side. Push each piston/connecting rod assembly up and remove it from the top of the bore making sure the connecting rod does not mark the cylinder bore walls **(see illustration)**.

 To ease removal of the pistons, carefully remove any ridge of carbon built up on the top of each cylinder bore using a scraper. If there is a pronounced wear ridge, remove it using a ridge reamer.

Caution: Do not try to remove the piston/connecting rod from the bottom of the cylinder bore. The piston will not pass the crankcase main bearing webs. If the piston is pulled right to the bottom of the bore the oil control ring will expand and lock the piston in position. If this happens it is likely the ring will be broken.

6 Fit the relevant bearing shells (if removed), bearing cap, and bolts on each piston/connecting rod assembly so that they are all kept together as a matched set.

7 Separate the pistons from the connecting rods if required (see Section 25).

Inspection

8 Check the connecting rods for cracks and other obvious damage.

9 Apply clean engine oil to the piston pin, insert it into its connecting rod small-end and check for any freeplay between the two **(see illustration)**. If freeplay is excessive, measure the pin external diameter **(see illustration 25.13b)**. Compare the result to the specifications at the beginning of the Chapter. Replace the pin with a new one if it is worn beyond its specified limits. If the pin diameter is within specifications, replace the connecting rod with a new one. Repeat the measurements for all the rods.

10 Refer to Section 22 and examine the connecting rod bearing shells. If they are scored, badly scuffed or appear to have seized, new shells must be installed. Always renew the shells in the connecting rods as a set. If they are badly damaged, check the corresponding crankpin. Evidence of extreme heat, such as discoloration, indicates that lubrication failure has occurred. Be sure to thoroughly check the oil pump and pressure regulator as well as all oil holes and passages before reassembling the engine.

24.4b . . . and pull the cap off the connecting rod

11 Have the rods checked for twist and bend by a Yamaha dealer if you are in doubt about their straightness.

Oil clearance check

12 Whether new bearing shells are being fitted or the original ones are being re-used, the connecting rod (big-end) bearing oil clearance should be checked prior to reassembly.

13 Remove the bearing shells from the rods and caps, keeping them in order **(see illustration)**. Clean the backs of the shells and the bearing locations in both the connecting rod and cap, and the crankpin journal.

14 Press the bearing shells into their locations, ensuring that the tab on each shell

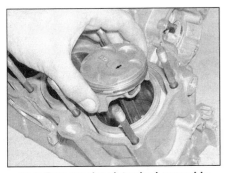

24.5 Remove the piston/rod assembly from the top of the bore

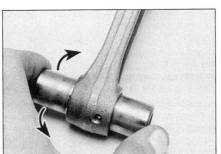

24.9 Slip the piston pin into the rod's small-end and rock it back and forth to check for looseness

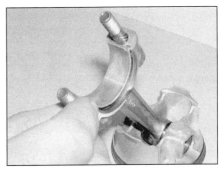

24.13 To remove a big-end bearing shell, push it sideways and lift it out

2•46 Engine, clutch and transmission

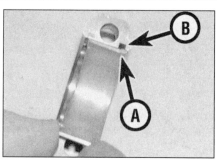

24.14 Make sure the tab (A) locates in the notch (B)

engages the notch in the connecting rod/cap **(see illustration)**. Make sure the bearings are fitted in the correct locations and take care not to touch any shell's bearing surface with your fingers.
15 Cut a length of the appropriate size Plastigauge (it should be slightly shorter than the width of the crankpin). Place a strand of Plastigauge on the crankpin journal **(see illustration 23.13)**.
16 Apply molybdenum disulphide grease to the bolt shanks and threads and to the seats of the nuts. Fit the connecting rod and cap onto the crankshaft **(see illustrations 24.30a and b)**. Make sure the cap is fitted the correct way around so the previously made markings align, and that the rod is facing the right way (see Step 3). Fit the nuts and tighten them finger-tight, making sure the connecting rod does not rotate on the crankshaft.
17 Tighten the cap nuts to the initial torque setting specified at the beginning of the Chapter, making sure the connecting rod does not rotate on the crankshaft **(see illustration 24.32)**. Now tighten each nut in turn and in one continuous movement to the final torque setting specified. If tightening is paused between the initial and final settings, slacken the nut to below the initial setting and repeat the procedure.
18 Slacken the cap nuts and remove the cap and rod, again taking great care not to rotate the rod or crankshaft.
19 Compare the width of the crushed Plastigauge on the crankpin to the scale printed on the Plastigauge envelope to obtain the connecting rod bearing oil clearance **(see illustration 23.19)**. Compare the reading to the specifications at the beginning of the Chapter.

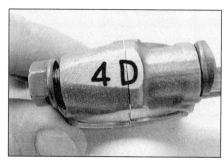

24.24 Connecting rod size number

20 On completion, carefully scrape away all traces of the Plastigauge material from the crankpin and bearing shells using a fingernail or other object which is unlikely to score the shells.
21 If the clearance is within the range listed in this Chapter's Specifications and the bearings are in perfect condition, they can be reused. If the clearance is beyond the service limit, replace the bearing shells with new ones (see Steps 24 and 25). Check the oil clearance once again (the new shells may be thick enough to bring bearing clearance within the specified range). Always renew all of the shells at the same time.
22 If the clearance is still greater than the service limit listed in this Chapter's Specifications, the big-end bearing journal is worn and the crankshaft should be replaced with a new one.
23 Repeat the procedure for the remaining connecting rods.

Bearing shell selection

24 Replacement bearing shells for the big-end bearings are supplied on a selected fit basis. Code numbers stamped on various components are used to identify the correct replacement bearings. The crankshaft journal size numbers are stamped on the outside of the crankshaft web on the left-hand end **(see illustration 23.23a)**. The right-hand block of four numbers are for the big-end bearing journals (the left-hand block of five numbers are for the main bearing journals). The first number of the four is for the left-hand (No. 1 cylinder) journal, and so on. Each connecting rod size number is marked in ink across the flat face of the connecting rod and cap **(see illustration)**.
25 A range of bearing shells is available. To select the correct bearing for a particular connecting rod, subtract the big-end bearing journal number (stamped on the crank web) from the connecting rod number (marked on the rod), and then subtract 2 from the result (i.e. connecting rod number minus crankweb number minus 2). Compare the bearing number calculated with the table below to find the colour coding of the replacement bearing required. The colour code is marked on the side of each bearing.

Number	Colour
-1	violet
0	white
1	blue
2	black

Installation

Note: *If new connecting rods are being fitted, note that modified rods will be supplied and that the new type rod requires new bolts and nuts to be fitted.*

26 If removed, fit the pistons onto the connecting rods (see Section 25).
27 Clean the backs of the bearing shells and the bearing locations in both cap and rod. If new shells are being fitted, ensure that all traces of the protective grease are cleaned off using paraffin (kerosene). Wipe the shells, cap and rod dry with a clean lint-free cloth. Install the bearing shells in the connecting rods and caps, making sure the tab on each shell locates in the notch in the connecting rod/cap **(see illustration 24.14)**. Make sure the bearings are fitted in their correct locations and take care not to touch any shell's bearing surface with your fingers. Lubricate the shells with clean engine oil.
28 Place the crankcase on blocks to prevent the connecting rod bolts hitting the bench before the piston is fully in the bore. Lubricate the pistons, rings and cylinder bore with clean engine oil. Insert the piston/connecting rod assembly into the top of its bore, taking care not to allow the connecting rod to mark the bore **(see illustration 24.5)**. Make sure the arrow marked on the piston crown points to the front of the bore and the 'Y' mark faces the left-hand side of the engine (see Step 3) **(see illustration 24.3)**, then carefully compress and feed each piston ring into the bore until the piston crown is flush with the top of the bore **(see illustration)**. If available, a piston ring compressor makes installation a lot easier **(see illustrations)**.

24.28a Carefully feed each ring into the bore using a screwdriver and/or fingernails

24.28b Clamp the rings in position by tightening the bands on the compressor . . .

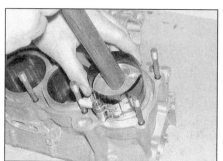

24.28c . . . then push the piston into its bore

Engine, clutch and transmission 2•47

24.30a Pull the rod onto the crankpin . . .

24.30b . . . then fit the cap

24.32 Tighten the nuts as described, first to the initial torque, then to the final torque

29 If the timing rotor is still on the crankshaft, fit the cam chain onto its sprocket. If the rotor has been removed, the cam chain can be installed later. Lower the crankshaft into position in the upper crankcase, feeding the cam chain into its tunnel if applicable, and making sure all bearings remain in place (see illustration 23.3).

30 Working on one connecting rod at a time, lubricate the crankpin and the shells in the connecting rod and cap with clean engine oil. Pull the rod onto the crankpin and fit the cap with its shell onto the rod (see illustrations). Make sure the cap is fitted the correct way around so the previously made markings align (see Step 3).

31 Apply molybdenum disulphide grease to the bolt shanks and threads and to the seats of the nuts. Fit the nuts and tighten them finger-tight. Check again to make sure all components have been returned to their original locations using the marks made on disassembly.

32 If refitting the original connecting rods, tighten the cap nuts to the initial torque setting specified at the beginning of this Chapter (see illustration). Now tighten each nut in turn and in one continuous movement to the final torque setting specified. If tightening is paused between the initial and final settings, slacken the nut to below the initial setting and repeat the procedure. Fit the remaining rods onto the crankshaft in the same way.

33 If installing new connecting rods (see Note at the beginning of this Section) note that the new bolts and nuts must be used. Tighten the cap nuts to the initial torque setting specified at the beginning of this Chapter (see illustration 24.32). Now tighten each nut in turn through 120° using an angle tightening gauge or degree disc (see Section 4 of Tools and Workshop Tips in the Reference Section). Do not exceed the tightening angle - if any nut is overtightened, the bolt and nut should be renewed.

34 Check that the crankshaft rotates smoothly and freely. If there are any signs of roughness or tightness, detach the rods and recheck the assembly, and if that is good, the oil main and big-end clearances (see above and Section 23). Sometimes tapping the bottom of the connecting rod cap will relieve tightness.

35 Reassemble the crankcase halves (see Section 20).

25 Pistons – removal, inspection and installation

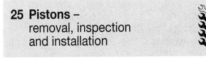

Note: *To remove the pistons the engine must be removed from the frame and the crankcase halves separated.*

Removal

1 Remove the connecting rods (see Section 24).

2 Before removing the piston from the connecting rod, use a sharp scriber or felt marker pen to write the cylinder identity on the crown of each piston (or on the inside of the skirt if the piston is dirty and going to be cleaned) as it must be installed in its original cylinder. Each piston also has an arrow on its crown which points to the exhaust (front) side of the bore (see illustration). If this is not visible, mark the piston accordingly so that it can be installed the correct way round.

3 Carefully prise out the circlip on one side of the piston using needle-nose pliers or a small flat-bladed screwdriver inserted into the notch (see illustration). Push the piston pin out from the other side to free the piston from the connecting rod (see illustration). Remove the other circlip and discard them both, as new ones must be used. When the piston has been removed, install its pin back into its bore so that related parts do not get mixed up. Rotate the crankshaft so that the best access is obtained for each piston.

 HAYNES HiNT *If a piston pin is a tight fit in the piston bosses, soak a rag in boiling water then wring it out and wrap it around the piston – this will expand the alloy piston sufficiently to release its grip on the pin. If the piston pin is particularly stubborn, extract it using a drawbolt tool, but be careful to protect the piston's working surfaces.*

Inspection

4 Before the inspection process can be carried out, the pistons must be cleaned and the old piston rings removed.

5 Using your thumbs or a piston ring removal and installation tool, carefully remove the rings from the pistons (see illustration). Do not nick or gouge the pistons in the process.

25.2 Note the arrow on the piston which must point forwards

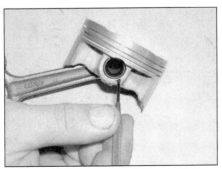

25.3a Prise out the circlip . . .

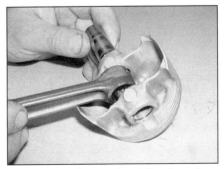

25.3b . . . then push out the pin and remove the piston

25.5a Removing the piston rings using a ring removal and installation tool

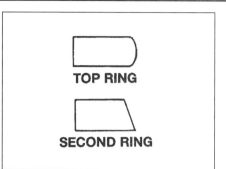

25.5b The rings can be identified by their different profiles

25.11 Measure the piston ring-to-groove clearance with a feeler gauge

Carefully note which way up each ring fits and in which groove, as they must be installed in their original positions if being re-used. The upper surface of each ring should have a manufacturer's mark or letter at one end – if the mark on each ring is different, note which mark is for the top ring and which is for the second. The rings can also be distinguished by their different profiles **(see illustration)**.

6 Scrape all traces of carbon from the tops of the pistons. A hand-held wire brush or a piece of fine emery cloth can be used once most of the deposits have been scraped away. Do not, under any circumstances, use a wire brush mounted in a drill motor to remove deposits from the pistons; the piston material is soft and will be eroded away by the wire brush.

7 Use a piston ring groove cleaning tool to remove any carbon deposits from the ring grooves. If a tool is not available, a piece broken off an old ring will do the job. Be very careful to remove only the carbon deposits. Do not remove any metal and do not nick or gouge the sides of the ring grooves.

8 Once the deposits have been removed, clean the pistons with solvent and dry them thoroughly. If the identification previously marked on the piston is cleaned off, be sure to re-mark it with the correct identity. Make sure the oil return holes below the oil ring groove are clear.

9 Carefully inspect each piston for cracks around the skirt, at the pin bosses and at the ring lands. Normal piston wear appears as even, vertical wear on the thrust surfaces of the piston and slight looseness of the top ring in its groove. If the skirt is scored or scuffed, the engine may have been suffering from overheating and/or abnormal combustion, resulting in excessively high operating temperatures. The oil pump should be checked thoroughly. Also check that the circlip grooves are not damaged.

10 A hole in the piston crown, although extreme, is an indication that abnormal combustion (pre-ignition) was occurring. Burned areas at the edge of the piston crown are usually evidence of spark knock (detonation). If any of the above problems exist, the causes must be corrected or the damage will occur again.

11 Measure the piston ring-to-groove clearance by laying each piston ring in its groove and slipping a feeler gauge in beside it **(see illustration)**. Make sure you have the correct ring for the groove (see Step 5). Check the clearance at three or four locations around the groove. If the clearance is greater than specified, renew both the piston and rings as a set. If new rings are being used, measure the clearance using the new rings. If the clearance is greater than that specified, the piston is worn and must be renewed.

12 Check the piston-to-bore clearance by measuring the bore (see Section 21) and the piston diameter. Make sure each piston is matched to its correct cylinder. Measure the piston 5.0 mm up from the bottom of the skirt and at 90° to the piston pin axis **(see illustration)**. Subtract the piston diameter from the bore diameter to obtain the clearance. If it is greater than the specified figure, check whether it is the bore or piston that is worn beyond its specified service limit. If the bores are good, install new pistons and rings. If the bores are worn, replace the crankcases with new ones and fit new rings onto the pistons. Yamaha recommend that the cases, pistons and rings should be renewed as a set if the clearance is too great.

13 Apply clean engine oil to the piston pin, insert it into the piston and check for any freeplay between the two **(see illustration)**. Measure the pin external diameter, and the pin bore in the piston **(see illustrations)**. Subtract the bore diameter from the pin diameter to obtain the clearance. If it is greater than the specified figure, check whether it is the pin bore or pin that is worn beyond its specified service limit and renew them as required. Repeat the checks between the pin and the connecting rod small-end (see Section 24).

25.12 Measure the piston diameter with a micrometer at the specified distance from the bottom of the skirt

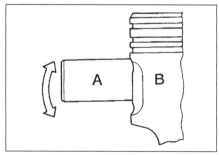

25.13a Slip the pin (A) into the piston (B) and try to rock it back and forth. If it is loose, renew the piston and pin

25.13b Measure the external diameter of the pin . . .

25.13c . . . and the internal diameter of the bore in the piston

Engine, clutch and transmission 2•49

Installation

14 Inspect and install the piston rings (see Section 26).
15 Lubricate the piston pin, the piston pin bore and the connecting rod small-end bore with clean engine oil.
16 Install a **new** circlip in one side of the piston (do not re-use old circlips). Line up the piston on its correct connecting rod so that the arrow on the piston crown will point to the front of the engine and the 'Y' mark will face the left-hand side of the engine when they are installed **(see illustration 24.3)**. Insert the piston pin from the side without the circlip **(see illustration 25.3b)**. Secure the pin with the other **new** circlip **(see illustration)**. When installing the circlips, compress them only just enough to fit them in the piston, and make sure they are properly seated in their grooves with the open end away from the removal notch.
17 Install the connecting rods (see Section 24).

26 Piston rings – inspection and installation

1 It is good practice to renew the piston rings when an engine is being overhauled. Before installing the new piston rings, the ring end gaps must be checked with the rings installed in the cylinder.

26.3 Measuring piston ring installed end gap

26.9a Fit the oil ring expander in its groove . . .

25.16 Do not over-compress the circlips and make sure they seat in the groove

Inspection

2 Lay out the pistons and the new ring sets so the rings will be matched with the same piston and cylinder during the end gap measurement procedure and engine assembly.
3 To measure the ring end gap, insert the top ring into the top of the first cylinder and square it up with the cylinder walls by pushing it in with the top of the piston. The ring should be about 20 mm below the top edge of the cylinder. Slip a feeler gauge between the ends of the ring and compare the measurement to the specifications at the beginning of the Chapter **(see illustration)**.
4 If the gap is larger or smaller than specified, double check to make sure that you have the correct rings before proceeding.
5 If the gap is too small, it must be enlarged or the ring ends may come in contact with

26.5 Ring end gap can be enlarged by clamping a file in a vice and filing the ring ends

26.9b . . . then fit the lower side rail . . .

each other during engine operation, which can cause serious damage. The end gap can be increased by filing the ring ends very carefully with a fine file. When performing this operation, file only from the outside in **(see illustration)**.
6 Excess end gap is not critical unless it exceeds the service limit. Again, double-check to make sure you have the correct rings for your engine and check that the bore is not worn.
7 Repeat the procedure for each ring that will be installed in the cylinders. When checking the oil ring, only the side-rails can be checked as the ends of the expander ring should contact each other. Remember to keep the rings, pistons and cylinders matched up.

Installation

8 Once the ring end gaps have been checked/corrected, the rings can be installed on the pistons.
9 The oil control ring (lowest on the piston) is installed first. It is composed of three separate components, namely the expander and the upper and lower side rails. Slip the expander into the groove, then install the lower side rail **(see illustrations)**. Do not use a piston ring installation tool on the oil ring side rails as they may be damaged. Instead, place one end of the side rail into the groove between the expander and the ring land. Hold it firmly in place and slide a finger around the piston while pushing the rail into the groove. Next, install the upper side rail in the same manner **(see illustration)**. Make sure the ends of the expander do not overlap.
10 After the three oil ring components have been installed, check to make sure that both the upper and lower side rails can be turned smoothly in the ring groove.
11 The upper surface of each compression ring should have a mark or letter at one end which must face up when the ring is installed on the piston. The rings can be distinguished by their different profiles **(see illustration 25.5b)**.
12 Fit the second ring into the middle groove in the piston. Do not expand the ring any more than is necessary to slide it into place **(see illustration)**. To avoid breaking the ring, use a

26.9c . . . and the upper side rail each side of it. The oil ring components must be installed by hand

26.12a Carefully feed the second ring into its groove

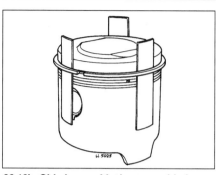

26.12b Old pieces of feeler gauge blade can be used to guide the ring over the piston

27 Transmission shafts – removal and installation

Note: *To remove the transmission shafts the engine must be removed from the frame and the crankcase halves separated.*

Removal

1 Remove the engine from the frame (see Section 5) and separate the crankcase halves (see Section 20).
2 Lift the output shaft out of the crankcase **(see illustration)**. If it is stuck, use a soft-faced hammer and gently tap on the ends of the shaft to free it. Remove the bearing half-ring retainer from the crankcase or bearing, noting how it fits **(see illustration 27.7a)**. Discard the oil seal from the left-hand end of the shaft and obtain a new seal for installation.
3 Remove the selector drum and forks (see Section 29).
4 Undo the Torx screws securing the input shaft bearing housing **(see illustration)**. Obtain two 6 mm bolts, 30 mm long excluding the bolt head, and with a 1 mm thread pitch, and screw them into the two holes in the bearing housing as shown **(see illustration)**. Tighten the bolts until they contact the surface of the crankcase, then continue tightening them evenly and a little at a time until the bearing housing is displaced. Withdraw the input shaft from the crankcase **(see illustration)**. Discard the Torx screws, as new ones should be used.

26.13 Finally install the top ring

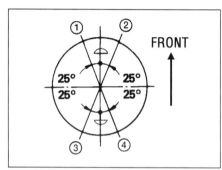

26.14 Stagger the ring end gaps as shown

1 Top compression ring
2 Oil ring lower rail
3 Oil ring upper rail
4 Second compression ring

piston ring installation tool **(see illustration 25.5a)**, or pieces of old feeler gauge blades **(see illustration)**.
13 Finally, install the top ring in the same manner into the top groove in the piston **(see illustration)**.
14 Once the rings are correctly installed, check they move freely without snagging and stagger their end gaps as shown **(see illustration)**.

Installation

5 Slide the input shaft into the crankcase, locating the shaft end into its bore **(see illustration)**. Make sure the bearing housing screw holes are correctly aligned, and tap the housing in with a soft-faced mallet if necessary. Install the **new** Torx screws and tighten them to the torque setting specified at the beginning of the Chapter **(see illustration)**. Stake the edge of each screw into the indent in the housing using a suitable punch **(see illustration)**.
6 Install the selector drum and forks (see Section 29).

27.2 Lift the output shaft out of the crankcase

27.4a Remove the Torx screws (arrowed)

27.4b Thread the specified bolts into the housing and tighten them as described to displace the housing . . .

27.4c . . . then withdraw the shaft from the crankcase

27.5a Slide the shaft into position . . .

Engine, clutch and transmission

27.5b ... then install the new Torx screws and tighten them to the specified torque

27.5c Stake the edge of each screw into the housing

27.7a Fit the retainer into the slot in the bearing ...

7 Fit the output shaft bearing half-ring retainer into its slot in the bearing (see illustration). Smear the lips of the output shaft seal with grease. Slide the seal onto the left-hand end of the shaft.

8 Lower the output shaft into position in the upper crankcase, making sure the groove in the bearing engages correctly with the half-ring retainer (see illustration 27.2).

9 Make sure both transmission shafts are correctly seated and their related pinions are correctly engaged.

Caution: If the half-ring retainer is not correctly engaged, the crankcase halves will not seat correctly.

10 Position the gears in the neutral position and check the shafts are free to rotate easily and independently (i.e. the input shaft can turn whilst the output shaft is held stationary) before proceeding further.

11 Reassemble the crankcase halves (see Section 20).

28 Transmission shafts – disassembly, inspection and reassembly

1 Remove the transmission shafts from the crankcase (see Section 27). Always disassemble the transmission shafts separately to avoid mixing up the components (see illustrations).

27.7b ... then slide on the new oil seal

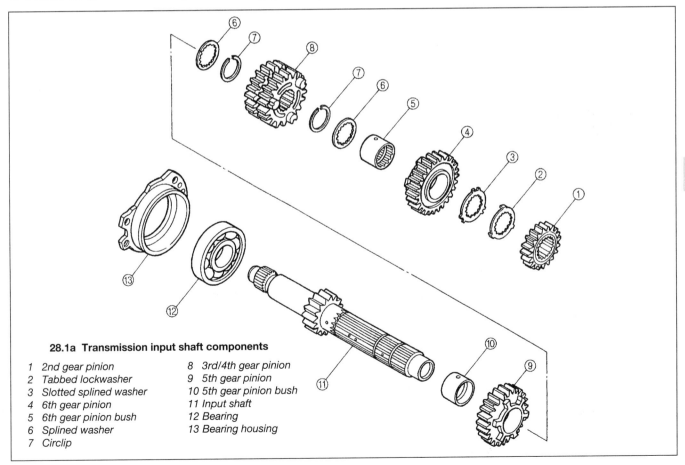

28.1a Transmission input shaft components

1. 2nd gear pinion
2. Tabbed lockwasher
3. Slotted splined washer
4. 6th gear pinion
5. 6th gear pinion bush
6. Splined washer
7. Circlip
8. 3rd/4th gear pinion
9. 5th gear pinion
10. 5th gear pinion bush
11. Input shaft
12. Bearing
13. Bearing housing

2•52 Engine, clutch and transmission

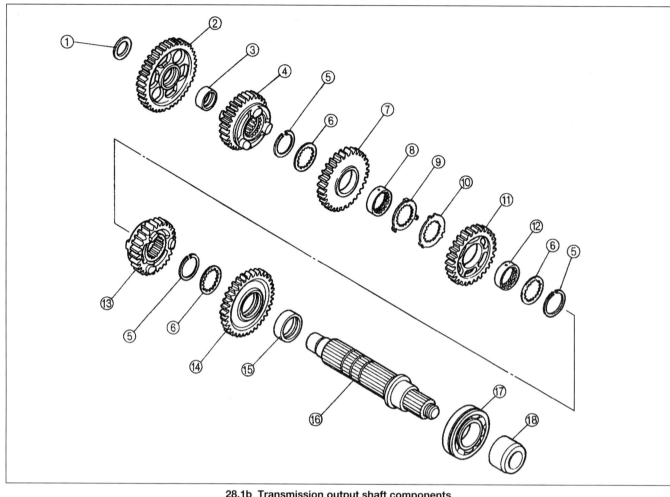

28.1b Transmission output shaft components

1 Thrust washer	5 Circlip	10 Slotted splined washer	14 2nd gear pinion
2 1st gear pinion	6 Splined washer	11 4th gear pinion	15 2nd gear pinion bush
3 1st gear pinion bush	7 3rd gear pinion	12 4th gear pinion bush	16 Output shaft
4 5th gear pinion	8 3rd gear pinion bush	13 6th gear pinion	17 Bearing
	9 Tabbed lockwasher		18 Spacer

Input shaft disassembly

 HAYNES HiNT *When disassembling the transmission shafts, place the parts on a long rod or thread a wire through them to keep them in order and facing the proper direction.*

2 Slide the 2nd gear pinion off the left-hand end of the shaft **(see illustration 28.21)**.
3 Slide the tabbed lockwasher off the shaft, then turn the slotted splined washer to offset the splines and slide it off the shaft, noting how they fit together **(see illustrations 28.20c and a)**. Slide the sixth gear pinion and its bush off the shaft, followed by the splined washer **(see illustrations 28.19c, b and a)**.
4 Remove the circlip securing the combined 3rd/4th gear pinion, then slide the pinion off the shaft **(see illustrations 28.18b and a)**.

5 Remove the circlip securing the 5th gear pinion, then slide the splined washer, the pinion and its bush off the shaft **(see illustrations 28.17d, c, b and a)**. The 1st gear pinion is integral with the shaft **(see illustration)**.
6 If required, remove the bearing and its housing from the right-hand end of the shaft, referring to *Tools and Workshop Tips* (Section 5) in the *Reference* section **(see illustration)**.

Input shaft inspection

7 Wash all of the components in clean solvent and dry them off.

28.5 The 1st gear pinion is integral with the shaft

28.6 Remove the housing and bearing if required

Engine, clutch and transmission 2•53

28.17a Slide the 5th gear pinion bush . . .

28.17b . . . the 5th gear pinion . . .

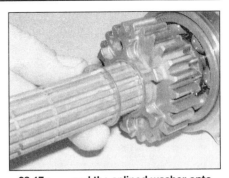

28.17c . . . and the splined washer onto the shaft . . .

8 Check the gear teeth for cracking, chipping, pitting and other obvious wear or damage. Any pinion that is damaged as such must be replaced.

9 Inspect the dogs and the dog holes in the gears for cracks, chips, and excessive wear especially in the form of rounded edges. Make sure mating gears engage properly. Renew the paired gears as a set if necessary.

10 Check for signs of scoring or bluing on the pinions, bushes and shaft. This could be caused by overheating due to inadequate lubrication. Check that all the oil holes and passages are clear. Replace any damaged pinions or bushes with new ones.

11 Check that each pinion moves freely on the shaft or bush but without undue freeplay. Check that each bush moves freely on the shaft but without undue freeplay.

12 The shaft is unlikely to sustain damage unless the engine has seized, placing an unusually high loading on the transmission, or the machine has covered a very high mileage. Check the surface of the shaft, especially where a pinion turns on it, and replace the shaft with a new one if it has scored or picked up, or if there are any cracks. Check the shaft runout using V-blocks and a dial gauge and replace the shaft with a new one if the runout exceeds the limit specified at the beginning of the Chapter.

13 Check the washers and circlips and renew any that are bent or appear weakened or worn. Use new ones if in any doubt. Note that it is good practice to renew all circlips when overhauling gearshafts.

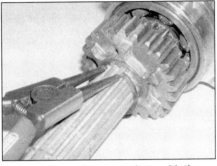

28.17d . . . and secure them with the circlip . . .

14 Referring to *Tools and Workshop Tips* (Section 5) in the *Reference* section, check the bearings and replace them with new ones, if necessary. Do not forget the input shaft left-hand bearing, which is housed in the crankcase.

Input shaft reassembly

15 During reassembly, apply engine oil or molybdenum disulphide oil (a 50/50 mixture of molybdenum disulphide grease and engine oil) to the mating surfaces of the shaft, pinions and bushes. When installing the circlips, do not expand their ends any further than is necessary. Install the stamped circlips and washers so that their chamfered side faces the pinion it secures (see illustration 28.1a).

16 If removed, fit the bearing and its housing onto the right-hand end of the shaft, referring to *Tools and Workshop Tips* (Section 5) in the *Reference* section (see illustration 28.6).

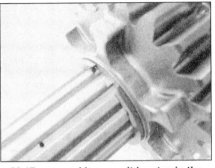

28.17e . . . making sure it locates in the groove

17 Slide the 5th gear pinion bush onto the left-hand end of the shaft, aligning the oil hole in the bush with the hole in the shaft (see illustration). Fit the 5th gear pinion with its dog slots facing away from the integral 1st gear (see illustration). Slide the splined washer onto the shaft, then fit the circlip, making sure that it locates correctly in the groove in the shaft (see illustrations).

18 Slide the combined 3rd/4th gear pinion onto the shaft with the smaller 3rd gear pinion facing the 5th gear pinion (see illustration). Fit the circlip, making sure it locates correctly in its groove in the shaft (see illustrations).

19 Slide the splined washer onto the shaft, followed by the 6th gear pinion bush, aligning the oil hole in the bush with the hole in the shaft, and the 6th gear pinion, making sure its dogs face the 3rd/4th gear pinion (see illustrations).

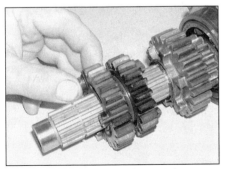

28.18a Slide the 3rd/4th gear pinion onto the shaft . . .

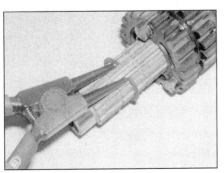

28.18b . . . and secure it with the circlip . . .

28.18c . . . making sure it locates in the groove

2•54 Engine, clutch and transmission

28.19a Slide the splined washer . . .

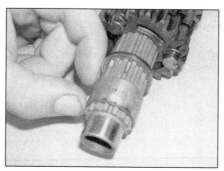

28.19b . . . the 6th gear pinion bush . . .

28.19c . . . and the 6th gear pinion onto the shaft

20 Slide the slotted splined washer onto the shaft and locate it in its groove, then turn it in the groove, so that the splines on the washer align with the splines on the shaft and secure the washer in the groove (see illustrations). Slide the tabbed lockwasher onto the shaft, so that the tabs locate into the slots in the outer rim of the splined washer (see illustrations).
21 Slide the 2nd gear pinion onto the shaft (see illustration).
22 Check that all components have been correctly installed (see illustration).

Output shaft disassembly

23 Slide the bearing off the right-hand end of the shaft (see illustration 28.38d).
24 Slide the thrust washer off the shaft, followed by the 1st gear pinion and its bush, and the 5th gear pinion (see illustrations 28.38c, b and a, and 28.37).
25 Remove the circlip securing the 3rd gear pinion, then slide the splined washer, the pinion and its splined bush off the shaft (see illustrations 28.36d, c, b and a).
26 Slide the tabbed lockwasher off the shaft, then turn the slotted splined washer to offset the splines and slide it off the shaft, noting how they fit together (see illustrations 28.35c and a).
27 Slide the 4th gear pinion and its splined bush, followed by the splined washer, off the shaft (see illustrations 28.34c, b and a).
28 Remove the circlip securing the 6th gear pinion, then slide the pinion off the shaft (see illustrations 28.33b and a).
29 Remove the circlip securing the 2nd gear pinion, then slide the splined washer, the pinion and its bush off the shaft (see illustrations 28.32d, c, b and a).

Output shaft inspection

30 Refer to Steps 7 to 13 above.

Output shaft reassembly

31 During reassembly, apply engine oil or molybdenum disulphide oil (a 50/50 mixture of molybdenum disulphide grease and engine oil) to the mating surfaces of the shaft, pinions and bushes. When installing the circlips, do not expand their ends any further than is necessary. Install the stamped circlips and washers so that their chamfered side faces the pinion it secures (see illustration 28.1b).
32 Slide the 2nd gear pinion bush onto the shaft, aligning the oil hole in the bush with the hole in the shaft, then slide on the 2nd gear pinion and the splined washer (see illustrations). Fit the circlip, making sure it is locates correctly in its groove in the shaft (see illustrations).

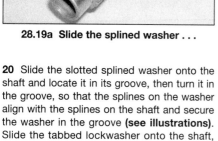

28.20a Slide the slotted splined washer onto the shaft . . .

28.20b . . . and position it as shown

28.20c Slide the tabbed lockwasher onto the shaft . . .

28.20d . . . and engage its tabs in the slots

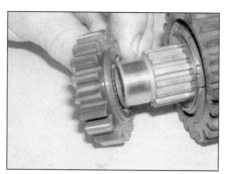

28.21 Slide the 2nd gear pinion onto the end of the shaft

28.22 The assembled shaft should be as shown

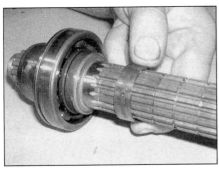

28.32a Slide the 2nd gear pinion bush ...

28.32b ... the 2nd gear pinion ...

28.32c ... and the splined washer onto the shaft ...

28.32d ... and secure them with the circlip ...

28.32e ... making sure it locates in the groove

28.33a Slide the 6th gear pinion onto the shaft ...

33 Slide the 6th gear pinion onto the shaft with its selector fork groove facing away from the 2nd gear pinion, then fit the circlip, making sure it locates correctly in its groove in the shaft (see illustrations).

34 Slide the splined washer and the 4th gear pinion bush onto the shaft, making sure the oil hole in the bush aligns with the hole in the shaft, then slide on the 4th gear pinion (see illustrations).

35 Slide the slotted splined washer onto the shaft and locate it in its groove, then turn it in the groove so that the splines on the washer align with the splines on the shaft and secure the washer in the groove (see illustrations).

28.33b ... and secure it with the circlip ...

28.33c ... making sure it locates in the groove

28.34a Slide the splined washer ...

28.34b ... the 4th gear pinion bush ...

28.34c ... and the 4th gear pinion onto the shaft

28.35a Slide the slotted splined washer onto the shaft ...

2•56 Engine, clutch and transmission

28.35b ... and position it as shown

28.35c Slide the tabbed lockwasher onto the shaft ...

28.35d ... and engage its tabs in the slots

28.36a Slide the 3rd gear pinion bush ...

28.36b ... the 3rd gear pinion ...

28.36c ... and the splined washer onto the shaft ...

Slide the lockwasher onto the shaft, so that the tabs on the lockwasher locate into the slots in the outer rim of the splined washer **(see illustrations)**.

36 Slide the 3rd gear pinion bush onto the shaft, making sure the oil hole in the bush aligns with the hole in the shaft, followed by the 3rd gear pinion and the splined washer. Fit the circlip, making sure it locates correctly in its groove in the shaft **(see illustrations)**.

37 Slide the 5th gear pinion onto the shaft with its selector fork groove facing the 3rd gear pinion **(see illustration)**.

38 Slide the 1st gear pinion bush onto the shaft, followed by the 1st gear pinion and the

28.36d ... and secure them with the circlip ...

28.36e ... making sure it locates in the groove

28.37 Slide the 5th gear pinion onto the shaft

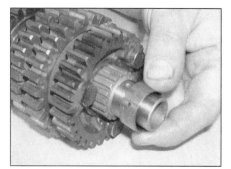

28.38a Slide the 1st gear pinion bush ...

28.38b ... and the 1st gear pinion onto the shaft ...

28.38c ... then fit the thrust washer ...

Engine, clutch and transmission 2•57

28.38d ... and the bearing

28.39 The assembled shaft should be as shown

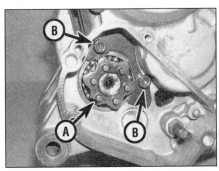

29.4 Note the position of the neutral detent (A). Unscrew the bolts (B) and remove the plate

thrust washer, then fit the bearing onto the end of the shaft **(see illustrations)**.

39 Check that all components have been correctly installed **(see illustration)**.

29 Selector drum and forks – removal, inspection and installation

Note: *To remove the selector drum and forks the engine must be removed from the frame and the crankcase halves separated.*

Removal

1 Position the transmission in neutral. Remove the engine from the frame (see Section 5) and separate the crankcase halves (see Section 20).

2 If not already done, remove the gearchange mechanism (see Section 17).

3 Remove the transmission output shaft (see Section 27).

4 Note the orientation of the selector drum as an aid for installation – the best way is to check which way the neutral detent in the end of the drum points, and to note how the neutral contact is located against the neutral switch contact **(see illustration)**. Unscrew the bolts securing the selector drum retainer plate and remove the plate, noting how it fits.

5 Before removing the selector forks, note that each fork is lettered for identification. The right-hand fork has an 'R', the centre fork a 'C', and the left-hand fork an 'L' **(see illustration)**. These letters face the right-hand side of the engine. If no letters are visible, mark them yourself using a felt pen.

6 Support the selector forks for the output shaft and withdraw the fork shaft from the crankcase, then remove the forks, noting how

they fit **(see illustration)**. Slide the forks back onto the shaft in the correct order and the right way round. Note the springs in the ends of the shaft and remove them for safekeeping, if loose.

7 Support the selector fork for the input shaft and withdraw the shaft from the casing **(see illustration)**. Move the fork guide pin out of its track in the selector drum, then withdraw the selector drum from the left-hand side of the casing **(see illustration)**. Move the selector fork round in its groove in the pinion and remove it, noting how it fits **(see illustrations)**. Slide the fork back onto the shaft. Note the springs in the ends of the shaft and remove them for safekeeping, if loose.

Inspection

8 Inspect the selector forks for any signs of wear or damage, especially around the fork

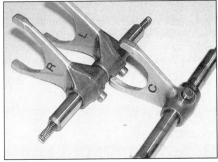

29.5 Note the letter on each fork denoting its position

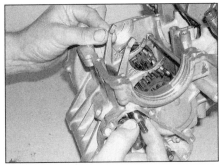

29.6 Withdraw the shaft and remove the forks

29.7a Withdraw the shaft ...

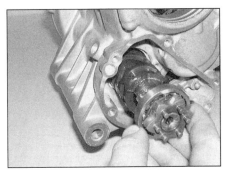

29.7b ... then displace the fork from its track and withdraw the selector drum

29.7c Slide the fork round ...

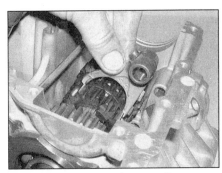

29.7d ... and remove it

2

2•58 Engine, clutch and transmission

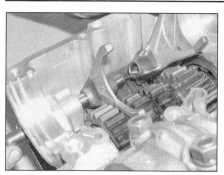

29.14 Locate the forks as described and slide in the shaft

ends where they engage with the groove in the pinion. Check that each fork fits correctly in its pinion groove. Check closely to see if the forks are bent. If the forks are in any way damaged they must be replaced with new ones.

9 Check that the forks fit correctly on their shaft. They should move freely with a light fit but no appreciable freeplay. Check that the fork shaft holes in the casing are not worn or damaged.

10 Check the selector fork shaft runout using V-blocks and a dial gauge and renew the shaft if the runout exceeds the limit specified at the beginning of the Chapter. A bent shaft will cause difficulty in selecting gears and make the gearchange action heavy. Replace the shafts with new ones if bent.

11 Inspect the selector drum grooves and selector fork guide pins for signs of wear or damage. If either show signs of wear or damage they must be replaced with new ones.

12 Check that the selector drum bearing rotates freely and has no signs of roughness or excessive freeplay between it and the drum or crankcase (when installed) (see *Tools and Workshop Tips* (Section 5) in the *Reference* section for more information on bearings). Replace the selector drum with a new one if necessary – the bearing is not available separately, though it would be worth checking with a bearing specialist before buying a new drum. Also check that the neutral switch contact in the drum is not damaged or worn away. If required, remove the contact and replace it with a new one.

Installation

13 Position the input shaft fork (marked C) in the crankcase, making sure the letter faces the right-hand side of the engine **(see illustration 29.7d)**. Locate the fork in the groove in the combined 3rd/4th gear pinion, then slide it round and support it clear so that it does not get in the way when installing the selector drum **(see illustration 29.7c)**. Align the selector drum so that the neutral detent points to the bottom of the engine and slide the drum into the crankcase **(see illustration 29.7b)**. Make sure the drum end locates in its bore in the crankcase, and that the neutral contact locates against the neutral switch. Apply clean engine oil to the fork shaft, and fit the springs into the ends of the shaft if removed. Locate the selector fork guide pin into its track in the drum, then slide the shaft into the crankcase and through the fork **(see illustration 29.7a)**.

14 Position the output shaft forks (marked 'R' and 'L') in the crankcase, making sure the letters face the right-hand side of the engine. Locate the fork marked 'R' in the groove in the 5th gear pinion, and the fork marked 'L' in the groove in the 6th gear pinion, and locate the fork guide pins in their tracks in the drum. Apply clean engine oil to the fork shaft, and fit the springs into the ends of the shaft if removed. Slide the shaft into the crankcase and through the forks **(see illustration)**.

15 Apply a suitable non-permanent thread locking compound to the selector drum retainer plate bolts. Install the plate and tighten the bolts to the torque setting specified at the beginning of the Chapter **(see illustration)**.

16 Install the transmission output shaft (see Section 27).

30 Initial start-up after overhaul

1 Make sure the engine oil level and coolant level are correct (see *Daily (pre-ride) checks*).
2 Make sure there is fuel in the tank, then turn the fuel tap to the 'ON' position, and set the choke.
3 Start the engine and allow it to run at a moderately fast idle until it reaches normal operating temperature.
4 As no oil pressure warning light is fitted, an oil pressure check must be carried out (see Chapter 1).
5 Check carefully that there are no oil and coolant leaks and make sure the transmission and controls, especially the brakes, function properly before road testing the machine. Refer to Section 31 for the recommended running-in procedure.

29.15 Install the retainer plate and tighten the bolts to the specified torque

6 Upon completion of the road test, and after the engine has cooled down completely, recheck the valve clearances (see Chapter 1) and check the engine oil and coolant levels (see *Daily (pre-ride) checks*).

31 Recommended running-in procedure

1 Treat the machine gently for the first few miles to make sure oil has circulated throughout the engine and any new parts installed have started to seat.
2 Even greater care is necessary if the engine has been extensively overhauled – the bike will have to be run in as when new. This means greater use of the transmission and a restraining hand on the throttle until at least 600 miles (1000 km) have been covered. There is no point in keeping to any set speed limit – the main idea is to keep from labouring the engine and to gradually increase performance up to the 600 mile (1000 km) mark. These recommendations apply less when only a partial overhaul has been done, though it does depend to an extent on the nature of the work carried out and which components have been renewed. Experience is the best guide, since it is easy to tell when an engine is running freely. If in any doubt, consult a Yamaha dealer. The following table shows the maximum engine speed limitations, which Yamaha provide for new motorcycles, can be used as a guide.
3 If a lubrication failure is suspected, stop the engine immediately and try to find the cause. If an engine is run without oil, even for a short period of time, severe damage will occur.

Maximum engine speed limitations		
Up to 600 miles (1000 km)	5000 rpm max	Vary throttle position/speed. Do not use full throttle
600 to 1000 miles (1000 to 1600 km)	6000 rpm max	Vary throttle position/speed. Do not use full throttle
Over 1000 miles (1600 km)		Do not exceed tachometer red line

Chapter 3
Cooling system

Contents

Coolant hoses, pipes and unions – removal and installation 9
Coolant level check .... see *Daily (pre-ride) checks*
Coolant reservoir – removal and installation 3
Coolant temperature display/warning light, and sender or thermo unit
 – check and renewal 5
Cooling fan, and cooling fan switch or relay – check and renewal .. 4
Cooling system checks .... see Chapter 1
Cooling system draining, flushing and refilling .... see Chapter 1
General information 1
Radiator – removal and installation 7
Radiator pressure cap – check 2
Thermostat housing and thermostat – removal, check and
 installation 6
Water pump – check and overhaul 8

Degrees of difficulty

| Easy, suitable for novice with little experience | Fairly easy, suitable for beginner with some experience | Fairly difficult, suitable for competent DIY mechanic | Difficult, suitable for experienced DIY mechanic | Very difficult, suitable for expert DIY or professional |

Specifications

Coolant
Mixture type and capacity .... see Chapter 1

Radiator
Cap valve opening pressure .... 13.5 to 18.0 psi (0.93 to 1.25 bars)

Cooling fan switch – 1998 and 1999 models
Cooling fan cut-in temperature .... 102 to 108°C
Cooling fan cut-out temperature .... 98°C

Coolant temperature sender – 1998 and 1999 models
Resistance
 @ 80°C .... 47.5 to 56.8 ohms
 @ 115°C .... 16.5 to 20.5 ohms

Thermo unit – 2000-on models
Resistance
 @ 50°C .... 9.7 to 11.4 K-ohms
 @ 80°C .... 3.4 to 4.0 K-ohms
 @ 105°C .... 1.6 to 1.9 K-ohms
 @ 120°C .... 1.1 to 1.2 K-ohms

Thermostat
Opening temperature .... 71 to 85°C
Valve lift .... 8 mm (min) @ 85°C

Water pump
Impeller shaft tilt (max.) .... 0.15 mm

Torque wrench settings
Cooling fan bolts .... 9 Nm
Cooling fan switch (1998 and 1999 models) .... 23 Nm
Coolant temperature sender (1998 and 1999 models) .... 15 Nm
Thermo unit (2000-on models) .... 23 Nm
Thermostat cover bolts .... 10 Nm
Radiator mounting bolts .... 20 Nm
Radiator earth wire bolt (1998 and 1999 models) .... 5 Nm
Water pump cover bolts .... 10 Nm
Coolant outlet pipes from valve cover .... 10 Nm
Coolant inlet and outlet pipes to/from water pump .... 10 Nm
Coolant inlet union to cylinder block bolts .... 10 Nm

3•2 Cooling system

1 General information

The cooling system uses a water/antifreeze coolant to carry away excess energy in the form of heat. The cylinders are surrounded by a water jacket, from which the heated coolant is circulated by thermo-syphonic action in conjunction with a water pump. The water pump is an integral assembly with the oil pump, and is driven by chain and sprockets off the clutch. The hot coolant passes upwards to the thermostat and through to the radiator. The coolant then flows across the radiator core, where it is cooled by the passing air, to the water pump and back to the engine, where the cycle is repeated.

A thermostat is fitted in the system to prevent the coolant flowing through the radiator when the engine is cold, therefore accelerating the speed at which the engine reaches normal operating temperature. A coolant temperature sender mounted in the radiator transmits to the temperature display on the instrument panel. A thermostatically-controlled cooling fan is fitted, to aid cooling in extreme conditions. The fan switch is also mounted in the radiator. On 1998 and 1999 models, the temperature sender and fan switch are independent components, while on 2000-on models they are combined into one thermo unit, with the circuit controlled by a relay.

The complete cooling system is partially sealed and pressurised, the pressure being controlled by a valve contained in the spring-loaded radiator cap. By pressurising the coolant the boiling point is raised, preventing premature boiling in adverse conditions. The overflow pipe from the system is connected to a reservoir, into which excess coolant is expelled under pressure. The discharged coolant automatically returns to the radiator when the engine cools.

Warning: *Do not remove the pressure cap from the radiator when the engine is hot. Scalding hot coolant and steam may be blown out under pressure and could cause serious injury. When the engine has cooled, place a thick rag such as a towel over the pressure cap; slowly rotate the cap anti-clockwise to the first stop. This procedure allows any residual pressure to escape. When the steam has stopped escaping, press down on the cap while turning it anti-clockwise, and remove it. Do not allow antifreeze to come into contact with your skin, or painted surfaces of the motorcycle. Rinse off any spills immediately with plenty of water. Antifreeze is highly toxic if ingested. Never leave antifreeze lying around in an open container or in puddles on the floor; children and pets are attracted by its sweet smell and may drink it. Check with the local authorities about disposing of used antifreeze. Many communities will have collection centres which will see that antifreeze is disposed of safely.*

Caution: *At all times use the specified type of antifreeze, and always mix it with distilled water in the correct proportion. The antifreeze contains corrosion inhibitors which are essential to avoid damage to the cooling system. A lack of these inhibitors could lead to a build-up of corrosion which would block the coolant passages, resulting in overheating and severe engine damage. Distilled water must be used as opposed to tap water to avoid a build-up of scale which would also block the passages.*

2 Radiator pressure cap – check

1 If problems such as overheating or loss of coolant occur, check the entire system as described in Chapter 1. The radiator cap opening pressure should be checked by a Yamaha dealer with the special tester required for the job. If the cap is defective, replace it with a new one.

3 Coolant reservoir – removal and installation

Removal

1 Remove the right-hand fairing side panel (see Chapter 8).
2 Release the clutch cable holder from reservoir **(see illustration)**. Release the clamp securing the breather/overflow hose (coming out of the back of the reservoir filler neck) and detach the hose. Remove the reservoir cap.
3 Place a suitable container underneath the reservoir. Unscrew the reservoir mounting bolts and tip the coolant out of the reservoir into the container **(see illustration)**.

Installation

4 Installation is the reverse of removal. Make sure the hoses are correctly installed and secured with their clamps. On completion, refill the reservoir as described in Chapter 1.

4 Cooling fan, and cooling fan switch or relay – check and renewal

Cooling fan

Check

1 If the engine is overheating and the cooling fan isn't coming on, first check the cooling fan circuit fuse (see Chapter 9) and then the fan switch or relay (according to model) as described below.
2 If the fan does not come on, (and the fan switch is good), the fault lies in either the

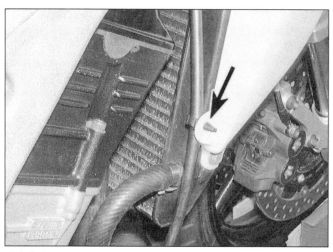

3.2 Release the clutch cable holder (arrowed) from the reservoir

3.3 Reservoir mounting bolts (arrowed)

Cooling system 3•3

4.3 Fan motor wiring connector

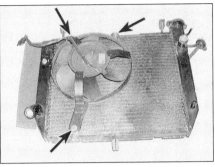

4.5 Fan assembly mounting bolts (arrowed)

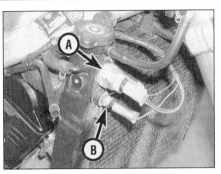

4.8 Cooling fan switch (A), temperature sender (B) – 1998 and 1999 models

cooling fan motor or the relevant wiring. Test all the wiring and connections as described in Chapter 9.

3 To test the cooling fan motor, remove the left-hand cockpit trim panel (see Chapter 8). If access is too restricted, also remove the fairing side panel. Trace the wiring from the fan motor and disconnect it at the connector located on the left-hand end of the radiator **(see illustration)**. Using a 12 volt battery and two jumper wires, connect the positive (+ve) battery lead to the blue wire terminal on the fan wiring connector and the negative (–ve) lead to the black wire terminal. Once connected, the fan should operate. If it does not, and the wiring is all good, then the fan motor is faulty. Replace the fan assembly with a new one – individual components are not available.

Renewal

 Warning: *The engine must be completely cool before carrying out this procedure.*

4 Remove the radiator (see Section 7).
5 Unscrew the three bolts securing the fan assembly to the radiator and remove it **(see illustration)**.
6 Installation is the reverse of removal. Tighten the bolts to the torque setting specified at the beginning of the Chapter.

Cooling fan switch – 1998 and 1999 models

Check

7 If the engine is overheating and the cooling fan isn't coming on, first check the cooling fan circuit fuse (see Chapter 9). If the fuse is blown, check the fan circuit for a short to earth (see *Wiring Diagrams* at the end of this book).
8 If the fuse is good, remove the right-hand fairing side panel (see Chapter 8). Disconnect the wiring connector from the fan switch, mounted in the front of the radiator **(see illustration)**. Turn the ignition switch ON. Using a jumper wire, connect between the terminals in the wiring connector. The fan should come on. If it does, the fan switch is defective and must be replaced with a new one. If it does not come on, test the fan motor (see Step 3).

9 If the fan stays on all the time, disconnect the switch wiring connector. The fan should stop. If it does, the switch is defective and must be replaced with a new one. If the fan doesn't stop, check the wiring between the switch and the fan, and the fan itself.
10 If the fan works but is suspected of cutting in at the wrong temperature, a more comprehensive test of the switch can be made as follows.
11 Remove the switch (see below). Fill a small heatproof container with coolant and place it on a stove. Connect the probes of an ohmmeter to the terminals of the switch, and using some wire or other support suspend the switch in the coolant so that just the sensing portion and the threads are submerged **(see illustration)**. Also place a thermometer capable of reading temperatures up to 110°C in the coolant so that its bulb is close to the switch. **Note:** *None of the components should be allowed to directly touch the container.*
12 Initially the ohmmeter reading should be very high indicating that the switch is open ('OFF'). Heat the coolant, stirring it gently.

 Warning: *This must be done very carefully to avoid the risk of personal injury.*

When the temperature reaches around 102 to 108°C the meter reading should drop to around zero ohms, indicating that the switch

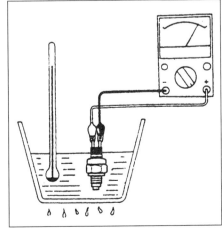

4.11 Fan switch testing set-up

has closed ('ON'). Now turn the heat off. As the temperature falls below 98°C the meter reading should show infinite (very high) resistance, indicating that the switch has opened ('OFF'). If the meter readings obtained are different, or they are obtained at different temperatures, then the switch is faulty and must be replaced with a new one.

Renewal

 Warning: *The engine must be completely cool before carrying out this procedure.*

13 Drain the cooling system (see Chapter 1).
14 Disconnect the wiring connector from the fan switch, mounted in the front of the radiator **(see illustration 4.8)**. Unscrew the switch and remove it from the radiator.
15 Apply a suitable sealant (Yamaha advise Three Bond Sealock 10) to the switch threads, then install the switch and tighten it to the torque setting specified at the beginning of the Chapter. Take care not to overtighten the switch as it or the radiator could be damaged.
16 Reconnect the switch wiring and refill the cooling system (see Chapter 1).

Cooling fan relay – 2000-on models

Check

17 If the engine is overheating and the cooling fan isn't coming on, first check the cooling fan circuit fuse (see Chapter 9). If the fuse is blown, check the fan circuit for a short to earth (see *Wiring Diagrams* at the end of this book).
18 If the fuse is good, disconnect the fan relay wiring connector. Set a multimeter to the ohms x 1 scale and connect the positive (+ve) probe to the brown wire terminal on the relay and the negative (–ve) probe to the blue wire terminal. Using a fully-charged 12 volt battery and two insulated jumper wires, connect the positive (+ve) terminal of the battery to the brown wire terminal on the relay, and the negative (–ve) terminal to the green/black wire terminal on the relay. At this point the relay should be heard to click and the multimeter read 0 ohms (continuity). If this is the case the relay is proved good. If the relay does not click when battery voltage is applied and indicates no continuity (infinite resistance)

3•4 Cooling system

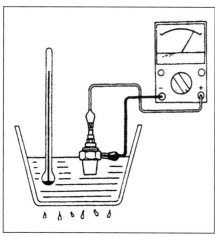

5.9 Temperature gauge sensor testing set-up

across its terminals, it is faulty and must be replaced with a new one. If the relay is good, test the thermo unit (see Section 5). If the unit is good, test the fan motor (see Step 3).

Renewal

19 The relay is located under the rider's seat. Refer to the wiring diagram at the end of this manual and use the wire colours to identify the relay. Disconnect the wiring connector from the relay, then release it from its mounting and remove it.

20 Install the new relay and connect the wiring connector.

5 Coolant temperature display/warning light, and sender or thermo unit – check and renewal

Temperature display and warning light

Check

1 The circuit consists of the sender (1998 and 1999 models) or thermo unit (2000-on models) mounted in the radiator, and the display mounted in the instrument cluster.

2 The temperature display should show 'LO' between 0°C and 40°C. Between 40°C and 117°C (105°F and 242°F) the display should show the actual temperature of the coolant. Between 117°C and 140°C (243°F and 284°F) the display should show the coolant temperature flashing, the temperature symbol should flash, and the warning light should come on. Above 140°C (284°F) the display should show 'HI', the temperature symbol should be flashing, and the warning light should be on.

3 If the system malfunctions, check first that the battery is fully charged and that the fuses are all good. Next check the temperature sender (1998 and 1999 models) or thermo unit (2000-on models) as described below.

4 If no problems are found, take the instrument cluster to a Yamaha dealer for further assessment – Yamaha provide no specific test data for the instruments themselves. If there are any faults, even if just the warning light LED is faulty, a new cluster will have to be fitted, as no individual components are available.

Renewal

5 See Chapter 9.

Temperature sender – 1998 and 1999 models

Check

6 The sender is mounted in the front of the radiator on the right-hand side, below the fan switch. Remove the right-hand fairing side panel (see Chapter 8).

7 Disconnect the sender wiring connector (see illustration 7.3). Using a continuity tester, check for continuity between the sender body and earth (ground). There should be continuity. If there is no continuity, check that the sender mounting and the earth wire below it are secure, and that the earth terminal is clean. With the ignition switch ON, earth the sender wire. The temperature display and warning light should come on. If they don't, check the wiring between the sender and the instrument cluster. If the wiring is good, refer to Step 4 above.

8 If the display and warning light do come on, remove the sender (see below).

9 Fill a small heatproof container with water and place it on a stove. Using an ohmmeter, connect the positive (+ve) probe of the meter to the terminal on the sender, and the negative (–ve) probe to the body of the sender. Using some wire or other support suspend the sender in the coolant so that just the sensing portion and the threads are submerged. Also place a thermometer capable of reading temperatures of up to 110°C in the water so that its bulb is close to the sender (see illustration). Note: *None of the components should be allowed to directly touch the container.* Heat the coolant, stirring it gently.

 Warning: *This must be done very carefully to avoid the risk of personal injury.*

When the temperature reaches around 80°C the meter reading should be as specified at the beginning of the Chapter. When the temperature reaches around 115°C the meter should again be as specified. If the meter readings obtained are different, or they are obtained at different temperatures, then the sender is faulty and must be renewed.

Renewal

 Warning: *The engine must be completely cool before carrying out this procedure.*

10 Drain the cooling system (see Chapter 1). The sender is mounted in the front of the radiator on the right-hand side, below the fan switch. Remove the right-hand fairing side panel (see Chapter 8).

11 Disconnect the sender wiring connector (see illustration 7.3). Unscrew the sender and remove it from the radiator.

12 Apply a suitable sealant (Yamaha advise Three Bond Sealock 10) to the switch threads, then install the switch and tighten it to the torque setting specified at the beginning of the Chapter. Take care not to overtighten the switch as it or the radiator could be damaged. Connect the sender wiring.

13 Refill the cooling system (see Chapter 1). Install the fairing side panel (see Chapter 8).

Thermo unit – 2000-on models

Check

14 The thermo unit is mounted in the front of the radiator on the right-hand side. Remove the right-hand fairing side panel (see Chapter 8).

15 Disconnect the wiring connector. With the ignition switch ON, connect a jumper wire between the terminals in the connector. The temperature display and warning light should come on. If they don't, check the wiring between the sender and the instrument cluster. If the wiring is good, refer to Step 4 above.

16 If the display and warning light do come on, remove the thermo unit (see below).

17 Fill a small heatproof container with water and place it on a stove. Using an ohmmeter, connect the probes to the terminals on the unit (see illustration 4.11). Using some wire or other support suspend the unit in the coolant so that just the sensing portion and the threads are submerged. Also place a thermometer capable of reading temperatures of up to 130°C in the water so that its bulb is close to the unit. Note: *None of the components should be allowed to directly touch the container.* Heat the coolant, stirring it gently.

 Warning: *This must be done very carefully to avoid the risk of personal injury.*

When the temperature reaches around 50°C the meter reading should be as specified at the beginning of the Chapter. When the temperature reaches around 80°C the meter should again be as specified, and so on as the listed temperatures are reached. If the meter readings obtained are different, or they are obtained at different temperatures, then the thermo unit is faulty and must be renewed.

Renewal

 Warning: *The engine must be completely cool before carrying out this procedure.*

18 Drain the cooling system (see Chapter 1). The thermo unit is mounted in the front of the radiator on the right-hand side. Remove the right-hand fairing side panel (see Chapter 8).

19 Disconnect the wiring connector. Unscrew the unit and remove it from the radiator.

20 Apply a suitable sealant (Yamaha advise Three Bond Sealock 10) to the thermo unit

Cooling system

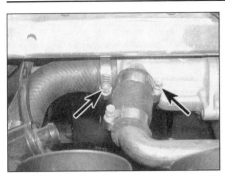

6.4a Slacken the clamp screws (arrowed) and detach the hoses from the thermostat housing . . .

6.4b . . . and manoeuvre it out of the front

6.4c Each outlet pipe is secured by a bolt (arrowed)

threads, then install the unit and tighten it to the torque setting specified at the beginning of the Chapter. Take care not to overtighten the unit as it or the radiator could be damaged. Connect the wiring connector.
21 Refill the cooling system (see Chapter 1). Install the fairing side panel (see Chapter 8).

6 Thermostat housing and thermostat – removal, check and installation

1 The thermostat is automatic in operation and should give many years' service without requiring attention. In the event of a failure, the valve will probably jam open, in which case the engine will take much longer than normal to warm up. Conversely, if the valve jams shut, the coolant will be unable to circulate and the engine will overheat. Neither condition is acceptable, and the fault must be investigated promptly.

Removal

⚠ **Warning: The engine must be completely cool before carrying out this procedure.**

Thermostat housing

2 Remove the radiator (see Section 7).
3 Remove the ignition HT coils (see Chapter 5).
4 Slacken the clamps securing the rear and left-hand hoses to the thermostat housing and detach the hoses, noting which fits where **(see illustration)**. Manoeuvre the housing out of the front **(see illustration)**. If required, for example when removing the valve cover, the housing can be removed along with the outlet pipes from the valve cover (or the pipes can be removed separately afterwards). Instead of detaching the rear and left-hand hoses, unscrew the bolt securing each pipe to the valve cover and pull the pipes out **(see illustration)**. Discard the O-rings, as new ones must be used.

Thermostat

5 Remove the thermostat housing (see above).
6 Unscrew the bolts securing the thermostat housing cover and separate it from the housing **(see illustrations)**. Withdraw the thermostat, noting how it fits **(see illustration)**.

Check

7 Examine the thermostat visually before carrying out the test. If it remains in the open position at room temperature, it should be replaced with a new one.
8 Suspend the thermostat by a piece of wire in a container of cold water. Place a thermometer in the water so that the bulb is close to the thermostat **(see illustration)**. Heat the water, noting the temperature when the thermostat opens, and compare the result with the specifications given at the beginning of the Chapter. Also check the amount the valve opens after it has been heated at 85°C for a few minutes and compare the measurement to the specifications. If the readings obtained differ from those given, the thermostat is faulty and must be replaced with a new one.
9 In the event of thermostat failure, as an emergency measure only, it can be removed and the machine used without it. **Note:** Take care when starting the engine from cold, as it will take much longer than usual to warm up. Ensure that a new unit is installed as soon as possible.

Installation

Thermostat housing

10 Installation is the reverse of removal. If the outlet pipes were removed, use new O-rings and smear them with grease **(see illustration)**. Tighten the outlet pipe bolts to the torque setting specified at the beginning

6.6a Unscrew the bolts (arrowed) . . .

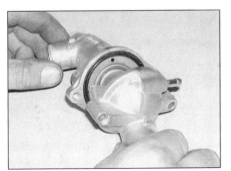

6.6b . . . and remove the cover . . .

6.6c . . . then withdraw the thermostat

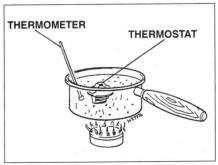

6.8 Thermostat testing set-up

3•6 Cooling system

6.10 Use new O-rings on the outlet pipes

of the Chapter. Make sure the hoses are pushed fully on to their unions and are secured by the clamps **(see illustrations 6.4b and a)**.

7.3 Disconnect the fan switch and temperature sender wiring connectors, and the earth wire (arrowed)

7.4a Release the clamps and detach the hoses (arrowed) from the right-hand side . . .

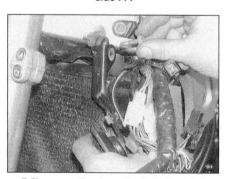

7.5b . . . noting how the horn bracket fits . . .

11 Refill the cooling system (see Chapter 1).

Thermostat

12 Fit the thermostat into the housing, making sure that it seats correctly and that the hole is at the top **(see illustration 6.6c)**. Fit the cover onto the housing, then install the bolts and tighten them to the torque setting specified at the beginning of the Chapter **(see illustrations 6.6b and a)**.
13 Install the housing (see above).

7 Radiator – removal and installation

Removal

 Warning: The engine must be completely cool before carrying out this procedure.

1 Remove the lower fairing and fairing side panels (see Chapter 8).
2 Drain the cooling system (see Chapter 1). Remove the reservoir (see Section 3).
3 Trace the fan motor wiring and disconnect it at the connector, located on the left-hand end of the radiator **(see illustration 4.3)**. On 1998 and 1999 models, disconnect the fan switch and temperature sender wiring connectors, and unscrew the bolt securing the earth wire **(see illustration)**. On 2000-on models, disconnect the thermo unit wiring connector.
4 Slacken the clamps securing all the radiator hoses and detach them from the radiator, noting which fits where **(see illustrations)**.

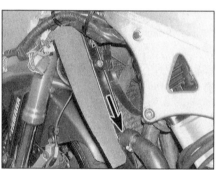

7.4b . . . and the left-hand side of the radiator

7.5c . . . then displace the radiator from the lug (arrowed) . . .

5 Unscrew the two bolts securing the radiator, noting the arrangement of the collars and rubber grommets and how the top bolt secures the horn bracket, and carefully manoeuvre the radiator to the right-hand side to free the other grommet from the lug, then remove the radiator **(see illustrations)**.
6 If necessary, separate the cooling fan from the radiator (see Section 4).
7 Check the radiator for signs of damage and clear any dirt or debris that might obstruct air flow and inhibit cooling. If the radiator fins are badly damaged or broken the radiator must be renewed. Also check the rubber mounting grommets, and replace them with new ones if necessary.

Installation

8 Installation is the reverse of removal, noting the following.
a) Make sure the collars, grommets and horn bracket are correctly installed with the mounting bolts **(see illustration 7.5b)**. Tighten the bolts to the torque setting specified at the beginning of the Chapter.
b) Make sure that the wiring connectors are correctly connected. On 1998 and 1999 models do not forget to attach the earth wire, and tighten the bolt to the specified torque setting **(see illustration 7.3)**.
c) Ensure the coolant hoses are in good condition (see Chapter 1), and are securely retained by their clamps, using new ones if necessary.
d) On completion refill the cooling system as described in Chapter 1.

7.5a Unscrew the bolts (arrowed) . . .

7.5d . . . and remove it

Cooling system 3•7

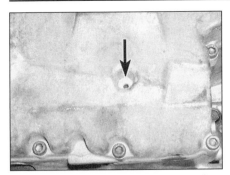

8.2 Water pump drain hole (arrowed)

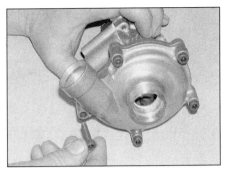

8.7a Unscrew the bolts and remove the cover

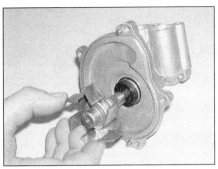

8.7b Withdraw the impeller from the pump

8 Water pump – check and overhaul

Check

1 The water pump is located inside the engine and is an integral assembly with the oil pump.
2 To prevent leakage of water from the cooling system to the lubrication system and vice versa, two seals are fitted on the pump shaft. The seal on the water pump side is of the mechanical type which bears on the rear face of the impeller. The second seal, which is mounted behind the mechanical seal, is of the normal feathered lip type. On the bottom of the sump there is a drain hole **(see illustration)**. If either seal fails, the drain allows the coolant or oil to escape and prevents them mixing. If on inspection there are signs of leakage, the seals must be removed and new ones installed (see below). If you are not sure about their condition, remove the pump (see below) and check them visually, looking for signs of damage and leakage.
3 Remove the pump, then remove the cover (see below). Wiggle the water pump impeller back-and-forth and in-and-out, and spin it by hand. If there is excessive movement, or the pump is noisy or rough when turned, the drive shaft bearing must be renewed. If you are not sure about its condition, disassemble the pump (see below) and check the bearing visually and mechanically, referring to *Tools and Workshop Tips* (Section 5) in the *Reference* Section. See below for bearing renewal.
4 Check the pump housing and cover, and the impeller and shaft for cracks, deformation and any other damage. Check that the shaft is straight – if it tilts by more than the specified limit, replace it with a new one. Also check for corrosion or a build-up of scale in the pump body and clean or renew the pump as necessary.
5 Check the condition of the rubber damper and its holder on the rear face of the impeller. Do not remove them from the shaft unnecessarily, as they cannot be reused. If they are damaged or deteriorated, fit a new impeller – Yamaha do not list the damper and holder as being available separately, though it is worth checking first. If they are available separately, lever off the old ones with a flat-bladed screwdriver. Apply coolant to the new ones and press them squarely down the shaft on to the back of the impeller.

Removal

6 The water pump is located inside the engine and is an integral assembly with the oil pump. Refer to Chapter 2, Section 19 and remove the pump.
7 To disassemble the water pump, first disassemble the oil pump (see Chapter 2). Unscrew the bolts securing the pump cover and remove it **(see illustration)**. Note the two locating pins and remove them if loose. Discard the O-ring, as a new one must be used. Withdraw the impeller from the housing **(see illustration)**.

Seal and bearing renewal

8 Remove the pump, then disassemble it (see above) **(see illustration)**.
9 To remove the mechanical seal, tap it out from the inside of the housing using a suitable punch, noting which way round it fits **(see illustration)**. Discard it, as a new one must be fitted.
10 To remove the oil seal, first remove the mechanical seal (see above). Tap the oil seal out from the inside of the housing using a suitable punch, noting which way round the seal fits. Discard it, as a new one must be fitted.

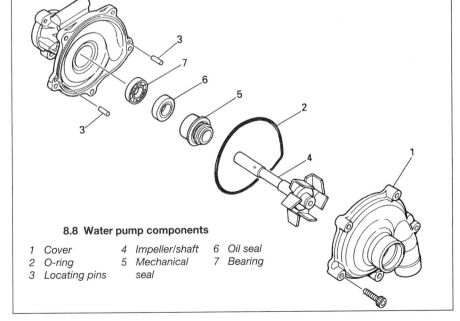

8.8 Water pump components

1 Cover
2 O-ring
3 Locating pins
4 Impeller/shaft
5 Mechanical seal
6 Oil seal
7 Bearing

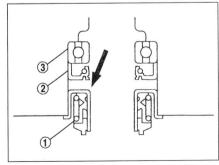

8.9 Mechanical seal (1), oil seal (2), bearing (3)

3•8 Cooling system

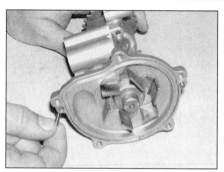

8.16a Fit the locating pins . . .

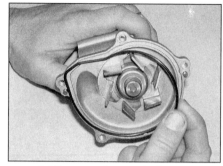

8.16b . . . and a new O-ring . . .

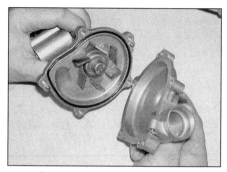

8.16c . . . then install the cover

11 To remove the bearing, first remove the seals (see above). Tap the bearing out from the inside of the housing using a bearing driver or suitable socket.
12 Press or drive the bearing into the housing, using a suitable sized socket or bearing driver. Drive the bearing in until it is properly seated **(see illustration 8.9)**.
13 Apply a smear of coolant to the lips of the new oil seal. Press or carefully drive the new oil seal into the housing, with its marked side facing out, so that it fits against the bearing – use a suitable sized socket or seal driver and make sure the seal is the correct way round **(see illustration 8.9)**.
14 Smear Yamaha Bond 1215 or a suitable equivalent to the mechanical seal housing. Press or carefully drive the new mechanical seal into the front of the pump body using a suitable sized socket or seal driver, shaped so that it bears only on the outer rim of the seal and not on the centre **(see illustration 8.9)**. Yamaha produce a special tool, Part No. 90890-04078 (European models) or YM-33221 (US models) for installing the seal if required.

Installation

15 Slide the drive shaft and impeller into the housing **(see illustration 8.7b)**.
16 If removed, fit the locating pins into the pump housing **(see illustration)**. Fit the new cover O-ring into its groove **(see illustration)**. Fit the cover, making sure it locates onto the pins, and tighten the bolts to the torque setting specified at the beginning of the Chapter **(see illustration)**.
17 Refer to Chapter 2, Section 19 and install the pump.

9 Coolant hoses, pipes and unions – removal and installation

Removal

1 Before removing a hose, pipe or union, drain the coolant (see Chapter 1).
2 Use a screwdriver to slacken the larger-bore hose clamps, then slide them back along the hose and clear of the union spigot. The smaller-bore hoses are secured by spring clamps which can be expanded by squeezing their ears together with pliers.
Caution: The radiator unions are fragile. Do not use excessive force when attempting to remove the hoses.
3 If a hose proves stubborn, release it by rotating it on its union before working it off. If all else fails, cut the hose with a sharp knife then slit it at each union so that it can be peeled off in two pieces. Whilst this means renewing the hose, it is preferable to buying a new radiator.
4 Remove the coolant pipes and the unions on the engine by detaching the hoses (see above), then unscrewing the retaining bolt(s). Discard the O-rings, as new ones must be used.

Installation

5 Slide the clamp onto the hose and then work it on to its respective union.

> **HAYNES HINT** *If the hose is difficult to push onto its union, it can be softened by soaking it in very hot water, or alternatively a little soapy water can be used as a lubricant.*

6 Rotate the hose on its union to settle it in position before sliding the clamp into place and tightening it securely.
7 If the water pipes or unions on the engine have been removed, fit new O-rings and smear them with grease, then install them and tighten their mounting bolts to the torque setting specified at the beginning of the Chapter. Apply a suitable non-permanent thread locking compound to the bolt for the water pump outlet pipe on the front of the engine.

Chapter 4
Fuel and exhaust systems

Contents

Air filter – cleaning and renewal .... see Chapter 1
Air filter housing – removal and installation 4
Air induction system (AIS) – function, disassembly and reassembly 17
Air/fuel mixture adjustment – general information 5
Carburettor overhaul – general information 6
Carburettor synchronisation .... see Chapter 1
Carburettors – disassembly, cleaning and inspection 8
Carburettors – reassembly and fuel level check 10
Carburettors – removal and installation 7
Carburettors – separation and joining 9
Choke cable – removal and installation 12
Exhaust system – removal and installation 13
EXUP system – cable renewal, check, removal and installation 14

EXUP system – check .... see Chapter 1
Fuel hoses – renewal .... see Chapter 1
Fuel pump and relay – check, removal and installation 15
Fuel system – check .... see Chapter 1
Fuel tank – cleaning and repair 3
Fuel tank and fuel tap – removal and installation 2
Fuel warning light and sensor – check and renewal 16
General information and precautions 1
Idle speed – check .... see Chapter 1
Throttle and choke cables – check and adjustment .... see Chapter 1
Throttle cables – removal and installation 11
Throttle position sensor – check and adjustment ...... see Chapter 5

Degrees of difficulty

| **Easy,** suitable for novice with little experience | **Fairly easy,** suitable for beginner with some experience | **Fairly difficult,** suitable for competent DIY mechanic | **Difficult,** suitable for experienced DIY mechanic | **Very difficult,** suitable for expert DIY or professional 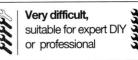 |

Specifications

Fuel
Grade .... Unleaded, minimum 91 RON (Research Octane Number)
Fuel tank capacity (including reserve) .... 18.0 litres
Reserve .... approx. 4.0 litres

Carburettors – European models
Type .... 4 x Mikuni BDSR40
Fuel level .... 4.1 to 5.1 mm below line on float chamber
Idle speed .... see Chapter 1
Pilot screw setting (no. of turns out)
  1998 and 1999 models 2½
  2000-on models 3⅛
Main jet 130
Main air jet
  Cylinders 1 and 4 60
  Cylinders 2 and 3 65
Jet needle 6DEY5-53-3
Needle jet
  1998 and 1999 models .... P-O
  2000-on models .... P-OM
Pilot air jet 120
Pilot jet
  1998 and 1999 models 17.5
  2000-on models 15
Starter jet 1 35
Starter jet 2 .... 0.7
Butterfly valve size 100

Carburettors – US models

Type	4 x Mikuni BDSR40
Fuel level	4.1 to 5.1 mm below line on float chamber
Idle speed	see Chapter 1
Pilot screw setting (no. of turns out)	3⅛
Main jet	130
Main air jet	
Cylinders 1 and 4	60
Cylinders 2 and 3	65
Jet needle	
1998 and 1999 models	6DEY5-53-3
2000-on models	6DEY8-53-1
Needle jet	
1998 and 1999 models	P-O
2000-on models	P-OM
Pilot air jet	120
Pilot jet	15
Starter jet 1	35
Starter jet 2	0.7
Butterfly valve size	100

EXUP system

Servo motor static resistance	5.3 to 9.8 K-ohms
Servo motor variable resistance	zero to approx. 7.5 K-ohms

Fuel pump

Resistance	4.0 to 30 ohms @ 20°C

Torque wrench settings

Fuel tank mountings	
Front mounting bolt nut	7 Nm
Rear mounting bolt	10 Nm
Fuel tap screws	7 Nm
Carburettor joining bolts	7 Nm
Silencer clamp bolt	20 Nm
Silencer mounting bolt	38 Nm
Downpipe assembly nuts	20 Nm
Downpipe assembly rear bolt	20 Nm
EXUP valve cover bolts	10 Nm
EXUP pulley cover bolts	10 Nm
Fuel level sensor screws	7 Nm

1 General information and precautions

General information

The fuel system consists of the fuel tank with internal level sensor, the fuel tap, fuel filter, fuel pump, fuel hoses, carburettors and control cables. The fuel tap has an integral strainer, and the fuel pump is mounted externally with an in-line filter.

The carburettors used are CV types. There is a carburettor for each cylinder. For cold starting, a choke lever is connected to the carburettors by a cable. The lever is housed in the left-hand switch housing on the handlebar.

Air is drawn into the carburettors via an air filter, which is housed under the fuel tank.

The exhaust is a four-into-one design, and incorporates Yamaha's EXUP system. 2000-on models have an air induction system (AIS) to reduce harmful emissions.

Many of the fuel system service procedures are considered routine maintenance items and for that reason are included in Chapter 1.

Precautions

Warning: *Petrol (gasoline) is extremely flammable, so take extra precautions when you work on any part of the fuel system. Don't smoke or allow open flames or bare light bulbs near the work area, and don't work in a garage where a natural gas-type appliance is present. If you spill any fuel on your skin, rinse it off immediately with soap and water. When you perform any kind of work on the fuel system, wear safety glasses and have a fire extinguisher suitable for a class B type fire (flammable liquids) on hand.*

Always perform service procedures in a well-ventilated area to prevent a build-up of fumes.

Never work in a building containing a gas appliance with a pilot light, or any other form of naked flame. Ensure that there are no naked light bulbs or any sources of flame or sparks nearby.

Do not smoke (or allow anyone else to smoke) while in the vicinity of petrol (gasoline), or of components containing petrol. Remember the possible presence of vapour from these sources and move well clear before smoking.

Check all electrical equipment belonging to the house, garage or workshop where work is being undertaken (see the *Safety First!* section of this manual). Remember that certain electrical appliances such as drills, cutters etc. create sparks in the normal course of operation and must not be used near petrol (gasoline) or any component containing it. Again, remember the possible presence of fumes before using electrical equipment.

Always mop up any spilt fuel and safely dispose of the rag used.

Any stored fuel that is drained off during servicing work must be kept in sealed containers that are suitable for holding petrol (gasoline), and clearly marked as such; the containers themselves should be kept in a safe place. Note that this last point applies equally to the fuel tank if it is removed from the machine; also remember to keep its filler cap closed at all times.

Read the *Safety first!* section of this manual carefully before starting work.

Fuel and exhaust systems 4•3

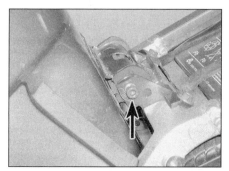

2.2a Slacken the rear bolt (arrowed) . . .

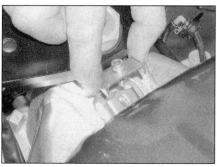

2.2b . . . then unscrew the front bolt and remove the bracket

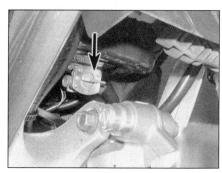

2.4a Turn the fuel tap (arrowed) to the 'OFF' position

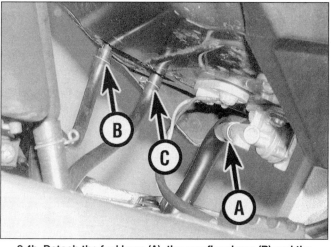

2.4b Detach the fuel hose (A), the overflow hose (B) and the breather hose (C)

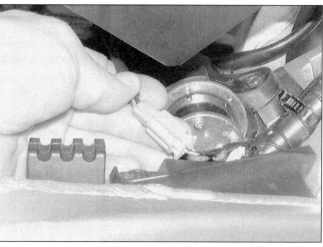

2.6 Disconnect the wiring connector

2 Fuel tank and fuel tap – removal and installation

Warning: *Refer to the precautions given in Section 1 before starting work.*

Fuel tank

Removal

1 Make sure the fuel filler cap is secure. Remove the rider's seat (see Chapter 8).
2 Slacken the bolt securing the rear of the tank **(see illustration)**. Unscrew the bolt securing the front of the tank and remove the bracket **(see illustration)**.
3 Raise the tank at the front and support it using a block of wood.
Caution: Do not raise the tank too high as the front could contact the fairing stay bolts and be damaged – place a rag between the tank and the stay, as a precaution.
4 Turn the fuel tap off – it is located on the base of the tank **(see illustration)**. Have a rag ready to catch any residual fuel from the hose and the fuel filter, then release the clamp securing the fuel hose to the fuel tap and detach the hose **(see illustration)**.
5 Release the clamp securing the overflow hose, and on all except California models the breather hose, to their unions and detach the hoses, noting which fits where **(see illustration 2.4b)**. On California models, detach the EVAP hose from its union.
6 Disconnect the fuel level sender wiring connector **(see illustration)**.
7 Remove the support and lower the tank. Unscrew the bolt securing the rear of the tank **(see illustration)**. Carefully lift the tank off the frame and remove it **(see illustrations)**.
8 Inspect the tank mounting rubbers for signs of damage or deterioration and replace them with new ones, if necessary.

Installation – all models

9 Installation is the reverse of removal, noting the following:
a) *Make sure the hoses are properly attached and secured by their clamps. Make sure the wiring connector is securely connected.*

2.7a Unscrew the rear bolt . . .

2.7b . . . and carefully remove the tank

4•4 Fuel and exhaust systems

2.9 Do not forget to turn the tap back on

b) Turn the fuel tap ON before lowering the tank *(see illustration)*.
c) Tighten the mounting bolts to the torque settings specified at the beginning of the chapter.
d) Start the engine and check that there is no sign of fuel leakage, then shut if off.

Fuel tap

Removal

10 The tap should not be removed unnecessarily from the tank, otherwise there is a possibility of damaging the O-ring or strainer.
11 Remove the fuel tank as described above.
12 Connect a drain hose to the fuel outlet union on the tap and insert its end in a container suitable and large enough for storing the fuel. Turn the fuel tap to the 'ON' position and allow the tank to drain. When the tank has drained, turn the tap to the 'OFF' position.
13 Remove the screws securing the tap to the tank and withdraw the tap assembly *(see illustration)*. Discard the O-ring, as a new one must be used.
14 Clean the gauze strainer to remove all traces of dirt and fuel sediment. Check the gauze for holes. If any are found, a new tap should be fitted, as the strainer is not available individually.

Installation

15 Installation is the reverse of removal.
16 Use a new O-ring on the tap, and tighten the screws to the torque setting specified at the beginning of the Chapter.
17 Install the fuel tank (see above).

3 Fuel tank – cleaning and repair

1 All repairs to the fuel tank should be carried out by a professional who has experience in this critical and potentially dangerous work. Even after cleaning and flushing of the fuel system, explosive fumes can remain and ignite during repair of the tank.
2 If the fuel tank is removed from the bike, it should not be placed in an area where sparks or open flames could ignite the fumes coming out of the tank. Be especially careful inside garages where a natural gas-type appliance is located, because the pilot light could cause an explosion.

4 Air filter housing – removal and installation

Removal

1 Remove the fuel tank (see Section 2).
2 Release the clamps securing the crankcase breather hose and the air filter housing breather hose and detach them from the filter housing, noting which fits where *(see illustration)*. On 2000-on models, also release the clamp securing the AIS hose and detach it.
3 Unscrew the bolt securing the front of the housing to the frame, noting the collar and grommet *(see illustration)*. Slacken the clamp screws securing the housing to the carburettor intakes on the underside *(see illustration)*.
4 Lift the housing up off the carburettors and remove it *(see illustration)*.

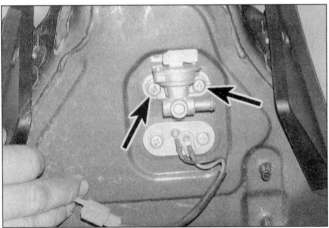

2.13 Fuel tap mounting screws (arrowed)

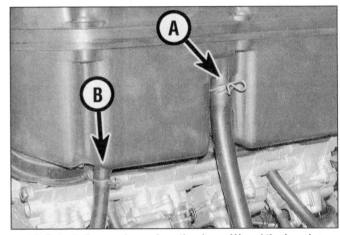

4.2 Detach the crankcase breather hose (A) and the housing breather hose (B)

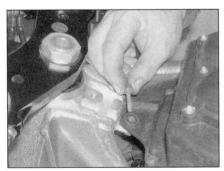

4.3a Unscrew the mounting bolt at the front . . .

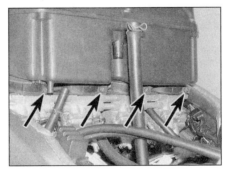

4.3b . . . then slacken the clamp screws (arrowed)

4.4 Lift the housing up off the carburettors

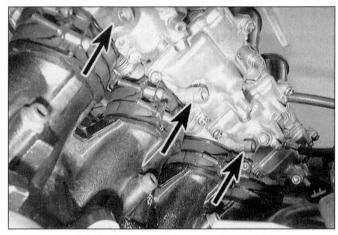
5.1 Carburettor pilot screws (arrowed)

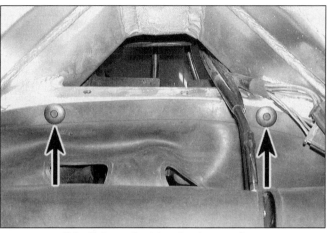
7.2a Release the trim clips (arrowed) . . .

Installation

5 Installation is the reverse of removal. Check the condition of the various hoses and their clamps and replace them with new ones, if necessary.

5 Air/fuel mixture adjustment – general information

UK market models

1 If the engine runs extremely rough at idle or continually stalls, and if a carburettor overhaul does not cure the problem (and it definitely is a carburation problem – see Section 6), the pilot screws may require adjustment. It is worth noting at this point that unless you have the experience to carry this out on a multi-cylinder machine it is best to entrust the task to a motorcycle dealer, tuner or fuel systems specialist. Note that you will need a long thin screwdriver with an angled end to access the pilot screws **(see illustration)**. Remove the fairing side panels (see Chapter 8).

2 Before adjusting the pilot screws, warm the engine up to normal working temperature. Stop the engine and remove the tank (see Section 2). Screw in all four pilot screws until they seat lightly, then back them out to the number of turns specified (see this Chapter's Specifications). This is the base position for adjustment.

3 Arrange a temporary fuel supply using an auxiliary tank and some hosing. Start the engine and reset the idle speed to the correct level (see Chapter 1). Working on one carburettor at a time, turn the pilot screw by a small amount either side of this position to find the point at which the highest consistent idle speed is obtained. When you've reached this position, reset the idle speed to the specified amount (see Chapter 1). Repeat on the other three carburettors in turn.

Other markets

4 Due to the increased emphasis on controlling exhaust emissions in certain world markets, regulations have been formulated which prevent adjustment of the air/fuel mixture. On such models the pilot screw positions are pre-set at the factory and in some cases have a limiter cap fitted to prevent tampering. Where adjustment is possible, it can only be made in conjunction with an exhaust gas analyser to ensure that the machine does not exceed the emissions regulations.

6 Carburettor overhaul – general information

1 Poor engine performance, hesitation, hard starting, stalling, flooding and backfiring are all signs that major carburettor maintenance may be required.
2 Keep in mind that many so-called carburettor problems are really not carburettor problems at all, but mechanical problems within the engine, or ignition system malfunctions. Try to establish for certain that the carburettors are in need of maintenance before beginning a major overhaul.
3 Check the fuel tap and filter, the fuel hoses, the fuel pump, the intake manifold joint clamps, the air filter, the ignition system, the spark plugs and carburettor synchronisation before assuming that a carburettor overhaul is required.
4 Most carburettor problems are caused by dirt particles, varnish and other deposits which build up in and block the fuel and air passages. Also, in time, gaskets and O-rings shrink or deteriorate and cause fuel and air leaks which lead to poor performance.
5 When overhauling the carburettors, disassemble them completely and clean the parts thoroughly with a carburettor cleaning solvent and dry them with filtered, unlubricated compressed air. Blow through the fuel and air passages with compressed air to force out any dirt that may have been loosened but not removed by the solvent. Once the cleaning process is complete, reassemble the carburettor using new gaskets and O-rings.
6 Before disassembling the carburettors, make sure you have all the necessary O-rings and other parts, some carburettor cleaner, a supply of clean rags, some means of blowing out the carburettor passages and a clean place to work. It is recommended that only one carburettor be overhauled at a time to avoid mixing up parts.

7 Carburettors – removal and installation

⚠ **Warning: Refer to the precautions given in Section 1 before starting work.**

Removal

1 Remove the fuel tank (see Section 2), the air filter housing (see Section 4), and the ignition coil assembly (see Chapter 5).
2 Pull the rubber cover sections up off the carburettor intakes and fold them forward out of the way. If required, release the trim clips securing the rubber cover and remove it, noting how it fits **(see illustrations)**. To

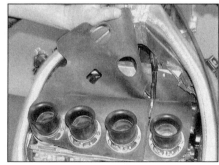
7.2b . . . and remove the cover

4•6 Fuel and exhaust systems

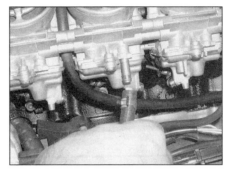

7.2c Push the centre into the body and remove the clip

7.4 Detach the fuel supply hose . . .

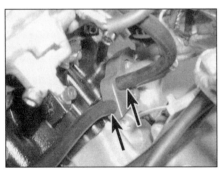

7.5 . . . and disconnect the throttle position sensor wiring connector

release the clips, push the centre into the body, then draw the clip out **(see illustration)**.
3 Detach the choke cable from the carburettors (see Section 12).
4 Release the clamp securing the fuel supply hose to the carburettors and detach the hose **(see illustration)**.
5 Disconnect the throttle position sensor wiring connector **(see illustration)**.
6 Detach the float chamber vent hoses from the unions on the back of the engine **(see illustration)**. On California models, detach the emission system hoses from the carburettors.
7 Free the idle speed adjuster from its holder **(see illustration)**.
8 Detach the throttle cables from the carburettors (see Section 11). If access is too restricted to access the cable ends, detach them after the carburettors have been lifted off the cylinder head intakes.
9 Fully slacken the upper clamp bolts on the cylinder head intake rubbers **(see illustration)**.
10 Ease the carburettors off the intakes and remove them **(see illustration)**. **Note:** *Keep the carburettors level to prevent fuel spillage from the float chambers and the possibility of the piston diaphragms being damaged.*
Caution: Stuff clean rag into each cylinder head intake after removing the carburettors, to prevent anything from falling in.
11 Place a suitable container below the float chambers, then slacken the drain bolt on each chamber in turn and drain all the fuel from the carburettors **(see illustration)**. Tighten the drain screws securely once all the fuel has been drained.

Installation

12 Installation is the reverse of removal, noting the following.
  a) Check for cracks or splits in the cylinder head intake rubbers, and replace them with new ones, if necessary – slacken the clamp bolts to release them. They are marked 'L' and 'R' for 'Left-' and 'Right-hand' side. Make sure the tabs on the underside locate around the lug on each intake duct **(see illustration)**.
  b) Make sure the carburettors are fully engaged with the cylinder head intake rubbers – they can be difficult to engage, so a squirt of WD40 or a smear of grease will ease entry. Make sure the clamps are positioned so the indents locate over the ridges on the rubber **(see illustration)**. Tighten them securely.

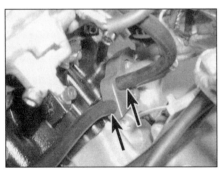

7.6 Detach the float chamber vent hoses (arrowed) . . .

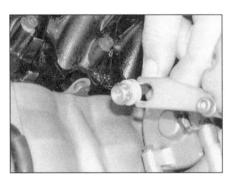

7.7 . . . and release the idle speed adjuster

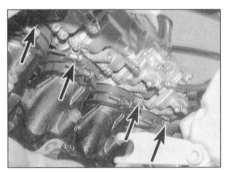

7.9 Fully slacken the clamp bolts (arrowed) . . .

7.10 . . . then ease the carburettors out of the intakes and remove them

7.11 Slacken the bolts (arrowed) to drain the carburettors

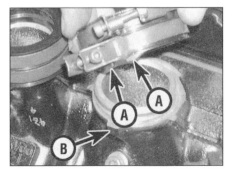

7.12a Make sure the tabs (A) locate on each side of the lug (B)

Fuel and exhaust systems 4•7

7.12b Position the clamps so that the ridges in the rubber locate in the indents in the clamp (arrowed)

7.12c Pull the centre out, then fit the clip into its socket. Push the centre into the body to secure it

c) Make sure all hoses are correctly routed and secured and not trapped or kinked.
d) Refer to Section 11 for installation of the throttle cables, and Section 12 for the choke cable. Check the operation of the cables and adjust them as necessary (see Chapter 1).
e) To install the clips securing the rubber cover, first pull the centre back out so that it protrudes from the top of the body. Fit the clip into its socket, then push the centre in so that it is flush with the top of the body **(see illustration)**.
f) Do not forget to connect the throttle position sensor wiring connector.
g) Check idle speed and carburettor synchronisation and adjust as necessary (see Chapter 1).

8 Carburettors – disassembly, cleaning and inspection

⚠️ **Warning:** Refer to the precautions given in Section 1 before starting work.

Disassembly

1 Remove the carburettors from the machine as described in the previous Section. **Note:** *Do not separate the carburettors unless absolutely necessary; each carburettor can be dismantled sufficiently for all normal cleaning and adjustments while in place on the mounting brackets. Dismantle the carburettors separately to avoid interchanging parts* **(see illustration)**.
2 Unhook the choke linkage bar return spring, noting how it fits **(see illustration)**. Remove the screws securing the choke linkage bar to the carburettors, noting the plastic washers **(see illustration)**. Lift off the bar, noting how it fits **(see illustration 10.9b)**. Unscrew the

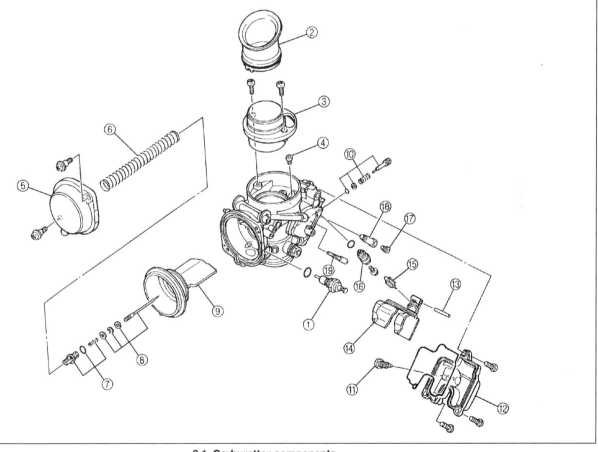

8.1 Carburettor components

1 Choke plunger assembly and O-ring
2 Intake duct
3 Intake duct seat
4 Pilot air jet
5 Top cover
6 Spring
7 Jet needle holder, O-ring and spring
8 Jet needle, spacer, E-clip and washer
9 Diaphragm/piston
10 Pilot screw
11 Drain bolt
12 Float chamber
13 Float pivot pin
14 Float
15 Needle valve
16 Needle valve seat and O-ring
17 Main jet
18 Main jet holder
19 Pilot jet

4•8 Fuel and exhaust systems

8.2a Unhook the return spring . . .

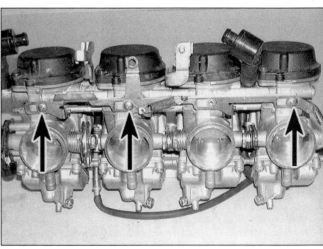

8.2b . . . then remove the screws (arrowed) and lift off the linkage bar . . .

8.2c . . . and unscrew the choke plunger nut

choke plunger nut, using a pair of thin-nosed pliers if access is too restricted for a spanner, and withdraw the plunger assembly from the carburettor body **(see illustration)**.

3 Unscrew and remove the top cover retaining screws and remove the cover **(see illustration)**.

4 Remove the spring from inside the piston **(see illustration)**. Carefully peel the diaphragm away from its sealing groove in the carburettor and withdraw the diaphragm/piston assembly **(see illustration)**. Using a pair of thin-nosed pliers, carefully withdraw the jet needle holder from inside the piston – it is a press fit, held by an O-ring **(see illustration)**. Note the spring fitted into the bottom of the holder. Push the jet needle up from the bottom of the piston and withdraw from the top, noting the washer on the top of the needle **(see illustration)**. If the E-clip and spacer are removed from the needle, note which notch the clip is fitted into.

Caution: Do not use a sharp instrument to displace the diaphragm, as it is easily damaged.

5 Remove the screws securing the float chamber to the base of the carburettor and remove the float chamber, noting how it fits **(see illustration)**. Remove the rubber gasket and discard it, as a new one must be used.

8.3 Undo the screws (arrowed) and remove the top cover

8.4a Remove the spring . . .

8.4b . . . and the piston/diaphragm assembly

8.4c Withdraw the needle holder using pliers . . .

8.4d . . . then push the needle up and remove it

8.5 Float chamber screws (arrowed)

Fuel and exhaust systems 4•9

8.6a Withdraw the pivot pin and remove the float assembly

8.6b Undo the screw (arrowed) and remove the needle valve seat

8.7 Remove the main jet (arrowed) . . .

6 Using a pair of thin-nosed pliers, carefully withdraw the float pivot pin **(see illustration)**. If necessary, carefully displace the pin using a small screwdriver or a nail. Remove the float assembly, noting how it fits **(see illustration 10.5d)**. Unhook the needle valve from the tab on the float, noting how it fits **(see illustration 10.5c)**. Remove the screw securing the needle valve seat and draw out the seat **(see illustration)**. Discard its O-ring, as a new one must be used.

7 Unscrew and remove the main jet **(see illustration)**.

8 Unscrew and remove the main jet holder **(see illustration)**.

9 Unscrew and remove the pilot jet **(see illustration)**.

10 The pilot screw can be removed from the carburettor, but note that its setting will be disturbed **(see Haynes Hint)**. Unscrew and remove the pilot screw along with its spring, washer and O-ring **(see illustration)**. Discard the O-ring, as a new one must be used.

 To record the pilot screw's current setting, turn the screw in until it seats lightly, counting the number of turns necessary to achieve this, then fully unscrew it. On installation, the screw is simply backed out the number of turns you've recorded.

11 A throttle position sensor is mounted on the outside of the right-hand carburettor. Do not remove the sensor from the carburettor unless it is known to be faulty and is being replaced with a new one. Refer to Chapter 5 for check and adjustment of the sensor.

Cleaning

Caution: Use only a dedicated carburettor cleaner or petroleum-based solvent for carburettor cleaning. Do not use caustic cleaners.

12 Remove the air intake duct, noting how it fits – it is held only by the tension of the spring around its base **(see illustration)**. Submerge the carburettor body and individual metal components in the solvent for approximately thirty minutes (or longer, if the directions recommend it).

13 After the carburettor has soaked long enough for the cleaner to loosen and dissolve most of the varnish and other deposits, use a nylon-bristle brush to remove the stubborn deposits. Rinse it again, then dry it with compressed air.

14 Use a jet of compressed air to blow out all of the fuel and air passages in the main and upper body. Do not forget to blow through the air jets and passages in the intake side of the carburettor **(see illustration)**. If required, undo the screws securing the intake duct seat and remove it, noting how it fits **(see illustration)**. The pilot air jet can now be removed, if required.

8.8 . . . the main jet holder (arrowed) . . .

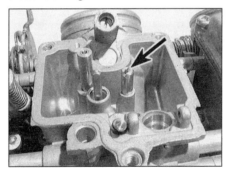

8.9 . . . and the pilot jet (arrowed)

8.10 Remove the pilot screw (arrowed) if required

8.12 Remove the intake duct, noting how it locates (arrowed)

8.14a Do not forget to clean the jet and passage on the intake side (arrowed)

8.14b The intake duct seat is held by two screws (arrowed)

4•10 Fuel and exhaust systems

8.15 Choke plunger assembly

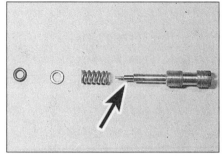

8.16 Check the tapered portion of the pilot screw (arrowed) for wear

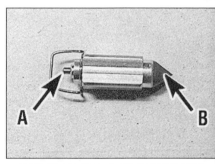

8.21 Check the valve's spring loaded rod (A) and tip (B) for wear or damage

Caution: *Never clean the jets or passages with a piece of wire or a drill bit, as they will be enlarged, causing the fuel and air metering rates to be upset.*

Inspection

15 Check the operation of the choke plunger assembly. If it doesn't move smoothly, inspect the needle on the end of the choke plunger, the spring and the plunger linkage bar **(see illustration)**. Replace the assembly with a new one if any component is worn, damaged or bent – individual parts are not available.

16 If removed from the carburettor, check the tapered portion of the pilot screw and the spring and O-ring for wear or damage **(see illustration)**. Replace the assembly with a new one if necessary – individual parts are not available.

17 Check the carburettor body, float chamber and top cover for cracks, distorted sealing surfaces and other damage. If any defects are found, replace the faulty component with a new one, although replacement of the entire carburettor may be necessary (check with a Yamaha dealer on the availability of separate components).

18 Check the piston diaphragm for splits, holes, creases and general deterioration. Holding it up to a light will help to reveal problems of this nature. Replace it with a new one if necessary – it is available separately from the piston.

19 Insert the piston or slide in the carburettor body and check that it moves up and down smoothly. Check the surface of the piston or slide for wear. If it is worn excessively or doesn't move smoothly in the guide, replace the components with new ones as necessary.

20 Check the jet needle for straightness by rolling it on a flat surface such as a piece of glass. Replace it with a new one if it is bent, or if the tip is worn.

21 Check the tip of the float needle valve and the valve seat. If either has grooves or scratches in it, or is in any way worn, they must be renewed as a set **(see illustration)**.

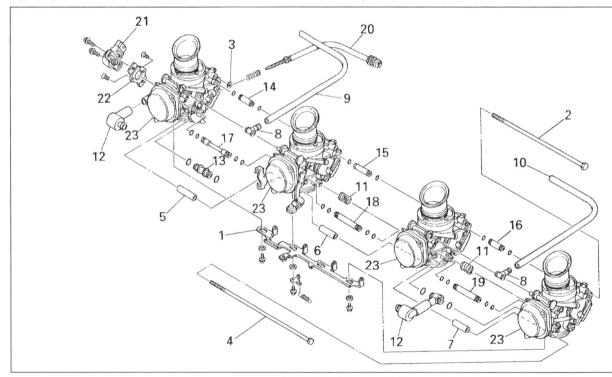

9.2 Carburettor assembly

1 Choke linkage bar	7 Spacer	13 Top cover pipe	19 Pipe
2 Connecting bolt	8 Hose joint	14 Fuel feed pipe	20 Throttle stop screw
3 Copper washer	9 Float chamber air vent hose	15 Fuel feed pipe	21 Throttle position sensor
4 Connecting bolt	10 Float chamber air vent hose	16 Fuel feed pipe	22 Throttle position sensor bracket
5 Spacer	11 Spring	17 Pipe	
6 Spacer	12 Top cover air vent hose	18 Pipe	23 Carburettor bodies

Fuel and exhaust systems 4•11

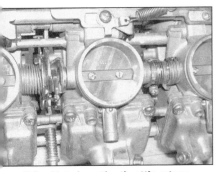
9.3a Note how the throttle return springs . . .

9.3b . . . linkage assembly . . .

9.3c . . . and carburettor synchronisation springs . . .

9.3d . . . hoses, unions . . .

9.3e . . . joint pieces and collars . . .

9.3f . . . and cable brackets fit, before separating the carburettors

22 Operate the throttle shaft to make sure the throttle butterfly valve opens and closes smoothly. If it doesn't, clean the throttle linkage, and also check the butterfly for distortion, or for any debris caught between its edge and the carburettor. Also check that the butterfly is central on the shaft – if the screws securing it to the shaft have come loose it may be catching.

23 Check the float for damage. This will usually be apparent by the presence of fuel inside the float. If it is damaged, replace it with a new one.

9 Carburettors – separation and joining

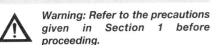

Warning: Refer to the precautions given in Section 1 before proceeding.

Separation

1 The carburettors do not need to be separated for normal overhaul. If you need to separate them (to replace a carburettor body with a new one, for example), refer to the following procedure.

2 Remove the carburettors from the machine (see Section 7). Mark the body of each carburettor with its cylinder location to ensure that it is positioned correctly on reassembly **(see illustration opposite)**.

3 Make a note of how the throttle return springs, linkage assembly and carburettor synchronisation springs are arranged, to ensure that they are fitted correctly on reassembly **(see illustrations)**. Also note the arrangement of the various hoses, unions, joint pieces and collars, and of the cable brackets **(see illustrations)**.

4 Unhook the choke linkage bar return spring, noting how it fits **(see illustration 8.2a)**. Remove the screws securing the choke linkage bar to the carburettors, noting the plastic washers **(see illustration 8.2b)**. Lift off the bar, noting how it fits **(see illustration 10.9b)**.

5 The carburettors are joined by two long bolts which pass through them **(see illustration)**. Remove the bolts, noting the copper washer with the bottom bolt, and remove the spacers, noting which fits where, as they are of different lengths.

6 Mark the position of each carburettor and gently separate them, noting how the throttle linkage is connected, and being careful not to

9.5 Carburettor joining bolts (arrowed)

lose any springs, or fuel and vent pipes and hose joints, that are present between the carburettors, and noting the O-rings fitted with them.

Joining

7 Fit new O-rings onto the fuel and vent pipes **(see illustration 9.2)**. Lubricate the O-rings with a light film of oil and install the fittings into their respective holes, making sure they seat completely. Also fit the hose joints.

8 Position the coil springs between the carburettors, gently push the carburettors together, then make sure the throttle linkages are correctly engaged **(see illustrations 9.3a, b and c)**. Check the fuel and vent fittings and hose joints to make sure they engage properly also **(see illustrations 9.3d, e and f)**.

9 Install the two long through-bolts with their spacers, but do not tighten them yet. Make sure the collars are correctly positioned, as they are of different lengths – the shortest (36 mm) fits between Nos. 1 and 2 carburettors (the carburettors are numbered 1 to 4 from left to right), the middle (38 mm) between Nos. 2 and 3, and the longest (43 mm) between Nos. 3 and 4. Set the carburettors on a sheet of glass, then align them with a straight-edge placed along the edges of the bores. When the centrelines of the carburettors are all in horizontal and vertical alignment, tighten the bolts to the torque setting specified at the beginning of the Chapter **(see illustration 9.5)**.

10 Fit the choke linkage bar onto the plungers, making sure the slots locate

4•12 Fuel and exhaust systems

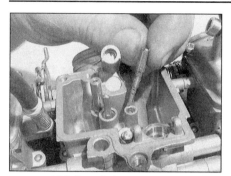

10.2 Install the pilot jet . . .

10.3 . . . the main jet holder . . .

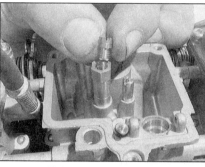

10.4 . . . and the main jet

correctly behind the nipple on the end of each choke plunger **(see illustration 10.9b)**. Fit the plastic washers and secure the linkage bar in place with the screws **(see illustration 10.9c)**. Hook up the linkage bar return spring **(see illustration 8.2a)**. Make sure the choke linkage operates smoothly, and returns quickly under spring pressure.

11 Install the throttle and carburettor synchronisation springs **(see illustrations 9.3a, b and c)**. Visually synchronise the throttle butterfly valves, turning the adjusting screws on the throttle linkage, if necessary, to equalise the clearance between the butterfly valve and throttle bore of each carburettor. Make sure the throttle operates smoothly, and returns quickly under spring pressure.

12 Install the carburettors (see Section 7) and check carburettor synchronisation and idle speed (see Chapter 1).

10 Carburettors – reassembly and fuel level check

Warning: Refer to the precautions given in Section 1 before proceeding.

Note: When reassembling the carburettors, be sure to use the new O-rings, seals and other parts supplied in the rebuild kit. Do not overtighten the carburettor jets and screws, as they are easily damaged.

Reassembly

1 Install the pilot screw (if removed) along with its spring, washer and O-ring, turning it in until it seats lightly **(see illustration 8.10)**. Now, turn the screw out the number of turns previously recorded, or as specified at the beginning of the Chapter.

2 Install the pilot jet **(see illustration)**.
3 Install the main jet holder **(see illustration)**.
4 Install the main jet **(see illustration)**.
5 If removed, fit a new O-ring onto the needle valve seat, then press it into place and secure it with the screw **(see illustrations)**. Hook the float needle valve onto the tab on the float assembly, then position the float assembly in the carburettor, making sure the needle valve locates in the seat **(see illustrations)**. Install the pivot pin, making sure it is secure **(see illustration)**.
6 Fit a new gasket onto the float chamber, making sure it is seated properly in its groove, then install the chamber on the carburettor and tighten its screws securely **(see illustration)**.
7 If removed, fit the spacer and E-clip onto the needle, making sure the clip is in the same groove from which it was removed, then fit the washer on top of the clip **(see illustration)**.

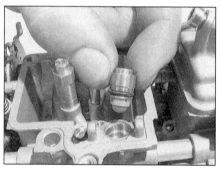

10.5a Fit the valve seat using a new O-ring . . .

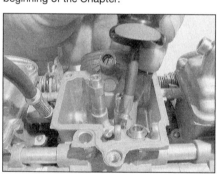

10.5b . . . and secure it with the screw

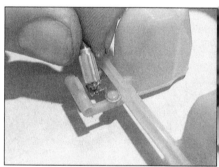

10.5c Hook the needle valve onto the tab on the float . . .

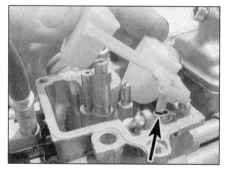

10.5d . . . then install the float assembly, locating the valve in the seat (arrowed) . . .

10.5e . . . and fit the pivot pin

10.6 Fit a new gasket into the groove and install the float chamber

Fuel and exhaust systems 4•13

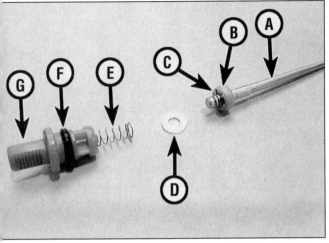

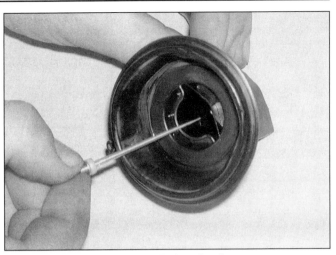

10.7a Needle (A), spacer (B), E-clip (C), washer (D), spring (E), O-ring (F), holder (G)

10.7b Fit the needle into the piston . . .

Check that the spring and O-ring are fitted to the needle holder – use a new O-ring if necessary. Fit the jet needle into the piston, then fit the needle holder onto it and press it gently down until it is secure **(see illustrations)**. Fit the piston/diaphragm assembly into the carburettor and lightly push the piston down, making sure the needle is correctly aligned with the needle jet **(see illustration 8.4b)**. Press the rim of the diaphragm into its groove, making sure it is correctly seated with the tab on the diaphragm locating correctly around the air passage in the carburettor **(see illustration)**.

If difficulty is encountered getting the rim to stay in the groove, fit the spring and locate the cover as described below, then feed the rim into the groove using a small screwdriver while holding the cover onto it, and fit the cover screws **(see illustration)**.

8 Fit the spring into the diaphragm assembly, making sure it locates correctly onto the needle holder, then fit the top cover onto the carburettor, locating the peg into the top of the spring, and tighten its screws securely **(see illustration)**. Check that the piston moves smoothly in the guide by pushing it up with your finger **(see illustration)**. Note that

the piston should descend slowly and smoothly as the diaphragm draws air into the chamber – it should not drop sharply under spring pressure.

9 Fit the choke plunger assembly into the carburettor body and tighten the nut to secure it **(see illustration)**. Fit the choke linkage bar onto the plungers, making sure the slots locate correctly behind the nipple on the end of each choke plunger **(see illustration)**. Fit the plastic washers and secure the linkage bar in place with the screws **(see illustration)**. Hook up the linkage bar return spring **(see illustration 8.2a)**.

10.7c . . . and secure it with the holder

10.7d Locate the tab around the air passage (arrowed)

10.7e Feed the rim into the groove with the cover located, as described

10.8a Insert the spring and fit the cover

10.8b Check the movement of the piston with your finger

10.9a Fit the choke plunger assembly into the carburettor . . .

4•14 Fuel and exhaust systems

10.9b ... and install the linkage bar, making sure the slots in the arms locate behind the nipples on the plungers (arrowed)

10.9c Fit the screws with their plastic washers and tighten them securely

10 Install the carburettors, but if the fuel level (float height) is to be checked do not yet install the air filter housing (see Section 4) or fuel tank (see Section 2).

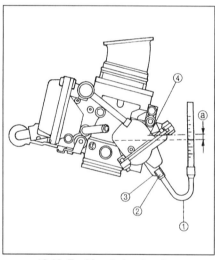

10.13 Fuel level check set-up

1 Gauge assembly
2 Union
3 Drain bolt
4 Float chamber line
a Fuel level measurement

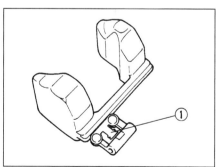

10.15 Adjust fuel level by bending the float tang (1)

Fuel level check

11 To check the fuel level, position the motorcycle on level ground and support it using an auxiliary stand so that it is vertical.
12 Arrange a temporary fuel supply, either by using a small temporary tank or by using an extra long fuel pipe to the now-remote fuel tank. Alternatively, position the tank on a suitable base on the motorcycle, taking care not to scratch any paintwork, and making sure that the tank is safely and securely supported. Connect the fuel line to the carburettors.
13 Yamaha produce a fuel level gauge (Pt. No. 90890-01312 for Europe, or YM-01312-A for USA), or alternatively a suitable length of clear plastic tubing can be used. Attach the gauge or tubing to the drain hose union on the bottom of the float chamber on the first carburettor and position its open end vertically alongside the carburettors (see illustration).
14 Slacken the drain screw and allow the fuel to flow into the tube. The level at which the fuel stabilises in the tubing indicates the level of the fuel in the float chamber. Refer to the Specifications at the beginning of the Chapter and measure the level relative to the line on the float chamber.

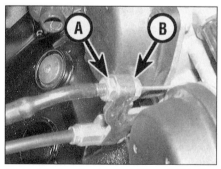

11.4a Slacken the top nut (A) and slide the cable in the bracket until the bottom nut (B) is clear of the lug, then slip it out of the bracket ...

15 If it is incorrect, detach the fuel supply and drain the carburettors, then remove the float from the chamber (see Section 8), and adjust the float height by carefully bending the float tab (see illustration) a little at a time until the correct height is obtained. Repeat the procedure for the other carburettors. Note: *Bending the tab up lowers the fuel level – bending it down raises the fuel level.*

11 Throttle cables – removal and installation

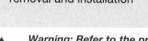 **Warning:** *Refer to the precautions given in Section 1 before proceeding.*

Removal

1 Remove the fuel tank (see Section 2), the air filter housing (see Section 4), and the ignition coil assembly (see Chapter 5).
2 Pull the rubber cover sections up off the carburettor intakes and fold them forward out of the way. If required, release the trim clips securing the rubber cover and remove it, noting how it fits (see illustrations 7.2a and b). To release the clips, push the centre into the body, then draw the clip out (see illustration 7.2c).
3 Mark each cable according to its location at both ends. If new cables are being fitted, match them to the old cables to ensure they are correctly installed.
4 Slacken the accelerator (opening) cable top nut and thread it up the elbow, then slide the cable down in the bracket until the bottom nut is clear of the small lug on the bracket (see illustration). Slip the cable out of the bracket and detach the nipple from the carburettors (see illustration). Slacken the decelerator (closing) cable hex, then slide the cable down in the bracket until the bottom nut is clear of the small lug on the bracket (see illustration). Slip the cable out of the bracket and detach

Fuel and exhaust systems 4•15

11.4b ... and detach the cable end from the carburettor

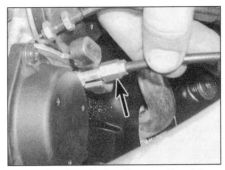

11.4c Slacken the hex (arrowed) and free the cable from the bracket ...

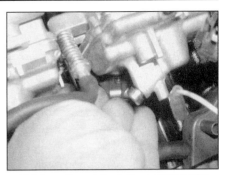

11.4d ... then detach the end from the carburettor

the nipple from the carburettors **(see illustration)**. Withdraw the cables from the machine, noting the correct routing of each cable. If you have difficulty disengaging the cables from the carburettors with them in place, displace them to improve access (see Section 7).

5 Pull the rubber boot back off the throttle housing on the handlebar **(see illustration)**. Remove the throttle housing screws and separate the halves **(see illustrations)**. Displace the cable elbows from the housing, noting how they fit, and detach the cable nipples from the pulley **(see illustration)**. Mark each cable to ensure it is connected correctly on installation.

Installation

6 Lubricate the cable nipples with multi-purpose grease and install them into the throttle pulley at the handlebar **(see illustration 11.5d)**. Fit the cable elbows into the housing, making sure they locate correctly **(see illustration 11.5c)**. Join the housing halves, making sure the pin locates in the hole in the handlebar, and tighten the screws **(see illustration)**. Fit the rubber boot **(see illustration 11.5a)**.

7 Feed the cables through to the carburettors, making sure they are correctly routed. The cables must not interfere with any other component and should not be kinked or bent sharply.

8 Lubricate the decelerator cable nipple with multi-purpose grease and fit it into the lower socket on the carburettor throttle cam **(see illustration 11.4d)**. Fit the decelerator cable into the bracket and draw it up into the bracket so that the bottom nut becomes captive against the small lug **(see illustration 11.4c)**. Tighten the hex down onto the bracket. Lubricate the accelerator cable nipple with multi-purpose grease and fit it into the upper socket on the carburettor throttle cam **(see illustration 11.4b)**. Fit the accelerator cable into the upper bracket and adjust as described in Chapter 1 to obtain the correct amount of freeplay **(see illustration 11.4a)**.

9 Operate the throttle to check that it opens and closes freely. Turn the handlebars back and forth to make sure the cable doesn't cause the steering to bind.

10 Install the rubber cover **(see illustration 7.2b)**, the ignition coil assembly (see Chapter 5), the air filter housing and the fuel tank (see Sections 4 and 2). To install the clips securing the rubber cover, first push the centre back out so that it protrudes from the top of the body. Fit the clip into its socket, then push the centre in so that it is flush with the top of the body **(see illustration 7.12c)**.

11 Start the engine and check that the idle speed does not rise as the handlebars are turned. If it does, the throttle cables are routed incorrectly. Correct the problem before riding the motorcycle.

12 Choke cable – removal and installation

Removal

1 Remove the fuel tank (see Section 2), the air filter housing (see Section 4), and the ignition coil assembly (see Chapter 5).

2 Pull the rubber cover sections up off the

11.5a Pull the boot back off the housing ...

11.5b ... and remove the housing screws (arrowed)

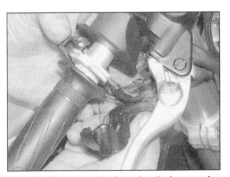

11.5c Separate the housing halves and detach the cable elbows ...

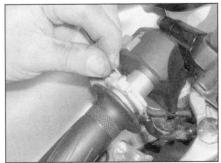

11.5d ... then free the cable ends from the throttle pulley

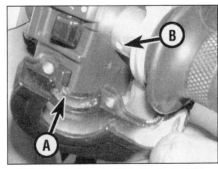

11.6 Locate the pin (A) in the hole (B) in the handlebar

4•16 Fuel and exhaust systems

12.3a Slacken the clamp screw (arrowed) . . .

12.3b . . . then free the outer cable from the clamp . . .

12.3c . . . and the cable nipple from the linkage bar

12.4a Remove the housing screws (arrowed) and separate the halves

12.4b Remove the choke lever and cable elbow from the housing . . .

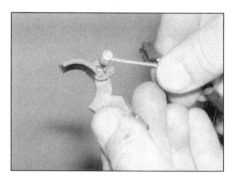

12.4c . . . and free the cable end from the lever

carburettor intakes and fold them forward out of the way. If required, release the trim clips securing the rubber cover and remove it, noting how it fits **(see illustrations 7.2a and b)**. To release the clips, push the centre into the body, then draw the clip out **(see illustration 7.2c)**.
3 Slacken the choke outer cable bracket screw and free the cable from the bracket on the front of the carburettors, then detach the inner cable nipple from the choke linkage bar arm **(see illustrations)**.
4 Unscrew the two handlebar switch/choke lever housing screws and separate the two halves **(see illustration)**. Lift the cable elbow and lever out of the housing, noting how they fit, and detach the cable nipple from the lever **(see illustrations)**.

Installation

5 Install the cable, making sure it is correctly

routed. The cable must not interfere with any other component and should not be kinked or bent sharply.
6 Lubricate the cable nipple with multi-purpose grease and attach it to the choke lever **(see illustrations 12.4c)**. Fit the lever and cable elbow into the housing **(see illustration 12.4b)**, then fit the two halves of the housing onto the handlebar, making sure the lever fits correctly, and the pin in the front half locates in the hole in the front of the handlebar **(see illustration)**. Install the screws and tighten them securely **(see illustration 12.4a)**.
7 Lubricate the cable nipple with multi-purpose grease and attach it to the choke linkage bar on the carburettor **(see illustration 12.3c)**. Fit the outer cable into its bracket, making sure there is a small amount of freeplay in the inner cable, and tighten the screw **(see illustrations 12.3b and a)**.

8 Check the operation of the choke cable (see Chapter 1).
9 Install the rubber cover **(see illustration 7.2b)**, the ignition coil assembly (see Chapter 5), the air filter housing and the fuel tank (see Sections 4 and 2). To install the clips securing the rubber cover, first push the centre back out so that it protrudes from the top of the body. Fit the clip into its socket, then push the centre in so that it is flush with the top of the body **(see illustration 7.12c)**.

13 Exhaust system – removal and installation

⚠ **Warning:** *If the engine has been running the exhaust system will be very hot. Allow the system to cool before carrying out any work.*

Silencer

Removal

1 Remove the lower fairing (see Chapter 8). Slacken the clamp bolt securing the silencer pipe to the downpipe assembly **(see illustration)**. Unscrew and remove the silencer mounting bolt, then release the silencer pipe from the downpipe assembly **(see illustration)**. Remove the sealing ring from the end of the silencer or downpipe assembly and discard it, as a new one should be used.

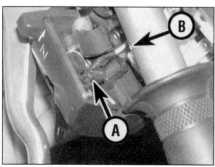

12.6 Make sure the pin (A) in the housing locates in the hole (B) in the handlebar

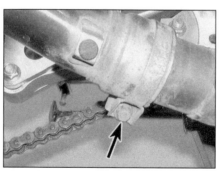

13.1a Slacken the clamp bolt (arrowed) . . .

Fuel and exhaust systems 4•17

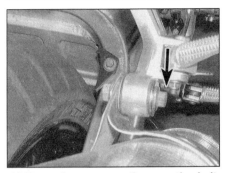

13.1b ... then unscrew the mounting bolt (arrowed) and remove the silencer

13.7 Unscrew the bolt (arrowed) ...

13.8 ... and the nuts (arrowed) ...

13.9 ... and remove the exhaust system

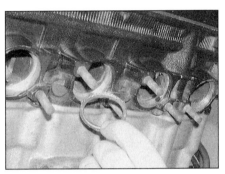

13.11 Fit a new gasket into each port

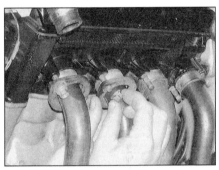

13.13 Fit the flanges onto the studs and secure them with the nuts

Installation

2 Check the condition of the silencer mounting rubber and replace it with a new one if it is damaged, deformed or deteriorated. Check that the collar is fitted into the inside of the rubber.
3 Fit the new sealing ring into the silencer pipe. Fit the silencer onto the downpipe assembly, making sure it is pushed fully home. Align the silencer mounting bracket at the rear and install the bolt with its washer, but do not tighten it yet. Tighten the clamp bolt first, then the silencer mounting bolt, to the torque settings specified at the beginning of the Chapter.
4 Run the engine and check the system for leaks. Install the lower fairing (see Chapter 8).

Complete system

Removal

5 Remove the lower fairing and the fairing side panels (see Chapter 8), and the radiator (see Chapter 3).
6 Detach the cables from the EXUP valve (see Section 14).
7 Slacken the silencer mounting bolt but do not remove it yet (see illustration 13.1b). Unscrew the bolt securing the rear of the downpipe assembly to the frame (see illustration).
8 Unscrew the eight downpipe flange nuts and draw the flanges off the studs (see illustration).
9 Supporting the system, remove the silencer mounting bolt, then detach the downpipes from the cylinder head and remove the system (see illustration).
10 Remove the gasket from each port in the cylinder head and discard it, as a new one must be fitted.

Installation

11 Check the condition of the silencer mounting rubber and replace it with a new one if it is damaged, deformed or deteriorated. Check that the collar is fitted into the inside of the rubber. Fit a new gasket into each of the cylinder head ports (see illustration). If necessary, apply a smear of grease to the gaskets to keep them in place whilst fitting the downpipe.
12 Manoeuvre the assembly into position so that the head of each downpipe is located in its port in the cylinder head (see

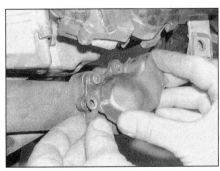

14.2 Unscrew the bolts and remove the cover

illustration 13.9), then install the silencer mounting bolt and the downpipe assembly rear bolt, but do not tighten them yet (see illustrations 13.1b and 13.7).
13 Locate the downpipe flanges onto the studs, then fit the nuts and tighten them to the torque setting specified at the beginning of the Chapter (see illustration). Now tighten the other bolts to the specified torque.
14 Attach the cables to the EXUP valve (see Section 14). Install the radiator (see Chapter 3).
15 Run the engine and check that there are no exhaust gas leaks. Install the fairing panels (see Chapter 8).

14 EXUP system – cable renewal, check, removal and installation

Cable renewal

1 Remove the fuel tank (see Chapter 4) and the lower fairing (see Chapter 8).
2 Unscrew the bolts securing the EXUP pulley cover on the exhaust and remove the cover (see illustration).
3 Unscrew the bolt securing the cable pulley and remove the washer, using a suitable punch or bar located as shown to stop the pulley turning (see illustration). Draw the pulley off the shaft and the cables out of the bracket, noting which fits where (see illustrations). Detach the cable ends from the

4•18 Fuel and exhaust systems

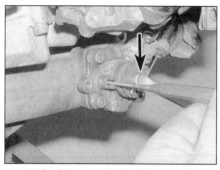

14.3a Unscrew the bolt (arrowed), securing the pulley as shown

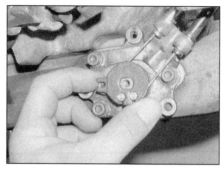

14.3b Draw the pulley off the shaft . . .

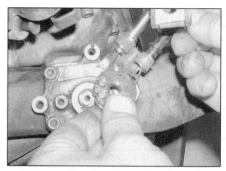

14.3c . . . then free the cables from the bracket . . .

14.3d . . . . and detach the cable nipples from the pulley

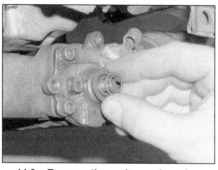

14.3e Remove the spring and washer

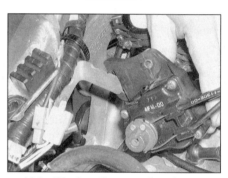

14.4a Displace the servo housing . . .

pulley, noting which fits where **(see illustration)**. Note the spring and washer fitted on the valve shaft and remove them for safekeeping, if required **(see illustration)**.

4 Displace the servo housing from its bracket

14.4b . . . then free the outer cables . . .

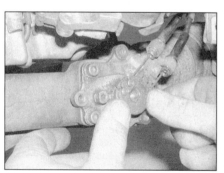

14.7a Fit the pulley onto the shaft . . .

(see illustration). Free the top ends of the outer cables from their holders on the servo housing, noting their location, and detach the cable nipples from the pulley **(see illustrations)**. Withdraw the cables from the

14.4c . . . and detach the cable ends

14.7b . . . and secure it with the bolt

machine, noting the correct routing of each cable.

5 Lubricate the cable nipples with multi-purpose grease and install them into the servo pulley **(see illustration 14.4c)**. Fit the outer cable ends into their holders, making sure they locate correctly **(see illustration 14.4b)**. Fit the servo onto its bracket **(see illustration 14.4a)**.

6 Feed the cables through to the exhaust, making sure they are correctly routed. The cables must not interfere with any other component and should not be kinked or bent sharply.

7 If removed, slide the washer and spring onto the shaft **(see illustration 14.3e)**. Fit the cable ends into the pulley, then fit the outer cables into the bracket **(see illustrations 14.3d and c)**. Fit the pulley onto the shaft, aligning the flats, and secure it with the bolt, not forgetting its washer **(see illustrations)**. Use a punch or rod as on removal to prevent the pulley turning while tightening the bolt **(see illustration 14.3a)**. Check and adjust cable freeplay (see Chapter 1), then install the cover and tighten the bolts to the torque setting specified at the beginning of the Chapter **(see illustration 14.2)**.

Check

8 When the ignition is first switched ON, or while the engine is running, the EXUP system performs its own self-diagnosis. If a fault occurs, the tachometer will be seen to display zero rpm for 3 seconds, then 7000 rpm for 2.5 seconds, then the actual engine speed for 3 seconds, whereupon it will repeat the cycle until the engine is switched off. Note that the

Fuel and exhaust systems 4•19

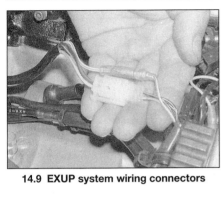

14.9 EXUP system wiring connectors

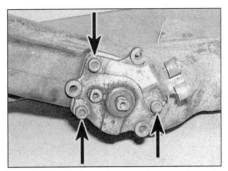

14.15a Unscrew the bolts (arrowed) and remove the bracket and cover

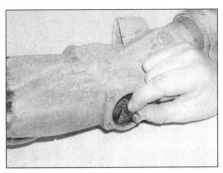

14.15b Draw the valve out of the exhaust (if it didn't come away with the cover)

motorcycle can be ridden even though a fault has been diagnosed, though a difference in performance will be noticed.

9 Trace the wiring from the EXUP system servo motor and disconnect it at the connectors (see illustration). Using a fully charged 12V battery and some jumper leads, connect the positive (+ve) terminal of the battery to the black/yellow wire terminal on the servo side of the connector, and the negative (–ve) terminal to the black/red wire terminal. When the battery is connected, the servo should operate. Disconnect the battery immediately after the test. If the servo does not operate, replace it with a new one.
Caution: *Disconnect the battery immediately after the test to prevent possible damage to the servo motor.*
10 If the servo now operates, yet did not beforehand, check for voltage at the wiring connector using a voltmeter – connect the positive (+ve) probe of the meter to the black/red wire terminal on the loom side of the connector, and the negative (–ve) terminal to the black/yellow wire terminal. Battery voltage should be present. If not, check the connector for loose or corroded terminals, then check the fuse and the battery, and then the wiring between the connector and the ignition control unit, referring to the test procedures and *Wiring Diagrams* in Chapter 9.
11 Check the static resistance between the white/black wire terminal and the yellow/blue wire terminal on the servo side of the connector, using an ohmmeter or multimeter set to the K-ohms scale. Compare the reading to that specified at the beginning of the Chapter. Now disconnect the single white/red wire bullet connector (see illustration 14.9). Check the variable resistance between the white/black wire terminal and the white/red wire terminal on the servo side of the connectors, while turning the servo pulley by hand. Compare the reading range to that specified at the beginning of the Chapter. If the readings from either test are not within the range specified, replace the servo with a new one.
12 If the operation of the valve has been sticky, remove it as described below, then check the valve shaft, face, cover and housing for any corrosion or dirt which may be causing the problem. Also check the shaft bushes in which the valve turns – one in the housing and one in the cover. Clean off all dirt and corrosion and check the operation of the valve. If the shaft ends or bushes are worn, replace them with new ones.

Removal

13 To access the servo, remove the fuel tank (see Chapter 4). To access the valve, remove the lower fairing (see Chapter 8). Before removing either the servo or the valve, detach the cables (see above). If only the servo is being removed, it may be necessary to detach the cables from the valve as well, to create enough slack.
14 To remove the servo, trace the wiring and disconnect it at the connectors (see illustration 14.9). Detach the cables as described above.
15 To remove the valve, detach the cables as described above and remove the spring and washer. Unscrew the bolts securing the valve cover and remove the cable bracket and cover (see illustration). The valve may come away with the cover. If not, draw it out of the exhaust, noting how it fits (see illustration). Note the bushes.

Installation

16 Installation is the reverse of removal. Lubricate the valve shaft ends, bushes and bolts with copper-based grease, as an anti-corrosion measure. Fit the inner bush onto the end of the shaft, then locate the shaft end into its bore in the exhaust (see illustrations). Fit the valve cover and tighten the bolts to the torque setting specified at the beginning of the Chapter (see illustration).
17 Attach the cables (see above), then check the operation of the system and the cable freeplay (see Chapter 1). Install the pulley cover and tighten the bolts to the specified torque (see illustration 14.2).

15 Fuel pump and relay – check, removal and installation

Warning: *Refer to the precautions given in Section 1 before starting work.*

Check

1 The fuel pump is mounted on the frame cross-member below the fuel tank (see

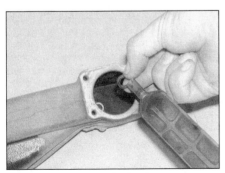

14.16a Fit the bush ...

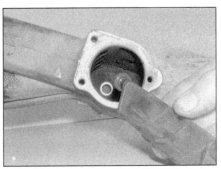

14.16b ... then locate the shaft end in its bore

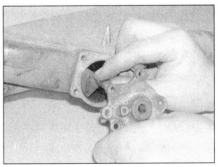

14.16c Fit the cover over the outer end of the shaft

4•20 Fuel and exhaust systems

15.1a Fuel pump assembly (arrowed)

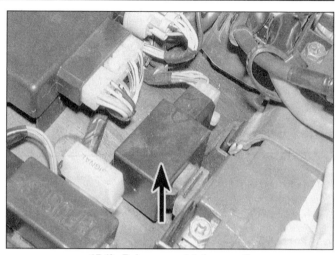

15.1b Relay assembly (arrowed)

illustration). The fuel pump relay is contained within the relay assembly, which is mounted behind the battery under the seat (see illustration). Remove the fuel tank to access all components (see Chapter 4).

2 The fuel pump is controlled through the relay so that it runs whenever the ignition is switched ON and the ignition is operative (i.e. only when the engine is turning over). As soon as the ignition is killed, the relay will cut off the fuel pump's electrical supply (so that there is no risk of fuel being sprayed out under pressure in the event of an accident).

3 It should be possible to hear or feel the fuel pump running whenever the engine is turning over – either place your ear close beside the fuel pump or feel it with your fingertips. If you can't hear or feel anything, check that the battery is fully charged and the ignition circuit fuse is good (see Chapter 9) and that all the wiring is good. Then check the pump and relay for loose or corroded connections or physical damage, and rectify as necessary.

4 If the circuit is fine so far, switch the ignition OFF. Displace the relay assembly from its mount and unplug the wiring connector (see illustration). Using an ohmmeter or continuity tester, connect the positive (+ve) probe to the relay's red/black terminal, and the negative (–ve) probe to the relay's blue/black terminal. There should be no continuity. Leaving the meter connected, now connect a fully charged 12 volt battery and two insulated jumper wires, connect the positive (+ve) terminal of the battery to the relay's red/black terminal, and the negative (–ve) terminal of the battery to the relay's blue/red terminal. There should now be continuity. If the relay assembly does not behave as described, replace it with a new one.

5 If the pump still does not work, trace the wiring from the pump and disconnect it at the connector (see illustration). Using an ohmmeter, connect the positive (+ve) probe to the pump's black/blue terminal, and the negative (–ve) probe to the black terminal, and measure the resistance. If the reading is not as specified at the beginning of the Chapter, replace the pump with a new one. A definitive check of the pump can be made by removing it (see below) and connecting a fully charged 12 volt battery and two insulated jumper wires, connecting the positive (+ve) terminal of the battery to the pump's black/blue terminal, and the negative (–ve) terminal of the battery to the pump's black terminal. The pump should work. If it doesn't, replace it with a new one.

Removal

6 Make sure the ignition is switched OFF. Remove the fuel tank (see Section 2).

7 Trace the wiring from the fuel pump and disconnect it at the connector (see illustration 15.5). Make a note or sketch of which fuel hose fits where, as an aid to installation. Using a rag to mop up any spilled fuel, disconnect the two fuel hoses from the fuel pump (see illustration). Unscrew the bolt securing the pump/filter sleeve to the frame and remove the pump and filter with its rubber mounting sleeve (see illustration 15.1a). Separate the pump from the sleeve.

8 To remove the relay assembly, displace it from its mounting and disconnect the wiring connector (see illustration 15.4).

Installation

9 Installation is a reverse of the removal procedure. Make sure the fuel hoses are correctly and securely fitted to the pump – the hose from the in-line filter attaches to the union marked 'INLET'; the hose to the carburettors attaches to the other union (see illustration 15.7). Start the engine and check carefully that there are no leaks at the pipe connections.

15.4 Displace the relay and disconnect the wiring connector

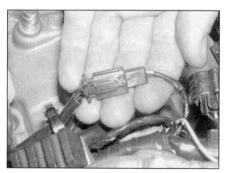

15.5 Fuel pump wiring connector

15.7 Detach the hoses (arrowed), noting which fits where

Fuel and exhaust systems 4•21

16 Fuel warning light and sensor – check and renewal

Check

1 When the ignition is first switched ON, or while the engine is running, the fuel warning circuit performs its own self-diagnosis. If a fault occurs, on 1998 and 1999 models, the tachometer will be seen to display zero rpm for 3 seconds, then 8000 rpm for 2.5 seconds, then the actual engine speed for 3 seconds, whereupon it will repeat the cycle until the engine is switched OFF. On 2000-on models, the warning light will be seen to flash eight times and then go out for three seconds, and this will repeat until the engine is switched OFF.

2 The circuit consists of the sensor mounted in the fuel tank and the warning light mounted in the instrument panel. If the system malfunctions check first that the battery is fully charged and that the fuses are good (see Chapter 9). If they are, remove the fuel tank and drain it (see Section 2).

3 Using an ohmmeter or continuity tester, check for continuity between the green and black wire terminals on the sensor side of the wiring connector coming from the fuel tank, with the tank the right way up. There should be continuity. If not, replace the sensor with a new one (see below).

4 If the sensor is good, install the fuel tank, but do not fill it with the fuel. Disconnect the instrument cluster wiring connector (see Chapter 9). With the ignition ON, check for voltage at the instrument cluster wiring connector by connecting the positive (+ve) probe of a voltmeter to the brown terminal on the loom side of the connector, and the negative (–ve) probe to the green/white terminal. If no voltage is present, the fault lies in the wiring. Check all the relevant wiring and wiring connectors (see Chapter 9), referring to the Wiring Diagrams at the end of Chapter 9.

5 If voltage is present, take the instrument cluster to a Yamaha dealer for further assessment – they provide no specific test data for the instruments themselves. If there are any faults, even if just the warning light LED is faulty, a new cluster will have to be fitted, as no individual components are available.

Renewal

6 Refer to Chapter 9 for removal of the instrument cluster.

7 To renew the sensor, remove the fuel tank (see Section 2) and drain it. Remove the screws securing the sensor and draw it out of the tank **(see illustration)**. Discard the O-ring. Fit a new O-ring onto the sensor and install it in the tank. Tighten the screws to the torque setting specified at the beginning of the Chapter.

8 Install the tank (see Section 2), and check carefully that there are no leaks, before using the bike.

17 Air induction system (AIS) – function, disassembly and reassembly

Function

1 The air induction system is fitted to 2000-on models. It uses exhaust gas pulses to suck fresh air into the exhaust ports, where it mixes with hot combustion gases **(see illustration overleaf)**. The extra oxygen causes continued combustion, allowing unburnt hydrocarbons to burn off, thereby reducing emissions. Reed valves control the flow of air into the ports, opening when there is negative pressure, and prevent exhaust gases flowing back into the AIS. An air cut-off valve shuts off the flow of air during deceleration, preventing backfiring.

2 Refer to Chapter 1, Section 6, for a check of the system.

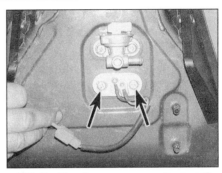

16.7 Fuel level sensor screws (arrowed)

Disassembly

3 Remove the fairing side panels (see Chapter 8).

4 To remove the AIS valve assembly, disconnect the hoses from it, then unscrew the bolts securing it to the bracket and remove it, noting the collars and grommets **(see illustration 17.1)**. No individual internal components are available for the valve assembly, so if there are any faults in its function the assembly must be replaced with a new one.

5 To access a reed valve, remove the valve assembly (see Step 4), then unscrew the reed valve cover bolts. Unscrew the bolts securing the reed itself and remove the reed, noting how it fits.

6 To renew any hoses, release the clamps securing them and disconnect them at each end. To renew the pipes, slacken the clamps and detach them from the unions on the cylinder head. Remove the sealing collars and replace them with new ones if they are deformed or damaged.

Reassembly

7 Reassemble the system, making sure all the hoses and pipes are securely connected at each end and held by their clamps, and are correctly routed **(see illustration 17.1)**.

4•22 Fuel and exhaust systems

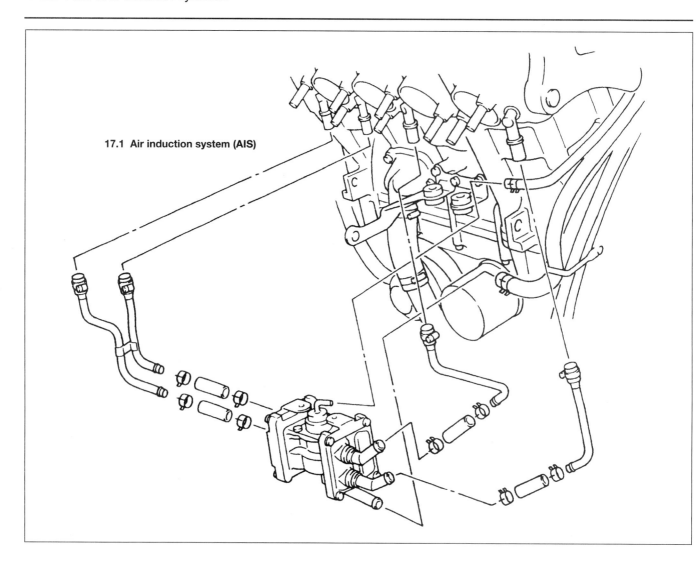

17.1 Air induction system (AIS)

Chapter 5
Ignition system

Contents

General information 1	Neutral switch – check and renewal .... see Chapter 9
Ignition (main) switch – check, removal and installation ..see Chapter 9	Pick-up coil – check and renewal 4
Ignition control unit – check, removal and installation 5	Relay assembly – check and renewal .... see Chapter 9
Ignition HT coil assembly – check, removal and installation 3	Sidestand switch – check and renewal .... see Chapter 9
Ignition system – check 2	Spark plugs – gap check/adjustment, and renewal ..... see Chapter 1
Ignition timing – general information and check 6	Throttle position sensor – check, adjustment and renewal 7

Degrees of difficulty

| Easy, suitable for novice with little experience | | Fairly easy, suitable for beginner with some experience | | Fairly difficult, suitable for competent DIY mechanic | | Difficult, suitable for experienced DIY mechanic | | Very difficult, suitable for expert DIY or professional | |

Specifications

General information
Cylinder numbering	1 to 4 from left to right
Spark plugs	see Chapter 1

Ignition timing
At idle
1998 and 1999 models	5° BTDC @ 1100 rpm
2000-on models	5° BTDC @ 1050 rpm
Full advance	55° BTDC @ 5000 rpm

Pick-up coil
Resistance	248 to 372 ohms @ 20°C

Ignition HT coils
Primary winding resistance	1.87 to 2.53 ohms @ 20°C
Secondary winding resistance (without plug caps)	12 to 18 K-ohms @ 20°C
Spark plug cap resistance	10 K-ohms @ 20°C
Minimum spark gap (see Section 2)	6 mm

Throttle position sensor
Maximum resistance	5.0 ± 1.0 K-ohms
Resistance range	Zero to 5.0 ± 1.0 K-ohms

Torque wrench settings
Pick-up coil bolts	10 Nm
Timing rotor/pick-up coil cover bolts	12 Nm

1 General information

All models are fitted with a fully transistorised electronic ignition system which, due to its lack of mechanical parts, is totally maintenance-free. The system comprises a trigger, pick-up coil, ignition control unit and ignition HT coils (refer to *Wiring Diagrams* at the end of Chapter 9 for details). All models are fitted with two HT coils. A throttle position sensor provides information for the ignition control unit.

The ignition trigger, which is on the rotor on the right-hand end of the crankshaft, magnetically operates the pick-up coil as the crankshaft rotates. The pick-up coil sends a signal to the ignition control unit, which then supplies the ignition HT coils with the power necessary to produce a spark at the plugs.

The ignition control unit incorporates an electronic advance system controlled by signals from the ignition triggers and pick-up coil and from the throttle position sensor.

The system also incorporates a safety interlock circuit which will cut the ignition if the sidestand is extended whilst the engine is running and in gear, or if a gear is selected whilst the engine is running and the sidestand is extended. It also prevents the engine from being started if the engine is in gear unless the clutch lever is pulled in.

Because of their nature, the individual ignition system components can be checked but not repaired. If ignition system troubles occur, and the faulty component can be isolated, the only cure for the problem is to replace the part with a new one. Keep in mind that most electrical parts, once purchased, cannot be returned. To avoid unnecessary expense, make very sure the faulty component has been positively identified before buying a replacement part.

Note that there is no provision for adjusting the ignition timing on these models.

5•2 Ignition system

2.2 Ground (earth) the spark plug and operate the starter – bright blue sparks should be visible

A simple spark gap testing tool can be made from a block of wood, a large alligator clip and two nails, one of which is fashioned so that a spark plug cap or bare HT lead end can be connected to its end. Make sure the gap between the two nail ends is the same as specified

2.5 Connect the tester as shown – when the starter is operated sparks should jump between the nails

2 Ignition system – check

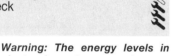

Warning: The energy levels in electronic systems can be very high. On no account should the ignition be switched on whilst the plugs or plug caps are being held. Shocks from the HT circuit can be most unpleasant. Secondly, it is vital that the engine is not turned over or run with any of the plug caps removed, and that the plugs are soundly earthed (grounded) when the system is checked for sparking. The ignition system components can be seriously damaged if the HT circuit becomes isolated.

1 As no means of adjustment is available, any failure of the system can be traced to failure of a system component or a simple wiring fault. Of the two possibilities, the latter is by far the most likely. In the event of failure, check the system in a logical fashion, as described below.

2 Displace the coil assembly (see Section 3), but disconnect the HT lead from one spark plug only, and leave the main wiring connector connected. Connect the HT lead to a spare spark plug and lay the plug on the engine with the threads contacting the engine **(see illustration)**. If necessary, hold the spark plug with an insulated tool.

Warning: Do not remove any of the spark plugs from the engine to perform this check – atomised fuel being pumped out of the open spark plug hole could ignite, causing severe injury!

3 Check that the kill switch is in the 'RUN' position and the transmission is in neutral, then turn the ignition switch ON and turn the engine over on the starter motor. If the system is in good condition a regular, fat blue spark should be evident at the plug electrodes. If the spark appears thin or yellowish, or is non-existent, further investigation will be necessary. Turn the ignition off and repeat the test for each spark plug in turn.

4 The ignition system must be able to produce a spark which is capable of jumping a particular size gap. Yamaha specify that a healthy system should produce a spark capable of jumping at least 6 mm. A simple testing tool can be made to test the minimum gap across which the spark will jump **(see Tool Tip)** or, alternatively, it is possible to buy an ignition spark gap tester tool and some of these are adjustable to alter the spark gap.

5 Connect one of the spark plug HT leads from one coil to the protruding electrode on the test tool, and clip the tool to a good earth (ground) on the engine or frame **(see illustration)**. Check that the kill switch is in the 'RUN' position, turn the ignition switch ON and turn the engine over on the starter motor. If the system is in good condition a regular, fat blue spark should be seen to jump the gap between the nail ends. Repeat the test for the other coil. If the test results are good the entire ignition system can be considered good. If the spark appears thin or yellowish, or is non-existent, further investigation will be necessary.

6 Ignition faults can be divided into two categories, namely those where the ignition system has failed completely, and those which are due to a partial failure. The likely faults are listed below, starting with the most probable source of failure. Work through the list systematically, referring to the appropriate Sections of this Chapter (or other Chapters, as indicated) for full details of the necessary checks and tests. **Note:** *Before checking the following items ensure that the battery is fully charged and that all fuses are in good condition.*

 a) *Loose, corroded or damaged wiring connections, broken or shorted wiring between any of the component parts of the ignition system (see Chapter 9).*
 b) *Faulty HT lead or spark plug cap, faulty spark plug, dirty, worn or corroded plug electrodes, or incorrect gap between electrodes.*
 c) *Faulty ignition (main) switch or engine kill switch (see Chapter 9).*
 d) *Faulty neutral, clutch or sidestand switch, or starter circuit cut-off relay (see Chapter 9).*
 e) *Faulty pick-up coil or damaged trigger.*
 f) *Faulty ignition HT coil(s).*
 g) *Faulty ignition control unit.*

7 If the above checks don't reveal the cause of the problem, have the ignition system tested by a Yamaha dealer.

3 Ignition HT coil assembly – check, removal and installation

Check

1 Check the coil visually for cracks and other damage.

2 Remove the rider's seat (see Chapter 8). Disconnect the battery negative (–ve) lead.

3 The coils are mounted under the air filter housing – remove the coil assembly as described below, as access is too restricted to test them in situ – there is no need to separate the coils themselves from the mounting plate to test them.

4 Measure the primary circuit resistance with a multimeter as follows. Disconnect the primary circuit electrical connectors from the coil **(see illustration)**. Set the meter to the ohms x 1 scale and measure the resistance between the terminals on the coil **(see**

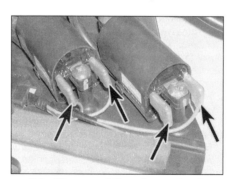

3.4a Disconnect the primary circuit connectors (arrowed) from the coil

Ignition system 5•3

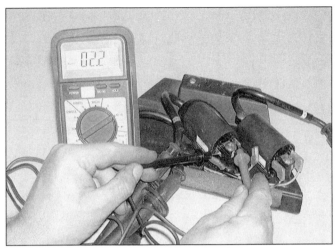

3.4b To test the coil primary resistance, connect the multimeter leads between the primary circuit terminals

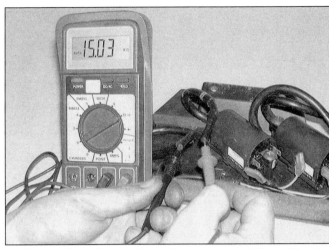

3.5 To test the coil secondary resistance, connect the multimeter leads between the spark plug lead ends

illustration). If the reading obtained is not within the range shown in the Specifications, it is likely that the coil is defective.

5 Measure the secondary circuit resistance with a multimeter as follows. Unscrew the spark plug caps from the HT leads and set the meter to the K-ohm scale. Connect one meter probe to one HT lead end and the other probe to the other lead end (see illustration). If the reading obtained is not within the range shown in the Specifications, it is likely that the coil is defective.

6 If the reading is as specified, measure the resistance of the spark plug cap by connecting the meter probes between the HT lead socket in the cap and the spark plug contact in the cap (see illustration). If the reading obtained is not as specified, replace the spark plug cap with a new ones.

7 If either coil is confirmed to be faulty, it must be replaced with a new one: the coils are sealed units and cannot therefore be repaired.

Removal

8 Remove the seat (see Chapter 8). Disconnect the battery negative (–ve) lead.

9 The coils are mounted under the air filter housing – remove the fuel tank and the air filter housing for access (see Chapter 4).

10 Trace the wiring from the coil mounting plate and disconnect it at the connector (see illustration).

11 Release the trim clips securing the coil mounting plate to the frame cross-member and displace the coil assembly, noting how it fits (see illustrations). To release the clips, pull the knob up, then draw the body out (see illustration). Disconnect the HT leads from the spark plugs (see illustration). Mark the locations of all wires and leads before disconnecting them. Note the routing of the HT leads.

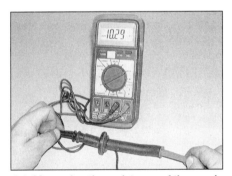

3.6 Measuring the resistance of the spark plug cap

3.10 Disconnect the coil assembly wiring connector

3.11a Release the trim clips (arrowed) . . .

3.11b . . . and displace the coil assembly

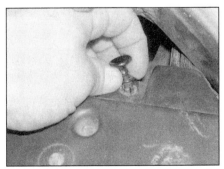

3.11c Pull the knob out then draw the body out of the socket

3.11d Disconnect the spark plug caps

5•4 Ignition system

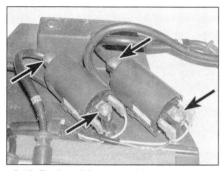

3.12 Each coil is secured by two screws (arrowed)

3.13 Fit the body into its socket then push the knob in to secure it

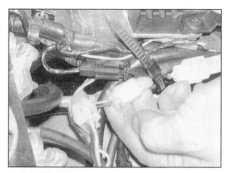

4.3 Disconnect the pick-up coil wiring connector

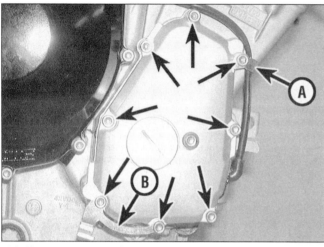

4.9 Unscrew the bolts (arrowed), noting the wiring clamp (A) and the clutch cable guide (B)

4.10a Unscrew the bolts and remove the pick-up coil . . .

12 To separate a coil from the mounting plate, disconnect the primary circuit electrical connectors from the coil **(see illustration 3.4a)**. Undo the screws securing the coil and remove the coil **(see illustration)**.

Installation

13 Installation is the reverse of removal. Make sure the wiring connectors and HT leads are securely connected. To install the coil mounting plate trim clips, fit the body into its panel, then push the centre in **(see illustration)**.

4 Pick-up coil – check and renewal

Check

1 Remove the rider's seat (see Chapter 8). Disconnect the battery negative (–ve) lead.
2 Remove the fuel tank (see Chapter 4).
3 Trace the wiring from the pick-up coil on the right-hand side of the engine and disconnect it at the 2-pin connector inside the frame on the right-hand side **(see illustration)**. Using a multimeter set to the ohms x 100 scale, measure the resistance between the terminals on the pick-up coil side of the connector.
4 Compare the reading obtained with that given in the Specifications at the beginning of this Chapter. The pick-up coil must be replaced with a new one if the reading obtained differs greatly from that given, particularly if the meter indicates a short circuit (no measurable resistance) or an open circuit (infinite, or very high resistance).
5 If the pick-up coil is thought to be faulty, first remove the lower fairing and the right-hand fairing side panel (see Chapter 8), and check that it is not due to a damaged or

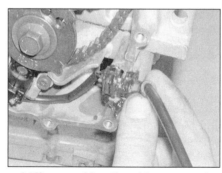

4.10b . . . and free the wiring grommet

broken wire from the coil to the connector: pinched or broken wires can usually be repaired.

Renewal

6 Remove the rider's seat (see Chapter 8). Disconnect the battery negative (–ve) lead.
7 Remove the fuel tank (see Chapter 4), the lower fairing and the right-hand fairing side panel (see Chapter 8).
8 Trace the wiring from the pick-up coil on the right-hand side of the engine and disconnect it at the 2-pin connector inside the frame on the right-hand side **(see illustration 4.3)**. Feed the wiring back to the coil, noting its routing and releasing it from any clips or ties.
9 Unscrew the bolts securing the timing rotor/pick-up coil cover on the right-hand side of the engine and remove the cover, noting the wiring clamp and clutch cable guide **(see illustration)**. Discard the gasket, as a new one must be used. Remove the dowels from either the crankcase or the cover if they are loose.
10 Unscrew the bolts securing the pick-up coil, then free its wiring grommet from the cutout and remove the coil from the engine **(see illustrations)**.
11 Fit the coil onto the engine and the wiring grommet into its cutout. Apply a suitable non-

Ignition system 5•5

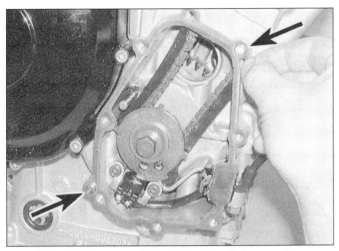

4.12a Fit the new gasket onto the dowels (arrowed) . . .

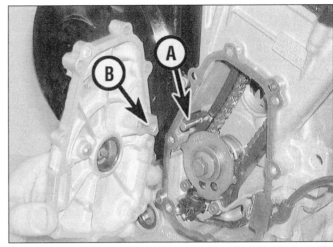

4.12b . . . then install the cover, locating the pin (A) in the hole (B)

permanent thread locking compound to the bolts and tighten them to the torque setting specified at the beginning of the chapter. Feed the wiring through to the connector, securing it with any clips or ties and making sure it is correctly routed, and reconnect it **(see illustration 4.3)**.

12 If removed, fit the timing rotor/pick-up coil cover dowels into the crankcase. Install the cover using a new gasket, making sure they locate correctly onto the dowels **(see illustration)**. Locate the cam chain tensioner blade pivot pin in the hole in the cover **(see illustration)**. Tighten the cover bolts to the torque setting specified at the beginning of the Chapter, not forgetting the wiring clamp and the cable guide **(see illustration 4.9)**.

13 Install the fuel tank (see Chapter 4) and the fairing panels (see Chapter 8). Reconnect the battery lead and install the seat.

5 Ignition control unit – check, removal and installation

Check

1 If the tests shown in the preceding or following Sections have failed to isolate the cause of an ignition fault, it is possible that the ignition control unit itself is faulty. No test details are available with which the unit can be tested. The best way to determine whether it is faulty is to substitute it with a known good one, if available. Otherwise, take the unit to a Yamaha dealer for assessment.

Removal

2 Remove the rider's seat (see Chapter 8). Disconnect the battery negative (–ve) lead.
3 Disconnect the wiring connectors from the ignition control unit **(see illustration)**.
4 Remove the screws securing the ignition control unit and remove the unit **(see illustration 5.3)**.

Installation

5 Installation is the reverse of removal. Make sure the wiring connectors are correctly and securely connected.

6 Ignition timing – general information and check

General information

1 Since no provision exists for adjusting the ignition timing and since no component is subject to mechanical wear, there is no need for regular checks: only if investigating a fault such as a loss of power or a misfire, should the ignition timing be checked.
2 The ignition timing is checked dynamically (engine running) using a stroboscopic lamp. The inexpensive neon lamps should be adequate in theory, but in practice may produce a pulse of such low intensity that the timing mark remains indistinct. If possible, one of the more precise xenon tube lamps should be used, powered by an external source of the appropriate voltage. **Note:** *Do not use the machine's own battery, as an incorrect reading may result from stray impulses within the machine's electrical system.*

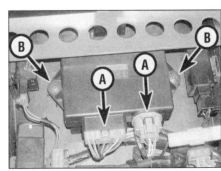

5.3 Ignition control unit wiring connectors (A) and mounting screws (B)

Check

3 Warm the engine up to normal operating temperature, then stop it. Remove the lower fairing and the left-hand fairing side panel (see Chapter 8).
4 Unscrew the timing inspection bolt (the small bolt, not the large slotted plug) from the centre of the timing rotor/pick-up coil cover on the right-hand side of the engine **(see illustration 4.9)**.
5 The mark on the timing rotor which indicates the firing point at idle speed for the No. 1 cylinder is an 'H' mark on its side **(see illustration)**. The static timing marks with which this should align are the cutouts in the inspection hole.

> **HAYNES HiNT** *The timing marks can be highlighted with white paint to make them more visible under the stroboscope light.*

6 Connect the timing light to the No. 1 cylinder HT lead as described in the manufacturer's instructions.
7 Start the engine and aim the light at the inspection hole.
8 With the machine idling at the specified speed, the static timing marks should point between the uprights of the 'H'.

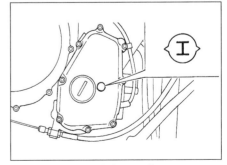

6.5 Timing mark

5•6 Ignition system

7.4 Disconnect the throttle position sensor wiring connector

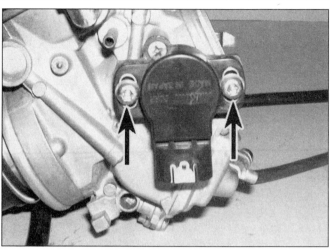

7.9 Slacken the throttle position sensor bolts (arrowed) and adjust as described

9 Slowly increase the engine speed whilst observing the 'H' mark. The mark should move anti-clockwise, increasing in relation to the engine speed until it reaches full advance (no identification mark).

10 As already stated, there is no means of adjustment of the ignition timing on these machines. If the ignition timing is incorrect, or suspected of being incorrect, one of the ignition system components is at fault, and the system must be tested as described in the preceding Sections of this Chapter.

11 When the check is complete, install the timing inspection bolt, using a new sealing washer if the old one is damaged or deformed, and tighten it securely. Install the fairing panels (see Chapter 8).

7 Throttle position sensor – check, adjustment and renewal

1 The throttle position sensor (TPS) is mounted on the outside of the right-hand (no. 4) carburettor and is keyed to the throttle shaft. The sensor provides the ignition control unit with information on throttle position and rate of opening or closing.

2 When the ignition is first switched ON, or while the engine is running, the throttle position sensor performs its own self-diagnosis. If a fault occurs, the tachometer will be seen to display zero rpm for 3 seconds, then 3000 rpm for 2.5 seconds, then the actual engine speed for 3 seconds, whereupon it will repeat the cycle until the engine is switched off. Note that the motorcycle can be ridden even though a fault has been diagnosed, though a difference in performance will be noticed.

Check

3 Remove the fuel tank and the air filter housing (see Chapter 4). The throttle sensor is mounted on the outside of the right-hand (no. 4) carburettor.

4 Make sure the ignition is switched OFF, then disconnect the sensor's wiring connector (see illustration). Using an ohmmeter or multimeter set to the K-ohms range, measure the sensor maximum resistance by connecting the meter probes between the blue and black/blue wire terminals on the sensor. Now measure the resistance range by connecting the meter probes between the yellow and black/blue wire terminals on the sensor, and slowly opening the throttle from fully closed to fully open. If the readings obtained differ greatly from those specified at the beginning of the Chapter, replace the sensor with a new one. When checking the resistance range, it is more important that there is a smooth and constant change in the resistance as the throttle is opened, than that the figures themselves are exactly as specified.

5 If the readings were as specified, using a multimeter set to resistance or a continuity tester, check for continuity between the terminals on the wiring loom side of the sensor wiring connector and the corresponding terminals on the ignition control unit connector, referring to Wiring Diagrams at the end of Chapter 9 (first disconnect it). There should be continuity between each terminal. If not, this is probably due to a damaged or broken wire between the connectors: pinched or broken wires can usually be repaired. Also check the connectors for loose or corroded terminals, and check the sensor itself for cracks and other damage. If the wiring and connectors are good, check the adjustment of the sensor as described below.

6 If the sensor is suspected of being faulty, take it to a Yamaha dealer for further testing. If it is confirmed to be faulty, it must be replaced with a new one – the sensor is a sealed unit and cannot therefore be repaired.

Adjustment

7 Before adjusting the sensor, check the idle speed and carburettor synchronisation (see Chapter 1).

8 Remove the fuel tank and the air filter housing (see Chapter 4). Turn the ignition switch ON, then disconnect and reconnect the sensor wiring connector (see illustration 7.4). This sets the tachometer to 'sensor adjustment mode'.

9 Slacken the sensor mounting bolts and rotate the sensor until the tachometer needle reads 5000 rpm (see Chapter 4) (see illustration). If the tachometer reads either 0 rpm or 10,000 rpm, the angle of the sensor is either too narrow or too wide. Adjust it as required until the reading is 5000 rpm, then tighten the bolts evenly and a little at a time. If it cannot be adjusted to within the range, or if no reading is obtained, check it as described above. Start the engine or turn the ignition switch OFF to reset the tachometer mode.

Renewal

10 Remove the fuel tank and the air filter housing (see Chapter 4). The throttle sensor is mounted on the outside of the right-hand (no. 4) carburettor.

11 Disconnect the wiring connector, then unscrew the sensor mounting bolts and remove the sensor, noting how it fits (see illustrations 7.4 and 7.9).

12 Install the sensor and lightly tighten the bolts, then connect the wiring connector and adjust the sensor as described above until the correct reading is obtained. On completion, tighten the bolts evenly and a little at a time.

13 Install the air filter housing and fuel tank (see Chapter 4).

Chapter 6
Frame, suspension and final drive

Contents

Drive chain – removal, inspection, cleaning and installation 15	Sidestand – check . . . . .see Chapter 1
Drive chain and sprockets – check, adjustment and lubrication . . . . .see Chapter 1	Sidestand – lubrication . . . . .see Chapter 1
Footrests, brake pedal and gearchange lever – removal and installation 3	Sidestand – removal and installation 4
Forks – disassembly, inspection and reassembly 7	Sidestand switch – check and renewal . . . . .see Chapter 9
Forks – oil change . . . . .see Chapter 1	Sprockets – check and renewal 16
Forks – removal and installation 6	Steering head bearings – freeplay check and adjustment . . . . .see Chapter 1
Frame – inspection and repair 2	Steering head bearings – inspection and renewal 9
General information 1	Steering head bearings – re-greasing . . . . .see Chapter 1
Handlebar switches – check . . . . .see Chapter 9	Steering stem – removal and installation 8
Handlebar switches – removal and installation . . . . .see Chapter 9	Suspension – adjustments 12
Handlebars and levers – removal and installation 5	Suspension – check . . . . .see Chapter 1
Rear shock absorber – removal, inspection and installation 10	Swingarm – inspection and bearing renewal 14
Rear sprocket coupling/rubber dampers – check and renewal 17	Swingarm – removal and installation 13
Rear suspension linkage – removal, inspection and installation 11	Swingarm and suspension linkage bearings – re-greasing . . . . .see Chapter 1

Degrees of difficulty

Easy, suitable for novice with little experience		Fairly easy, suitable for beginner with some experience		Fairly difficult, suitable for competent DIY mechanic		Difficult, suitable for experienced DIY mechanic		Very difficult, suitable for expert DIY or professional

Specifications

Front forks
Fork oil type	Yamaha suspension oil '01' or equivalent
Fork oil capacity (per leg)	
1998 and 1999 models	477 cc
2000-on models	482 cc
Fork oil level*	
1998 and 1999 models	78 mm
2000-on models	74 mm
Fork spring free length	255 mm

*Oil level is measured from the top of the tube with the fork spring removed and the leg fully compressed.

Rear suspension
Shock absorber spring free length	176 mm

Final drive
Chain type	DAIDO 50ZVM (114 links)
Chain freeplay and stretch limit	see Chapter 1

Torque wrench settings
Footrest bracket bolts	28 Nm
Steering stem nut	115 Nm
Fork clamp bolts (top yoke)	23 Nm
Handlebar positioning bolts	13 Nm
Handlebar clamp bolts	17 Nm
Handlebar end-weight bolts	4 Nm
Fork clamp bolts (bottom yoke)	23 Nm
Damper cartridge Allen bolt	40 Nm
Damper rod to top bolt locknut	15 Nm
Fork top bolt	23 Nm
Rear shock absorber nuts	40 Nm
Rear suspension linkage plate and linkage arm nuts	40 Nm
Swingarm pivot bolt nut	125 Nm
Front sprocket nut	85 Nm
Front sprocket cover bolts	10 Nm
Rear sprocket nuts	69 Nm

6•2 Frame, suspension and final drive

3.1a Remove the split pin (A) and withdraw the clevis pin (B)

3.1b Unhook the springs (arrowed) from the lug

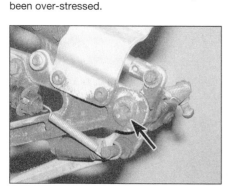

3.1c Unscrew the bolts (arrowed) and displace the bracket . . .

1 General information

All models use a twin spar box-section aluminium frame, which uses the engine as a stressed member.

Front suspension is by a pair of oil-damped upside-down telescopic forks, which use a cartridge-type damper. The forks are adjustable for spring pre-load and both rebound and compression damping.

At the rear, an aluminium alloy swingarm acts on a single shock absorber via a three-way linkage. The shock absorber is adjustable for spring pre-load and for both rebound and compression damping.

The drive to the rear wheel is by chain and sprockets.

2 Frame – inspection and repair

1 The frame should not require attention unless accident damage has occurred. In most cases, frame renewal is the only satisfactory remedy for such damage. A few frame specialists have the jigs and other equipment necessary for straightening the frame to the required standard of accuracy, but even then there is no simple way of assessing to what extent the frame may have been over-stressed.

2 After the machine has accumulated a lot of miles, the frame should be examined closely for signs of cracking or splitting at the welded joints. Loose engine mount bolts can cause ovaling or fracturing of the mounts themselves. Minor damage can often be repaired by welding, depending on the extent and nature of the damage, but this is a task for an expert.

3 Remember that a frame which is out of alignment will cause handling problems. If misalignment is suspected as the result of an accident, it will be necessary to strip the machine completely so that the frame can be thoroughly checked.

3 Footrests, brake pedal and gearchange lever – removal and installation

Brake pedal and rider's right-hand footrest

Removal

1 Remove the split pin and washer from the clevis pin securing the brake pedal to the master cylinder pushrod **(see illustration)**. Remove the clevis pin and separate the pushrod from the pedal. Unhook the pedal return spring and brake light switch spring from the lug on the pedal **(see illustration)**. Unscrew the bolts securing the footrest bracket and displace it enough to provide access to the footrest bolt on the back **(see illustration)**. Make sure no strain is placed on the brake switch wiring or brake hose. Unscrew the footrest bolt, then withdraw the footrest from the bracket and remove the pedal **(see illustration)**. Note the washer fitted between the pedal and the footrest bracket.

Installation

2 Installation is the reverse of removal. Apply grease to the brake pedal pivot. Use a new split pin on the clevis pin securing the brake pedal to the master cylinder pushrod. Tighten the footrest bracket bolts to the torque setting specified at the beginning of the Chapter. Check the operation of the rear brake light switch (see Chapter 1, Section 9).

Gearchange lever and rider's left-hand footrest

Removal

3 Slacken the gearchange lever linkage rod locknuts, then unscrew the rod and separate it from the lever and the arm (the rod is reverse-threaded on one end, so will unscrew from both lever and arm simultaneously when turned in the one direction) **(see illustration)**. Note how far the rod is threaded into the lever and arm, as this determines the height of the lever relative to the footrest. Withdraw the rod from the frame **(see illustration)**. Unscrew the bolts securing the footrest bracket and displace it **(see illustration 3.1c)**. Unscrew the footrest bolt, then withdraw the footrest from the bracket and remove the lever **(see illustration)**. Note the washer fitted between the lever and the footrest bracket.

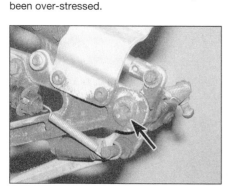

3.1d . . . then unscrew the footrest bolt (arrowed)

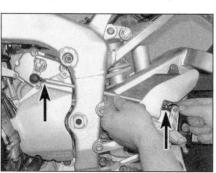

3.3a Slacken the locknuts (arrowed), then thread the rod off the lever and arm . . .

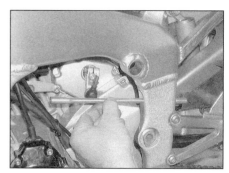

3.3b . . . and withdraw it from the frame

Frame, suspension and final drive

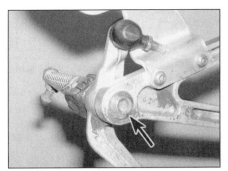

3.3c Unscrew the bolt (arrowed) and remove the footrest and lever

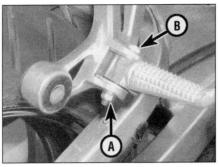

3.5 Unscrew the nut (A) and withdraw the bolt (B)

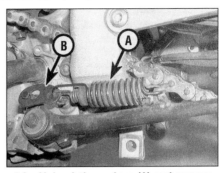

4.2a Unhook the springs (A) and remove the link plate (B)

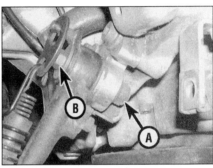

4.2b Unscrew the nut (A) and remove the pivot bolt (B)

Installation

4 Installation is the reverse of removal. Apply grease to the lever pivot. Tighten the footrest bracket bolts to the torque setting specified at the beginning of the Chapter. Slide the linkage rod into the frame and thread it onto the lever and arm (the rod is reverse-threaded on one end, so will screw onto both lever and arm simultaneously when turned in the one direction) **(see illustration 3.3b)**. Set it as noted on removal to give the same gearchange lever height, then tighten the locknuts securely **(see illustration 3.3a)**. To adjust the lever height, slacken the locknuts and thread the rod in or out as required, then retighten the nuts.

Passenger footrests

Removal

5 Unscrew the nut from the bottom of the footrest pivot bolt, then withdraw the bolt and remove the footrest, noting the fitting of the detent plates, ball and spring – take care that they do not spring out when removing the footrest **(see illustration)**. Also note the collar for the pivot bolt.

Installation

6 Installation is the reverse of removal.

4 Sidestand – removal and installation

1 Support the motorcycle securely in an upright position using an auxiliary stand.
2 Unhook the stand springs and remove the link plate, noting how it fits **(see illustration)**. Counter-hold the pivot bolt and unscrew the nut on the inside of the bracket **(see illustration)**. Remove the pivot bolt and the stand, noting how the contact plate on the back locates against the switch plunger.
3 On installation, apply grease to the pivot and to the contact surfaces of the stand and bracket. Apply a suitable non-permanent thread locking compound to the bolt threads. Fit the washer and tighten the nut securely. Fit the link plate, then install the bolt with its

washer and tighten it securely. Reconnect the sidestand spring. Check that the spring holds the stand securely up when not in use – an accident is almost certain to occur if the stand extends while the machine is in motion.
4 Check the operation of the sidestand switch (see Chapter 1).

5 Handlebars and levers – removal and installation

Handlebars

Removal

Note: *The handlebars can be displaced from the forks without having to remove any of the lever or switch assemblies.*

1 Remove the fairing (see Chapter 8). **Note:** *The fairing can remain on the bike if required, though it is advisable to remove it to prevent the possibility of damaging it.*
2 If removing the right handlebar, displace the front brake master cylinder and reservoir (see Chapter 7). There is no need to disconnect the hydraulic hose. Keep the reservoir upright to prevent possible fluid leakage and make sure no strain is placed on the hydraulic hose(s). If the handlebar is just being displaced with the master cylinder/reservoir still attached, release the brake hoses from the clamp on the bottom yoke to provide some slack, and unscrew the bolt securing the reservoir to the top yoke **(see illustrations)**. Displace the throttle cable housing from the handlebars (see Chapter 4). There is no need to detach the cables from the carburettors. Displace the handlebar switch (see Chapter 9). There is no need to disconnect the loom wiring connector.
3 If removing the left handlebar, detach the clutch cable from the lever (see Chapter 2). Displace the handlebar switch (see Chapter 9) and the choke cable (see Chapter 4). There is no need to disconnect the loom wiring connector.
4 If required, unscrew the handlebar end-weight bolts and remove them from the ends of the handlebars **(see illustration)**. Slide the throttle twistgrip off the right-hand bar. Remove the grip from the left-hand bar – it may be necessary to slit it open using a sharp blade as it may have been glued in place, although a screwdriver between the grip and the handlebar and some compressed air or

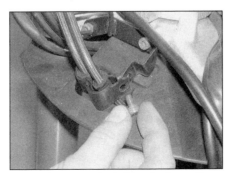

5.2a Unscrew the bolt and release the hoses from the clamp . . .

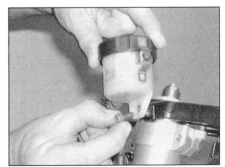

5.2b . . . and displace the reservoir from the top yoke

5.4a Handlebar end-weight bolt (arrowed)

5.4b Clutch lever bracket clamp bolt (arrowed)

5.5a Remove the blanking caps . . .

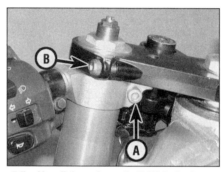

5.5b . . . then unscrew the positioning bolts

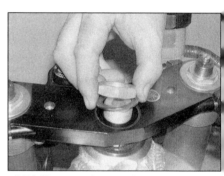

5.5c Handlebar clamp bolt (A) fork clamp bolt (B)

5.5d Unscrew the steering stem nut and remove the washer . . .

spray lubricant directed into the grip, will usually work. Depending on your removal method and its success, a new grip may be required on assembly. Slacken the clamp bolt securing the clutch lever bracket and slide it off the bar **(see illustration)**.

5 Remove the blanking caps from the heads of the handlebar positioning bolts using a small flat-bladed screwdriver, then unscrew the bolts **(see illustrations)**. Slacken the fork clamp bolts in the top yoke **(see illustration)**. Unscrew the steering stem nut and remove it along with its washer **(see illustration)**. Gently ease the top yoke upwards off the fork tubes and position it clear, using a rag to protect the tank or other components **(see illustration)**.

6 Slacken the handlebar clamp bolt **(see illustration 5.5c)**. Ease the handlebar up and off the fork **(see illustration)**.

Installation

7 Installation is the reverse of removal, noting the following.
 a) Tighten the steering stem nut, fork clamp bolts, handlebar positioning bolt, handlebar clamp bolts, and the end-weight bolts if removed, to the torque settings specified at the beginning of the Chapter, in that order.
 b) Apply some rubber adhesive to the left-hand bar before fitting the grip. Apply some grease to the right-hand bar before sliding on the throttle twistgrip. Apply grease to the clutch, throttle and choke cable ends.
 c) Refer to the relevant Chapters (as directed) for the installation of the handlebar mounted assemblies. Align the slit in the clutch lever bracket with the punchmark on the underside of the handlebar.
 d) Do not forget to reconnect the front brake light switch and clutch switch wiring connectors.
 e) Adjust throttle and clutch cable freeplay (see Chapter 1).
 f) Check the operation of all switches and the front brake and clutch before taking the machine on the road.

Clutch lever

Removal

8 Thread the clutch cable adjuster fully into the bracket to provide maximum freeplay in the cable **(see illustration)**. Unscrew the lever pivot bolt locknut, then unscrew the pivot bolt

5.5e . . . then ease the yoke up and off the forks

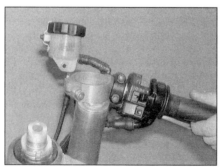

5.6 Slide the handlebar up and off the fork

5.8a Thread the adjuster fully in . . .

Frame, suspension and final drive 6•5

and remove the lever, detaching the cable nipple as you do (see illustration).

Installation

9 Installation is the reverse of removal. Apply grease to the pivot bolt shaft and the contact areas between the lever and its bracket, and to the clutch cable nipple. Adjust the clutch cable freeplay (see Chapter 1).

Front brake lever

Removal

10 Unscrew the lever pivot bolt locknut, then unscrew the pivot bolt and remove the lever (see illustration).

Installation

11 Installation is the reverse of removal. Apply grease to the pivot bolt shaft and the contact areas between the lever and its bracket.

6 Forks – 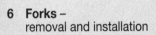 removal and installation

Removal

Caution: *Although not strictly necessary, before removing the forks it is recommended that the fairing side panels and fairing are removed (see Chapter 8). This will prevent accidental damage to the paintwork should a tool slip.*

1 Remove the fairing and the fairing side panels (see Chapter 8).
2 Displace the front brake calipers (see Chapter 7). There is no need to disconnect the hydraulic hoses.
3 Remove the front wheel (see Chapter 7).
4 Remove the front mudguard (see Chapter 8).
5 Work on each fork individually. Note the routing of the various cables and hoses around the forks.
6 Slacken the handlebar clamp bolt and the fork clamp bolt in the top yoke (see illustration 5.5c). If the forks are to be disassembled, or if the fork oil is being changed, slacken the fork top bolt now.
7 Note the alignment or amount of protrusion of the tops of the fork tube with the top yoke. Slacken but do not remove the fork clamp bolts in the bottom yoke (see illustration 6.9a). Remove the fork by twisting it and pulling it downwards (see illustration). Note which fork fits on which side.

If the fork legs are seized in the yokes, spray the area with penetrating oil and allow time for it to soak in before trying again.

Installation

8 Remove all traces of corrosion from the fork tubes and the yokes. Slide the fork up through

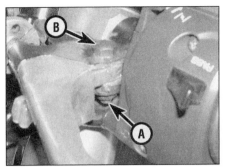
5.8b ... then unscrew the nut (A), remove the bolt (B) and slide the lever out of the bracket

the bottom yoke and the handlebars and up into the top yoke, making sure the wiring, cables and hoses are the correct side of the fork as noted on removal (see illustration 6.7). Make sure that the fork with the threaded section for the axle fits on the left, and the fork with the axle clamp bolt fits on the right. Check that the amount of protrusion of the fork tube above the top yoke is as noted on removal and equal on both sides – the tops of the tubes should be flush with the top of the top yoke (see illustration 5.5c).
9 Tighten the fork clamp bolts in the bottom yoke to the torque setting specified at the beginning of the Chapter (see illustration). If the fork leg has been dismantled or if the fork oil has been changed, tighten the fork top bolt to the specified torque setting (see illustration). Now tighten the fork clamp bolt in the top yoke, and the handlebar clamp bolt, to

6.7 Draw the fork down and out of the yokes – twisting it as you go eases removal

6.9b ... the fork top bolt ...

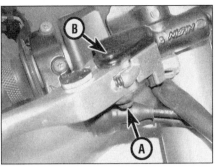
5.10 Unscrew the nut (A), remove the screw (B) and slide the lever out of the bracket

the specified torque settings (see illustration).
10 Install the front wheel (see Chapter 7), the front mudguard (see Chapter 8), and the brake calipers (see Chapter 7).
11 Install the fairing and the fairing side panels (see Chapter 8).
12 Check the operation of the front forks and brakes before taking the machine out on the road.

7 Forks – disassembly, inspection and reassembly

Disassembly

1 Always dismantle the fork legs separately to avoid interchanging any parts and thus causing an accelerated rate of wear. Store all

6.9a Tighten the bottom yoke clamp bolts ...

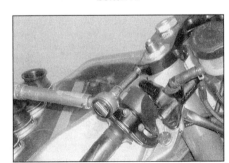
6.9c ... and the top yoke clamp bolt and the handlebar clamp bolt to the specified torque

6•6 Frame, suspension and final drive

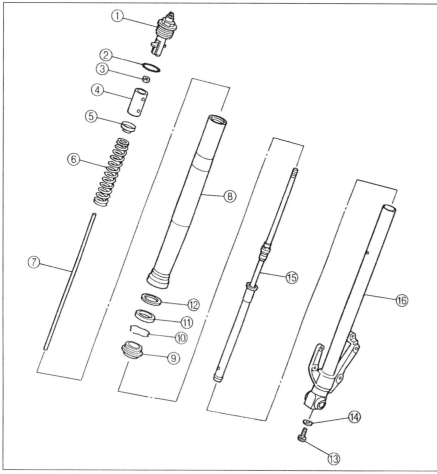

7.1 Front fork components

1 Top bolt assembly
2 O-ring
3 Locknut
4 Spacer
5 Spring seat
6 Spring
7 Damping adjuster rod
8 Fork outer tube
9 Dust seal
10 Retaining clip
11 Oil seal
12 Washer
13 Damper rod bolt
14 Sealing washer
15 Damper cartridge
16 Fork inner tube

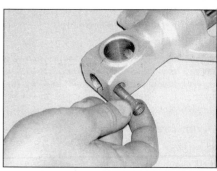

7.2a Remove the axle clamp bolt from the right-hand fork

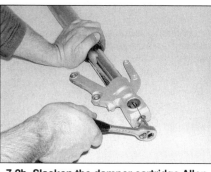

7.2b Slacken the damper cartridge Allen bolt

components in separate, clearly marked containers **(see illustration)**. Due to the construction of these forks, an assistant and a special tool are required (see Step 5).

2 Before dismantling the fork, it is advised that the damper cartridge bolt be slackened at this stage – when working on the right-hand fork, first remove the axle clamp bolt **(see illustration)**. **Note:** *If the fork is being dismantled for seal renewal or to change the oil rather than for full disassembly, there is no need to separate the damper cartridge from the inner tube, so ignore this Step and Step 9 onwards.* Compress the outer fork tube over the inner tube so that the spring exerts maximum pressure on the damper cartridge head, then have an assistant slacken the damper cartridge Allen bolt in the base of the inner tube **(see illustration)**. If an assistant is not available, clamp the brake caliper lugs on the inner tube between the padded jaws of a vice. If the bolt won't undo (usually because the damper cartridge turns with it), use an air wrench if available, or wait until the spring has been removed, then use a holding tool to prevent the damper cartridge turning with the bolt.

3 If the fork top bolt was not slackened with the fork in situ, carefully clamp the outer tube in a vice equipped with soft jaws, taking care not to overtighten or score its surface.

4 Unscrew the fork top bolt from the top of the outer tube **(see illustration)**. The bolt will remain threaded on the damper cartridge.

5 Carefully clamp the brake caliper lugs on the inner tube between the padded jaws of a vice and slide the outer tube fully down onto the inner tube (wrap a rag around the top of the outer tube to minimise oil spillage) while keeping the damper cartridge fully extended (with the aid of an assistant if necessary). Obtain either the Yamaha service tool (Pts. Nos. 90890-01441 and -01434 (Europe) or YM-01441 and –01434 (USA)), or construct a home-made equivalent from a piece of steel strap bent into a U-shape, some threaded rod and nuts, and a suitably-sized washer with a slot cut into it, as shown **(see illustration)**. Fit the tool onto the spacer, locating it into the

7.4 Unscrew the top bolt from the fork tube

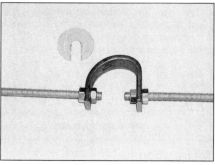

7.5a A home-made tool can be made as shown

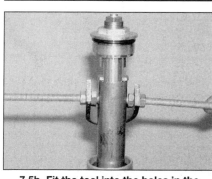

7.5b Fit the tool into the holes in the spacer as shown

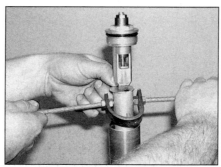

7.5c Press down on the tool and slide the slotted washer between the top of the spacer and the locknut

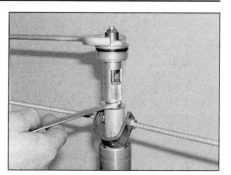

7.6a Counter-hold the locknut and unscrew the top bolt . . .

holes **(see illustration)**. Push down on the spacer using the tool and have an assistant insert the slotted washer between the top of the spacer and the base of the locknut on the damper cartridge **(see illustration)**. This will keep the spacer and spring compressed while the top bolt assembly is removed.

6 Using two spanners, one on the locknut and one on the fork top bolt, counter-hold the locknut and release the top bolt from it **(see illustration)**. Now hold the rod and thread the top bolt assembly off the damper cartridge **(see illustration)**. Allow the damper rod to settle inside the fork.

7 Push down on the spacer using the tool and have an assistant remove the fabricated slotted washer **(see illustration 7.5c)**, then slowly release the spring pressure and remove the spacer and the spring seat **(see illustrations)**. Withdraw the spring from the tube, noting which way up it fits **(see illustration)**. Draw the damper rod out of the fork using a pair of thin-nosed pliers, then withdraw the damping adjuster rod from inside it **(see illustrations)**. Remove the holding tool from the spacer if required, but note that it will be needed for installation.

8 Invert the fork leg over a suitable container and pump the fork and damper cartridge vigorously to expel as much fork oil as possible **(see illustration)**. **Note:** *If you are only changing the oil and not carrying out any other work on the fork, ignore the following Steps and proceed to Step 23, though it is always worth checking through Steps 14 to 18.*

9 If required (for full disassembly), remove the previously slackened damper cartridge bolt and its sealing washer from the bottom of the inner tube **(see illustration)**. If the bolt wasn't slackened earlier, insert the Yamaha holding tool (Pts. Nos. 90890-01423 (Europe) or

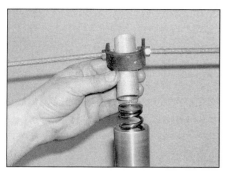

7.6b . . . then thread it off the rod

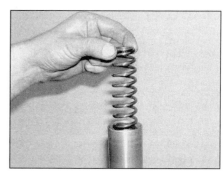

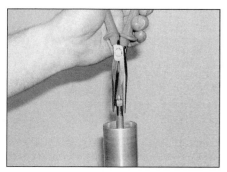

7.7a Remove the spacer . . .

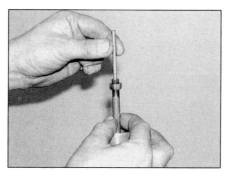

7.7b . . . and the spring seat

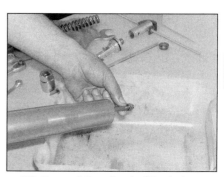

7.7c Withdraw the spring . . .

7.7d . . . then lift the damper rod . . .

7.7e . . . and withdraw the damping adjuster rod

7.8 Pump all the oil out of the fork as described

7.9a Unscrew and remove the damper cartridge bolt . . .

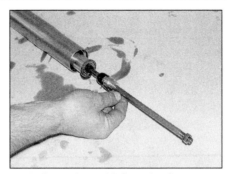

7.9b . . . then withdraw the damper cartridge from the fork

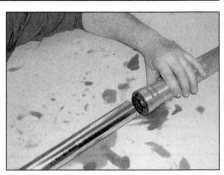

7.10 Slide the inner and outer tubes apart

YM-01423 (USA)). Discard the sealing washer, as a new one must be used on reassembly. Invert the fork and withdraw the damper cartridge **(see illustration)**.

10 Separate the inner and outer tubes by sliding them apart, taking care not to scratch the inside of the outer tube with the damper cartridge end if it has not been removed **(see illustration)**.

11 Carefully prise out the dust seal from the bottom of the outer tube to gain access to the oil seal retaining clip **(see illustration)**. Discard the dust seal, as a new one must be used.

12 Carefully remove the retaining clip, taking care not to scratch the surface of the tube **(see illustration)**.

13 Hook or lever the oil seal out using a seal removing tool or a large flat-bladed screwdriver, noting which way up it fits **(see illustration)**. Remove the washer **(see illustration)**. Discard the oil seal, as a new one must be used.

Inspection

14 Clean all parts in solvent and blow them dry with compressed air, if available. Check the inner tube for score marks, scratches, flaking of the chrome finish and excessive or abnormal wear. Renew the inner tube in both forks if any dents are found. Check the fork seal seat in the bottom of the outer tube for nicks, gouges and scratches. If damage is evident, leaks will occur. Also check the oil seal washer for damage or distortion, and renew it if necessary.

15 Check the fork tube for runout (bending) using V-blocks and a dial gauge, or have it done by a Yamaha dealer or suspension specialist. Yamaha do not specify a runout limit, but if the tube is bent beyond the generally accepted limit of 0.2 mm seek the advice of a suspension specialist.

 Warning: If the tube is bent, it should not be straightened; replace it with a new one.

16 Check the spring for cracks and other damage. Measure the spring free length and compare the measurement to the specifications at the beginning of the Chapter. If it is defective or sagged below the service limit, replace the springs in both forks with new ones. Never renew only one spring.

17 Examine the working surfaces of the two bushes inside the outer tube **(see illustration)**. Note that the bushes are available separately on 2001 models, but only complete with the outer tube on earlier models. If the bushes require renewal, and can be obtained separately, seek the advice of a suspension specialist on how best to extract the old bushes and install the new ones.

18 Check the damper cartridge assembly for damage and wear, and renew it if necessary. Holding the outside of the damper, pump the rod in and out. If the rod does not move smoothly in the damper, it must be replaced with a new one.

Reassembly

19 Fit the washer into the bottom of the outer tube **(see illustration)**. Apply a smear of lithium grease oil to the lips of the oil seal, then fit it into the outer tube, making sure that

7.11 Prise off the dust seal . . .

7.12 . . . then lever out the retaining clip using a flat bladed screwdriver

7.13a Remove the oil seal as described . . .

7.13b . . . then remove the washer

7.17 The bushes (arrowed) are inside the outer tube, one at the bottom and one about half way up

Frame, suspension and final drive 6•9

7.19a Install the washer ...

7.19b ... then fit the oil seal ...

7.19c ... and drive it in as described

the side of the oil seal with the spring faces out **(see illustration)**. Drive the seal in using a suitable socket or piece of tubing, making sure it doesn't have a sharp edge, and locates on the outer rim of the seal, not on the spring

7.20a Fit the retaining clip into its groove ...

or lips **(see illustration)**. Drive it in squarely until it is fully seated. The seal is in place when the groove for the retaining clip is fully exposed and the seal is square to it.

20 Fit the retaining clip into its groove in the outer tube, making sure it locates correctly **(see illustration)**. Apply a smear of lithium grease to the lips of the dust seal, then press it into the tube **(see illustration)**.

21 Apply a smear of the specified clean fork oil to the outer surface of the inner tube, then carefully insert it squarely into the outer tube, twisting it as you do so that it passes through the seals easily **(see illustration)**.

22 Lay the fork flat and push the inner tube fully into the outer tube. If removed, insert the damper cartridge in the top of the fork until it seats on the bottom of the inner tube **(see illustration)**. Apply a few drops of a suitable

non-permanent thread locking compound onto the damper cartridge Allen bolt and fit a new sealing washer, then install the bolt into the bottom of the inner tube and tighten it to the torque setting specified at the beginning of the Chapter **(see illustrations)**. If the damper cartridge rotates inside the tube, either use the holding tool if you have one (see Step 9), or wait until the fork is fully reassembled before tightening the bolt.

23 Stand the fork upright. Slowly pour in the specified quantity and grade of fork oil and pump the fork and damper cartridge slowly at least ten times each to distribute it evenly and expel all the air. Be careful not to extend the outer tube by more than 100 mm when pumping it, as this can cause air to enter the system – in which case the process must be repeated. Wait ten minutes, then fully

7.20b ... then press in the dust seal

7.21 Slide the inner tube into the outer tube

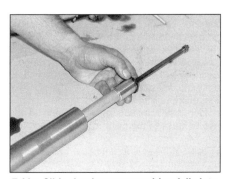

7.22a Slide the damper cartridge fully into the fork

7.22b Apply a threadlock to the bolt threads ...

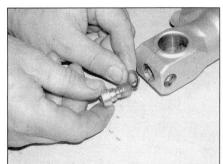

7.22c ... and fit a new washer ...

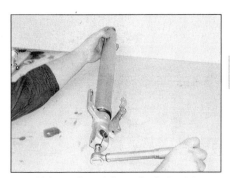

7.22d ... then tighten the bolt to the specified torque

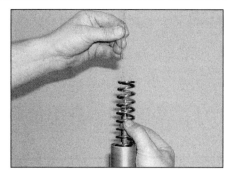

7.23 Measure the oil level as described

7.24a Install the spring...

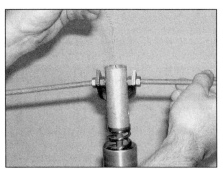

7.24b ... the spring seat...

7.24c ... and the spacer

7.26a Fit a new O-ring onto the top bolt...

7.26b ... then thread the bolt onto the damper rod

compress the fork and measure the fork oil level from the top of the tube **(see illustration)**. Add or subtract fork oil until it is at the level specified at the beginning of the Chapter.

24 Withdraw the damper cartridge fully from the fork, using a pair of thin-nosed pliers to grab it if necessary **(see illustration 7.7d)**. Fit the adjuster rod into the damper cartridge **(see illustration 7.7e)**. As the damper cartridge will have to be kept extended out of the tube, tie a piece of wire around the base of the locknut on the top of the rod to use as a holder. Install the spring, with its tapered end upwards **(see illustration)**. If the holding tool used on disassembly was removed from the spacer, fit it back on (see Step 5). Install the spring seat, making sure the lip on the bottom fits into the top of the spring, then install the spacer **(see illustrations)**. Push down on the

7.28 Thread the top bolt into the fork tube

spacer using the tool, and insert the slotted washer (as used on removal) between the top of the spacer and the base of the locknut on the damper cartridge **(see illustration 7.5c)**. This will keep the spacer and spring compressed while the top bolt assembly is installed. You can now remove the piece of wire.

25 Position the locknut so that the distance between the top of the nut and the top of the rod is 11 mm.

26 Fit a new O-ring onto the damping adjuster/pre-load adjuster/top bolt assembly and thread it down onto the damper cartridge until it seats against the locknut, holding the rod to prevent it from turning **(see illustrations)**. Counter-hold the top bolt and tighten the locknut securely against it, to the specified torque if the correct tools are available **(see illustration 7.6a)**.

27 Push down on the spacer using the tool and have an assistant remove the fabricated slotted washer, then slowly release the spring pressure and allow the spacer to settle against the pre-load adjuster arms **(see illustration 7.5c)**. Remove the holding tool from the spacer.

28 Apply a smear of the specified clean oil to the top bolt O-ring. Fully extend the outer tube and carefully screw the top bolt into the tube, making sure it is not cross-threaded **(see illustration)**. Note: *The top bolt can be tightened to the specified torque setting at this stage if the tube is held between the padded jaws of a vice, but do not risk distorting the tube by doing so. A better method is to tighten*

the top bolt when the fork leg has been installed and is securely held in the yokes.

 TOOL TIP *Use a ratchet-type tool when installing the fork top bolt. This makes it unnecessary to remove the tool from the bolt whilst threading it in.*

29 If the damper cartridge Allen bolt requires tightening, clamp the fork between the padded jaws of a vice and have an assistant compress the fork so that maximum spring pressure is placed on the damper cartridge head – tighten the bolt to the specified torque setting **(see illustration 7.22d)**.

30 Install the forks as described in Section 6. Set the spring pre-load and rebound damping as required (see Section 12).

8 Steering stem – removal and installation

Removal

1 Remove the fairing and the fairing side panels (see Chapter 8). It is also advisable to remove the fuel tank to avoid the possibility of scratching it (see Chapter 4).

2 Remove the front forks (see Section 6).

3 Unscrew the bolt securing the front brake master cylinder reservoir bracket to the top yoke **(see illustration 5.2b)**. Keep the reservoir upright.

Frame, suspension and final drive 6•11

8.4a Unscrew the bolts (arrowed) and displace the shield . . .

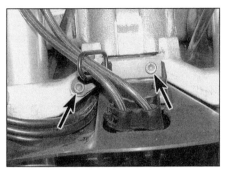

8.4b . . . and the hose bracket

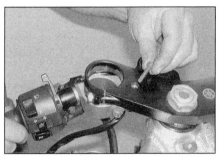

8.5 Remove the blanking caps and positioning bolts and displace the handlebars

8.6a Unscrew the steering stem nut and remove the washer . . .

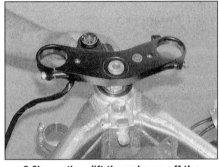

8.6b . . . then lift the yoke up off the steering stem

8.7a Remove the lockwasher . . .

4 Unscrew the bolts securing the shield to the bottom yoke and displace it **(see illustration)**. Unscrew the bolts securing the front brake hose bracket to the bottom yoke and displace it **(see illustration)**. If the top yoke is being removed from the bike rather than just being displaced, trace the wiring from the ignition switch and disconnect it at the connector.

5 Remove the blanking caps from the heads of the handlebar positioning bolts using a small flat-bladed screwdriver, then unscrew the bolts and displace the handlebars from under the yoke **(see illustration)**. Secure the handlebar assemblies so that the master cylinder is upright, making sure there is no strain on the hoses, cables or wiring.

6 Unscrew the steering stem nut and remove it along with its washer, where fitted **(see illustration)**. Lift the top yoke up off the steering stem and position it clear, using a rag to protect the tank or other components **(see illustration)**.

7 Remove the tabbed lockwasher, noting how it fits, then unscrew and remove the locknut using, if necessary, either a C-spanner, a peg spanner or a drift located in one of the notches (though it shouldn't be tight and can probably be undone with your fingers) **(see illustrations)**. Remove the rubber washer **(see illustration)**.

8 Supporting the bottom yoke, unscrew the adjuster nut using either a C-spanner, a peg-spanner or a drift located in one of the notches, then remove the adjuster nut and the bearing cover from the steering stem **(see illustrations)**.

9 Gently lower the bottom yoke and steering stem out of the frame **(see illustration)**.
10 Remove the inner race and bearing from the top of the steering head (see illustrations 8.12b and a). Remove the bearing and rubber dust seal from the base of the steering stem **(see illustrations 8.11b and a)**. Remove all traces of old grease from

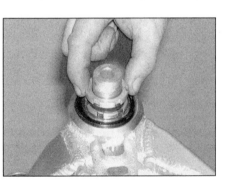

8.7b . . . the locknut . . .

8.7c . . . and the rubber washer

8.8a Unscrew the adjuster nut . . .

8.8b . . . and remove the bearing cover

6•12 Frame, suspension and final drive

8.9 Gently lower the steering stem out of the frame

8.11a Fit the dust seal . . .

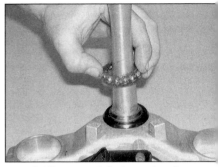

8.11b . . . and the lower bearing

the bearings and races and check them for wear or damage as described in Section 9. **Note:** *Do not attempt to remove the races from the steering head or the steering stem unless they are to be replaced with new ones.*

Installation

11 Smear a liberal quantity of lithium-based grease onto the bearing races. Also work some grease well into both the upper and lower bearings. Fit the rubber dust seal over the lower bearing inner race on the steering stem, then fit the bearing **(see illustrations)**.

12 Carefully lift the steering stem/bottom yoke up through the steering head **(see illustration 8.9)**. Fit the upper bearing then the inner race into the top of the steering head, then install the bearing cover **(see illustrations and 8.8b)**. Thread the adjuster nut onto the steering stem and adjust the bearings as described in Chapter 1, noting

that you may need to carry out the procedure several times if new bearings have been fitted to allow them to settle in **(see illustration)**.

13 Install the rubber washer and the locknut **(see illustrations 8.7c and b)**. Tighten the locknut finger-tight, then tighten it further until its notches align with those in the adjuster nut. If necessary, counter-hold the adjuster nut and tighten the locknut using a C-spanner or drift until the notches align, but make sure the adjuster nut does not turn as well. Install the tabbed lockwasher so that the tabs fit into the notches in both the locknut and adjuster nut **(see illustration 8.7a)**.

14 Fit the top yoke onto the steering stem **(see illustration 8.6b)**, then install the washer and steering stem nut and tighten it finger-tight **(see illustration 8.6a)**. Temporarily install one of the forks to align the top and bottom yokes, and secure it by tightening the bottom yoke clamp bolts only **(see**

illustration). Now tighten the steering stem nut to the torque setting specified at the beginning of the Chapter **(see illustration)**.

15 Install the remaining components in a reverse of the removal procedure, referring to the relevant Sections or Chapters, and to the torque settings specified at the beginning of the Chapter. Tighten the handlebar positioning bolts before the clamp bolts.

16 Carry out a check of the steering head bearing freeplay as described in Chapter 1 and if necessary re-adjust.

9 Steering head bearings – inspection and renewal

Inspection

1 Remove the steering stem (see Section 8).
2 Remove all traces of old grease from the bearings and races and check them for wear or damage.
3 The outer races should be polished and free from indentations. Inspect the bearing balls for signs of wear, damage or discoloration, and examine their retainer cage for signs of cracks or splits. Spin the bearing balls by hand. They should spin freely and smoothly. If there are any signs of wear on any of the above components, both upper and lower bearing assemblies must be renewed as a set. Only remove the outer races in the steering head and the lower bearing inner race on the steering stem if they need to be renewed – do not re-use them once they have been removed.

8.12a Fit the upper bearing . . .

8.12b . . . and its inner race into the steering head

8.12c Tighten the adjuster nut to set the bearings as described in Chapter 1

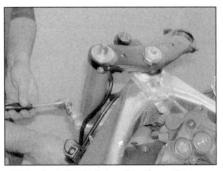

8.14a Install a fork to align the yokes . . .

8.14b . . . then tighten the steering stem nut to the specified torque

Frame, suspension and final drive 6•13

9.4a Locate the end of the drift in the cutouts (arrowed) . . .

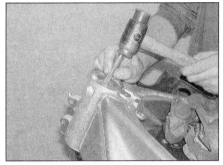

9.4b . . . and drive the races out

Renewal

4 The outer races are an interference fit in the steering head and can be tapped from position with a suitable drift located in the cutouts in the head **(see illustrations)**. Alternate between the cutouts so that the race is driven out squarely. It may prove advantageous to curve the end of the drift slightly to improve access.

5 Alternatively, the races can be removed using a slide-hammer type bearing extractor – these can often be hired from tool shops.

6 The new outer races can be pressed into the head using a drawbolt arrangement **(see illustration)**, or by using a large diameter tubular drift. Ensure that the drawbolt washer or drift (as applicable) bears only on the outer edge of the race and does not contact the working surface.

 Installation of new bearing outer races is made much easier if the races are left overnight in the freezer. This causes them to contract slightly making them a looser fit. Alternatively, use a freeze spray.

7 The lower bearing inner race should only be removed from the steering stem if a new one is being fitted. To remove the race from the steering stem, use two screwdrivers placed on opposite sides of the race to work it free, using blocks of wood to improve leverage and protect the yoke, or tap under it using a cold chisel **(see illustration)**. If the race is firmly in place it will be necessary to use a puller, or in extreme circumstances to split the race using an angle grinder **(see illustration)**. Take the steering stem to a Yamaha dealer if required.

8 Fit the new lower race onto the steering stem. A length of tubing with an internal diameter slightly larger than the steering stem will be needed to tap the new race into position **(see illustration)**.

9 Install the steering stem (see Section 8).

10 Rear shock absorber – removal, inspection and installation

> **Warning:** Do not attempt to disassemble this shock absorber. It is nitrogen-charged under high pressure. Improper disassembly could result in serious injury. Instead, take the shock to a Yamaha dealer or suspension specialist to do the job.

Removal

1 Support the motorcycle securely in an upright position using an auxiliary stand. Position a support under the rear wheel so that it does not drop when the shock absorber is removed, while also making sure that the weight of the machine is off the rear suspension so that the shock is not compressed.

2 Remove the lower fairing (see Chapter 8).

3 Unscrew the nut and withdraw the bolt securing the linkage plate to the swingarm **(see illustration)**.

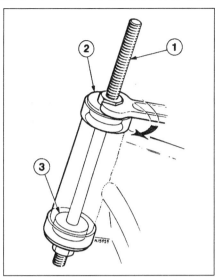

9.6 Drawbolt arrangement for fitting steering stem bearing races

1 Long bolt or threaded bar
2 Thick washer
3 Guide for lower race

9.7a Remove the lower bearing inner race as described . . .

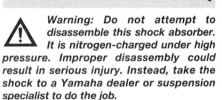

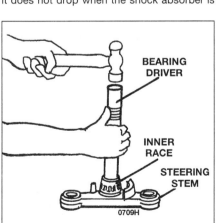

9.7b . . . you may have to use a puller

9.8 Drive the new race on, using a suitable driver or a length of pipe or tubing

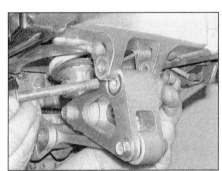

10.3 Unscrew the nut and withdraw the bolt securing the linkage plate to the swingarm

6•14 Frame, suspension and final drive

10.4 Unscrew the nut and withdraw the shock lower mounting bolt

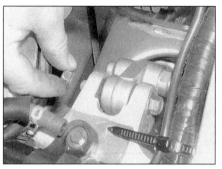

10.5a Unscrew the nut . . .

10.5b . . . then withdraw the upper mounting bolt and spacer . . .

10.5c . . . and lower the shock

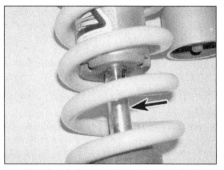

10.7 Look for cracks, pitting and oil leakage on the damper rod (arrowed)

11 Rear suspension linkage – removal, inspection and installation

Removal

1 Support the motorcycle securely in an upright position using an auxiliary stand. Position a support under the rear wheel so that it does not drop when the shock absorber lower mounting bolt is removed, while also making sure that the weight of the machine is off the rear suspension so that the shock is not compressed.
2 Remove the lower fairing (see Chapter 8).
3 Unscrew the nuts and withdraw the bolts securing the linkage plate to the swingarm and the shock absorber **(see illustrations 10.3 and 10.4)**. Note which bolts fit where.
4 Unscrew the nut and withdraw the bolt securing the linkage plate to the linkage arm and remove the plate, noting which way round it fits **(see illustration)**.
5 Unscrew the nut and withdraw the bolt securing the linkage arm to the frame and remove the linkage arm, noting the spacer with the bolt, and which way round the arm fits **(see illustrations)**.

Inspection

6 Withdraw all the spacers from the linkage plate and arm, noting their different sizes. Lever the grease seals out **(see illustration)**.

4 Unscrew the nut and withdraw the bolt securing the bottom of the shock absorber to the linkage plate, then swing the plate and the arm down to provide clearance for the shock **(see illustration)**.
5 Unscrew the nut on the shock absorber upper mounting bolt, then support the shock absorber and withdraw the bolt, noting the spacer **(see illustrations)**. Raise the swingarm and lower the shock out of the frame **(see illustration)**.

Inspection

6 Inspect the shock absorber for obvious physical damage and the coil spring for looseness, cracks or signs of fatigue.
7 Inspect the damper rod for signs of bending, pitting and oil leakage **(see illustration)**.
8 Inspect the pivot hardware at the top and bottom of the shock for wear or damage.

Yamaha do not list the needle bearing in the bottom of the shock as being available separately. However, if it needs renewing it is worth checking with Yamaha or a bearing specialist about fitting a new one, rather than having to fit a new shock.
9 Check the reservoir for damage, cracks or leakage.
10 Individual components are not available for the shock absorber. If it is worn or damaged, it must be replaced with a new one.

Installation

11 Installation is the reverse of removal. Apply molybdenum lithium-based grease to the shock absorber and linkage plate pivot points. Install the bolts and nuts finger-tight only until all components are in position, then tighten the nuts to the torque settings specified at the beginning of the Chapter.

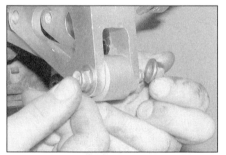

11.4 Unscrew the nut and withdraw the bolt securing the plate to the arm and remove the plate

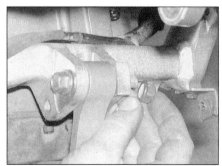

11.5a Unscrew the nut . . .

11.5b . . . then withdraw the bolt and spacer and remove the arm

Frame, suspension and final drive

11.6a Remove the spacers and lever out the grease seals

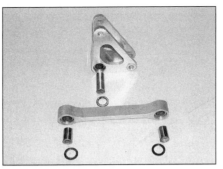

11.6b Suspension linkage components

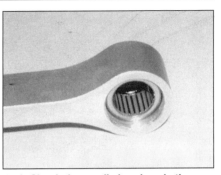

11.8 Check the needle bearings in the arm and plate

Thoroughly clean all components, removing all traces of dirt, corrosion and grease **(see illustration)**.

7 Inspect all components closely, looking for obvious signs of wear such as heavy scoring, or for damage such as cracks or distortion. Slip each spacer back into its bearing and check that there is not an excessive amount of freeplay between the two components. Renew any components as required.

8 Check the condition of the needle roller bearings in the linkage arm and plate **(see illustration)**. Refer to *Tools and Workshop Tips* (Section 5) in the *Reference* section for more information on bearings.

9 Worn bearings can be drifted out of their bores, but note that removal will destroy them; new bearings should be obtained before work commences. The new bearings should be pressed or drawn into their bores rather than driven into position. In the absence of a press,

a suitable drawbolt tool can be made up as described in *Tools and Workshop Tips* (Section 5) in the *Reference* section.

10 Lubricate the needle roller bearings and the spacers with lithium-based grease and install the spacers.

11 Check the condition of the grease seals, and renew them if they are damaged or deteriorated. Press the seals squarely into place.

Installation

12 Installation is the reverse of removal. Apply lithium-based grease to the pivot points. Install the bolts and nuts finger-tight only until all components are in position, then tighten the nuts to the torque settings specified at the beginning of the Chapter.

12 Suspension – adjustments

Note: *Refer to the Owner's Handbook supplied with the machine for recommended front and rear suspension settings to suit loading.*

Front forks

1 The front forks are adjustable for spring pre-load, rebound damping, and compression damping.

2 Spring pre-load is adjusted using a suitable spanner on the adjuster flats on the top of the forks **(see illustration)**. The amount of pre-load is indicated by lines on the adjuster **(see illustration)**. There are eight lines. The standard position is with the 6th line just visible above the top bolt hex. Turn the adjuster clockwise to increase pre-load, and anti-clockwise to decrease it. Always make sure both adjusters are set equally.

3 Rebound damping is adjusted using a screwdriver in the slot in the adjuster protruding from the pre-load adjuster **(see illustration)**. The amount of damping is indicated by the number of clicks when turned anti-clockwise from the fully screwed-in position. There are thirteen positions on 1998 and 1999 models, and eleven on 2000-on models. The standard position is five clicks out on all models. Turn the adjuster clockwise to increase damping and anti-clockwise to decrease it. To establish the current setting, turn the adjuster in (clockwise) until it stops, counting the number of clicks, then reset it as required by turning it out. Always make sure both adjusters are set equally.

4 Compression damping is adjusted using a screwdriver in the slot in the adjuster on the base of each fork **(see illustration)**. The amount of damping is indicated by the number of clicks when turned anti-clockwise from the fully screwed-in position. There are eleven positions on 1998 and 1999 models, and nine on 2000-on models. The standard position is five clicks out on all models. Turn the adjuster clockwise to increase damping and anti-clockwise to decrease it. To establish the current setting, turn the adjuster in (clockwise) until it stops, counting the number of clicks, then reset it as required by turning it out. Always make sure both adjusters are set equally.

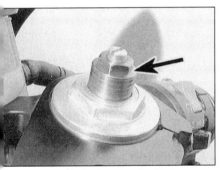

12.2a Spring pre-load adjuster (arrowed)

12.2b The amount of pre-load is indicated by the lines

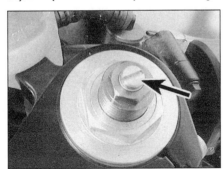

12.3 Rebound damping adjuster (arrowed)

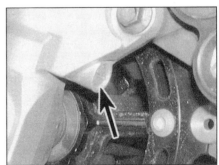

12.4 Compression damping adjuster (arrowed)

6•16 Frame, suspension and final drive

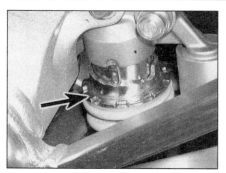

12.6a Spring pre-load adjuster (arrowed)

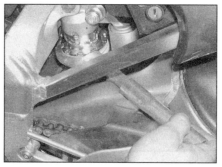

12.6b Adjusting the pre-load using the Yamaha tool supplied

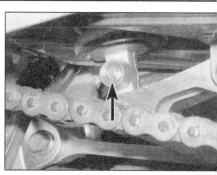

12.7 Rebound damping adjuster (arrowed)

Rear shock absorber

5 The rear shock absorber is adjustable for spring pre-load, rebound damping, and compression damping.

6 Spring pre-load is adjusted using a suitable C-spanner (one is provided in the toolkit) to turn the spring seat on the top of the shock absorber **(see illustrations)**. There are nine positions. Position 1 is the softest setting, position 4 is the standard, position 9 is the hardest. Align the setting required with the adjustment stopper. Turn the spring seat anti-clockwise to increase pre-load and clockwise to decrease it.

7 Rebound damping is adjusted using a screwdriver in the slot in the adjuster on the base of the shock absorber **(see illustration)**. Turn the adjuster clockwise to increase damping and anti-clockwise to decrease it. There are twelve positions on 1998 and 1999 models, and eleven on 2000-on models. The standard position is six clicks out on 1998 and 1999 models, and seven on 2000-on models. To establish the current setting, turn the adjuster in (clockwise) until it stops, counting the number of clicks, then reset it as required by turning it out (anti-clockwise).

8 Compression damping is adjusted using a screwdriver in the slot in the adjuster on the top of the shock absorber **(see illustration)**. Turn the adjuster clockwise to increase damping and anti-clockwise to decrease it. There are twelve positions on 1998 and 1999 models, and eleven on 2000-on models. The standard position is eight clicks out on 1998 and 1999 models, and nine on 2000-on models. To establish the current setting, turn the adjuster in (clockwise) until it stops, counting the number of clicks, then reset it as required by turning it out (anti-clockwise).

13 Swingarm – removal and installation

Removal

Note: *Before removing the swingarm, it is advisable to perform the rear suspension checks described in Chapter 1 to assess the extent of any wear.*

1 Remove the front sprocket (see Section 16), the rear wheel (see Chapter 7), and the rear shock absorber (see Section 10).

12.8 Compression damping adjuster (arrowed)

2 Unscrew the bolts securing the rear hugger to the swingarm and remove it, noting how it fits **(see illustrations)**. If required, unscrew the bolts securing the chainguard to the swingarm and remove it, noting how it fits.

3 Unscrew the bolts securing the brake hose guides/clamps to the swingarm **(see illustration)**. Feed the brake caliper through the swingarm and position it clear, making sure no strain is placed on the hose.

4 Before removing the swingarm it is advisable to re-check for play in the bearings (see Chapter 1). Any problems which may have been overlooked with the other suspension components attached to the frame are highlighted with them loose.

5 Unscrew the nut on the end of the swingarm pivot bolt and remove the washer **(see illustration)**. Support the swingarm, then

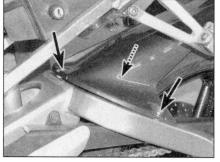

13.2a The hugger is secured by three bolts (arrowed) . . .

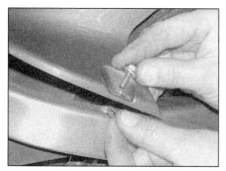

13.2b . . . note the arrangement of the washers

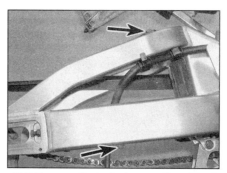

13.3 Remove the bolts (arrowed) and feed the caliper assembly through the swingarm

13.5a Unscrew the nut (arrowed) and remove the washer . . .

Frame, suspension and final drive 6•17

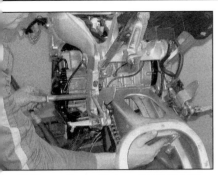

13.5b ... then withdraw the bolt and remove the swingarm

13.6 Remove the chain slider if required

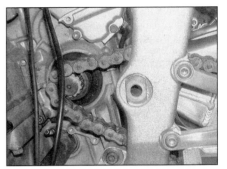

13.10a Locate the bolt head flats correctly in the frame

13.10b Fit the washer and the nut . . .

13.10c . . . and tighten the nut to the specified torque

13 Install the shock absorber (see Section 10), the rear wheel (see Chapter 7) and the front sprocket (see Section 16).
14 Check and adjust the drive chain slack (see Chapter 1). Check the operation of the rear suspension before taking the machine on the road.

14 Swingarm – inspection and bearing renewal

Inspection

1 Thoroughly clean the swingarm, removing all traces of dirt, corrosion and grease. In particular, wipe all chain grease from the chain slider so that it can be inspected for wear.
2 Remove the bearing cover from each side of the swingarm, then withdraw the collar (see illustrations). Check the condition of the seal in each cover and replace the cover with a new one, if necessary.
3 Inspect all components closely, looking for obvious signs of wear such as heavy scoring, and cracks or distortion due to accident damage. Check the bearings for roughness, looseness and any other damage, referring to *Tools and Workshop Tips* (Section 5) in the *Reference* section (see illustration). Any damaged or worn component must be renewed.
4 Check the swingarm pivot bolt for straightness by rolling it on a flat surface such

withdraw the pivot bolt and remove the swingarm, along with the drive chain, manoeuvring it around the EXUP cable and rear brake light switch (see illustration). Knock the pivot bolt through using a drift if required.
6 Remove the chain slider from the front of the swingarm if necessary, noting how it fits (see illustration). If it is badly worn or damaged, it should be replaced with a new one.
7 Inspect all components for wear or damage as described in Section 14.

Installation

8 If removed, install the chain slider and tighten the bolt securely (see illustration 13.6).
9 If not already done, remove the bearing cover from each side of the swingarm and withdraw the collar (see illustrations 14.2a and b). Lubricate the bearings with lithium-

based grease. Also grease the collar and the covers. Check the condition of the seal in each cover and replace the cover with a new one, if necessary. Re-install the collar and covers.
10 Offer up the swingarm and have an assistant hold it in place – don't forget to loop the drive chain over the chain slider on the pivot. Slide the pivot bolt through from the left-hand side (see illustration 13.5b), locating the flats on head correctly in the frame (see illustration). Install the nut with its washer, and tighten the nut to the torque setting specified at the beginning of the Chapter (see illustrations).
11 Feed the caliper through the swingarm, then fit the brake hose guides/clamps (see illustration 13.3).
12 Install the rear hugger and chainguard (if removed) and tighten their bolts securely (see illustration 13.2b and a).

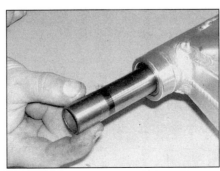

14.2a Remove the cover . . .

14.2b . . . then withdraw the collar

14.3 Check the needle bearing in each side as described

as a piece of plate glass (first wipe off all old grease and remove any corrosion using fine emery cloth). If the equipment is available, place the axle in V-blocks and measure the runout using a dial gauge. If the axle is bent, replace it with a new one.

5 Check the chain slider for wear, particularly at its contact points with the chain, and renew the slider if worn; it is retained to the swingarm by a single screw with washer and spacer and its front end clips around the swingarm pivot.

Bearing renewal

6 Remove the bearing cover from each side of the swingarm, then withdraw the collar **(see illustrations 14.2a and b)**. Refer to *Tools and Workshop Tips* (Section 5) in the *Reference* section for more information on bearing checks and replacement methods.

7 Needle bearings can be drawn or drifted out of their bores, but note that removal will make them unusable; new bearings should be obtained before work commences **(see illustration 14.3)**. Pass a long drift with a hooked end through one side of the swingarm and locate it on the inner edge of the bearing on the other side. Tap the drift around the bearing's inner edge to ensure that it leaves its bore squarely. Use the same method to extract the other bearing. If available, a slide-hammer with knife-edged bearing puller can be used, and is preferable to using a drift, to extract the bearings.

8 The new bearings should be pressed or drawn into their bores, rather than driven into position. In the absence of a press, a suitable drawbolt arrangement can be made up as described in *Tools and Workshop Tips* (Section 5) in the *Reference* section. Lubricate the bearings with lithium-based grease. Install the collar, then fit the bearing cover onto each side **(see illustrations 14.2b and a)**.

15 Drive chain –
removal, inspection, cleaning and installation

Removal

Note: *All models are fitted with a staked-type master (joining) link which can be disassembled using one of several commercially-available drive chain cutting/staking tools. Such chains can be recognised by the master link side plate's identification marks (and usually its different colour), as well as by the staked ends of the link's two pins which look as if they have been deeply centre-punched, instead of peened over as with all the other pins.*

⚠ **Warning:** *Use ONLY the correct service tools to disassemble the staked-type of master link – if you do not have access to such tools or do not have the skill to operate them correctly, have the chain removed by a dealer service department or bike repair shop.*

1 Remove the front sprocket cover (see Section 16).
2 Locate the joining link in a suitable position for working on, by rotating the back wheel.
3 Slacken the drive chain as described in Chapter 1.
4 Split the chain at the joining link using the chain cutter, following the manufacturer's operating instructions carefully (see also Section 8 in *Tools and Workshop Tips* in the *Reference* Section). Remove the chain from the bike, noting its routing through the swingarm.

Inspection and cleaning

5 Soak the chain in paraffin (kerosene) for approximately five or six minutes, then clean it with a brush to remove any remaining dirt and wipe it dry.
Caution: Don't use gasoline (petrol), solvent or other cleaning fluids. Don't use high-pressure water. Remove the chain, wipe it off, then blow-dry it with compressed air immediately. The entire process shouldn't take longer than ten minutes – if it does, the O-rings in the chain rollers could be damaged.
6 When dry, check the chain for damage to the rollers, sideplates and sealing O-rings.
7 The chain can be checked for wear (stretch) by measuring 10-link sections of the chain at three different points along its length and comparing this to the service limit – see Chapter 1, Section 1 for details.
8 If the chain is in good condition, lubricate it with engine oil or an aerosol lube for O-ring chains, and refit it as follows.

Installation

⚠ **Warning: NEVER install a drive chain which uses a clip-type master (split) link. Use ONLY the correct service tools to secure the staked-type of master link – if you do not have access to such tools or do not have the skill to operate them correctly, have the chain installed by a dealer service department or bike repair shop to be sure of having it securely installed.**

9 Slip the drive chain through the swingarm sections and around the front sprocket, leaving the two ends in a convenient position to work on.

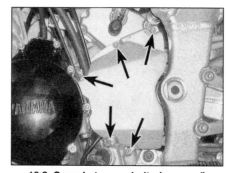

16.3 Sprocket cover bolts (arrowed)

10 Refer to Section 8 in *Tools and Workshop Tips* in the *Reference* Section. Fit an O-ring onto each pin on the new joining link, then slide the link through from the inside and fit the other two O-rings. Install the new side plate with its identification marks facing out. Stake the new link using the drive chain cutting/staking tool, following the instructions of both the chain manufacturer and the tool manufacturer carefully. DO NOT re-use old joining link components.

11 After staking, check the joining link and staking for any signs of cracking. If there is any evidence of cracking, the joining link, O-rings and side plate must be renewed. Measure the diameter of the staked ends in two directions and check that it is evenly staked.

12 Install the sprocket cover (see Section 16).
13 On completion, adjust and lubricate the chain following the procedures described in Chapter 1.
Caution: Use only the recommended lubricant.

16 Sprockets –
check and renewal

Check

1 Remove the lower fairing and the left-hand fairing side panel (see Chapter 8).
2 Slacken the gearchange lever linkage rod locknuts, then unscrew the rod and separate it from the lever and the arm (the rod is reverse-threaded on one end, so will unscrew from both lever and arm simultaneously when turned in the one direction) **(see illustration 3.3a)**. Note how far the rod is threaded into the lever and arm as this determines the height of the lever relative to the footrest. Withdraw the rod from the frame **(see illustration 3.3b)**.
3 Unscrew the bolts securing the front sprocket cover and remove it, noting how the hose and wiring are routed **(see illustration)**.
4 Check the wear pattern on both sprockets **(see illustration 1.7 in Chapter 1)**. Whenever the sprockets are inspected, the drive chain should also be inspected. If the sprocket teeth are worn excessively, or you are fitting a new chain, renew the chain and sprockets as a set.
5 Adjust and lubricate the chain following the procedures described in Chapter 1.
Caution: Use only the recommended lubricant.

Renewal

Front sprocket

6 Remove the lower fairing and the left-hand fairing side panel (see Chapter 8).
7 Slacken the gearchange lever linkage rod locknuts, then unscrew the rod and separate it from the lever and the arm (the rod is reverse-threaded on one end, so will unscrew from both lever and arm simultaneously when

Frame, suspension and final drive 6•19

16.9a Bend back the lockwasher tab(s) . . .

16.9b . . . then unscrew the nut (arrowed) and remove the washer

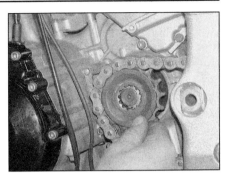

16.10 Slide the sprocket off the shaft and remove it

16.11 Fit the sprocket into the chain and slide it onto the shaft

16.12a Fit the new lockwasher . . .

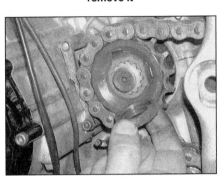

16.12b . . . and the nut . . .

turned in the one direction) **(see illustration 3.3a)**. Note how far the rod is threaded into the lever and arm as this determines the height of the lever relative to the footrest. Withdraw the rod from the frame

16.12c . . . and tighten it to the specified torque

(see illustration 3.3b).
8 Unscrew the bolts securing the front sprocket cover and remove it **(see illustration 16.3)**.
9 Bend down the tab(s) on the sprocket nut lockwasher **(see illustration)**. Have an assistant apply the rear brake hard, then unscrew the nut and remove the washer **(see illustration)**. Discard the washer, as a new one should be used. Refer to Chapter 1 and adjust the chain so that it is fully slack.
10 If the sprocket is not being replaced with a new one, mark the outside with a scratch, or dab of paint, so that it can be installed the same way round. Slide the sprocket and chain off the shaft and slip the sprocket out of the chain **(see illustration)**. If there is not enough slack on the chain to remove the sprocket, disengage the chain from the rear wheel.
11 Engage the new sprocket with the chain and slide it on the shaft **(see illustration)**.

Take up the slack in the chain (see Chapter 1).
12 Slide on the new lockwasher, then fit the nut with the shouldered side facing in and tighten it to the torque setting specified at the beginning of the Chapter, applying the rear brake to prevent the sprocket from turning **(see illustrations)**. Bend up the tabs of the lockwasher against the nut **(see illustration)**.
13 Install the front sprocket cover, making sure the hose and wiring are correctly routed **(see illustration)**. Apply a suitable non-permanent thread locking compound to the front bolt and tighten all the bolts to the specified torque setting **(see illustration)**.
14 Slide the linkage rod into the frame and thread it onto the lever and arm (the rod is reverse-threaded on one end, so will screw simultaneously onto both lever and arm when turned in the one direction) **(see illustration 3.3b)**. Set it as noted on removal to give the same gearchange lever height,

16.12d Bend the tabs up against the nut

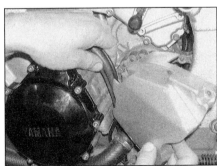

16.13a Make sure everything is correctly routed . . .

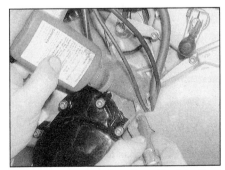

16.13b . . . and apply a threadlock to the front bolt

6•20 Frame, suspension and final drive

16.16 Unscrew the nuts (arrowed) and remove the sprocket

17.2 Lift the sprocket coupling out of the wheel . . .

17.3 . . . and remove the rubber dampers

then tighten the locknuts securely **(see illustration 3.3a)**. To adjust the lever height, slacken the locknuts and thread the rod in or out as required, then retighten the nuts.

Rear sprocket

15 Remove the rear wheel (see Chapter 7).
16 Unscrew the nuts securing the sprocket to the hub assembly **(see illustration)**. Remove the sprocket, noting which way round it fits.
17 Install the sprocket onto the hub with the stamped mark facing out. Tighten the nuts evenly and in a criss-cross sequence to the torque setting specified at the beginning of the Chapter.
18 Install the rear wheel (see Chapter 7).

17 Rear sprocket coupling/ rubber dampers – check and renewal

1 Remove the rear wheel (see Chapter 7).
Caution: Do not lay the wheel down on the disc, as it could become warped. Lay the wheel on wooden blocks so that the disc is off the ground.
2 Lift the sprocket coupling away from the wheel, leaving the rubber dampers in position in the wheel **(see illustration)**. Check the coupling for cracks or any obvious signs of damage. Also check the sprocket studs for looseness, wear or damage. The coupling was modified on all 1998 models after isolated cases of cracking were found on machines tested under extreme conditions. If the modification has been carried out, there will be no logo marking on the new coupling (the old component is marked FUSHIMA).
3 Lift the rubber damper segments from the wheel and check them for cracks, hardening and general deterioration **(see illustration)**. Renew the rubber dampers as a set, if necessary.
4 Checking and renewal procedures for the sprocket coupling bearing are described in Chapter 7.
5 Installation is the reverse of removal.
6 Install the rear wheel (see Chapter 7).

Chapter 7
Brakes, wheels and tyres

Contents

Brake fluid level check . . . . .see *Daily (pre-ride) checks*	Rear brake caliper – removal, overhaul and installation 7
Brake hoses and unions – inspection and renewal 10	Rear brake disc – inspection, removal and installation 8
Brake light switches – check and renew . . . . .see Chapter 9	Rear brake master cylinder – removal, overhaul and installation . . . 9
Brake pad wear check . . . . .see Chapter 1	Rear brake pads – renewal 6
Brake system bleeding 11	Rear wheel – removal and installation 15
Brake system check . . . . .see Chapter 1	Tyres – general information and fitting 17
Front brake calipers – removal, overhaul and installation 3	Tyres – pressure, tread depth and condition see *Daily (pre-ride) checks*
Front brake discs – inspection, removal and installation 4	Wheel bearings – check . . . . .see Chapter 1
Front brake master cylinder – removal, overhaul and installation . . . 5	Wheel bearings – removal, inspection and installation 16
Front brake pads – renewal 2	Wheels – alignment check 13
Front wheel – removal and installation 14	Wheels – general check . . . . .see Chapter 1
General information 1	Wheels – inspection and repair 12

Degrees of difficulty

Easy, suitable for novice with little experience	Fairly easy, suitable for beginner with some experience	Fairly difficult, suitable for competent DIY mechanic 	Difficult, suitable for experienced DIY mechanic	Very difficult, suitable for expert DIY or professional

Specifications

Brakes
Brake fluid type	DOT 4
Brake pad minimum thickness	0.5 mm
Front caliper bore ID	
Upper bore	30.20 mm
Lower bore	27.00 mm
Front disc thickness	
Standard	5.0 mm
Service limit	4.5 mm
Front disc maximum runout	0.1 mm
Front master cylinder bore ID	14.0 mm
Rear caliper bore ID	38.2 mm
Rear disc thickness	
Standard	5.0 mm
Service limit	4.5 mm
Rear disc maximum runout	0.1 mm
Rear master cylinder bore ID	12.7 mm

Wheels
Rim size	
Front	17 x MT3.50
Rear	17 x MT6.00
Wheel runout (max)	
Axial (side-to-side)	0.5 mm
Radial (out-of-round)	1.0 mm

Tyres
Tyre pressures	see *Daily (pre-ride) checks*
Tyre sizes*	
Front	120/70-ZR17 (58W)
Rear	190/50-ZR17 (73W)

*Refer to the owners handbook, the tyre information label on the swingarm, or your dealer for approved tyre brands.

7•2 Brakes, wheels and tyres

Torque wrench settings

Front brake caliper mounting bolts	40 Nm
Brake hose banjo bolts	30 Nm
Front brake disc bolts	18 Nm
Front brake master cylinder clamp bolts	13 Nm
Rear brake caliper mounting bolts	40 Nm
Rear brake disc bolts	18 Nm
Rear brake master cylinder mounting bolts	23 Nm
Bleed valves	6 Nm
Front wheel axle	72 Nm
Front axle clamp bolt	23 Nm
Rear axle nut	150 Nm

1 General information

All models are fitted with cast alloy wheels designed for tubeless tyres only. Both front and rear brakes are hydraulically-operated disc brakes.

The front brakes are twin opposed-piston calipers. The rear brake is a single opposed-piston caliper.

Caution: Disc brake components rarely require disassembly. Do not disassemble components unless absolutely necessary. If a hydraulic brake line is loosened, the entire system must be disassembled, drained, cleaned and then properly filled and bled upon reassembly. Do not use solvents on internal brake components. Solvents will cause the seals to swell and distort. Use only clean brake fluid or denatured alcohol for cleaning. Use care when working with brake fluid as it can injure your eyes and it will damage painted surfaces and plastic parts.

2 Front brake pads – renewal

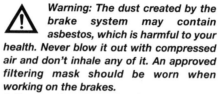

Warning: The dust created by the brake system may contain asbestos, which is harmful to your health. Never blow it out with compressed air and don't inhale any of it. An approved filtering mask should be worn when working on the brakes.

1 If new pads are being installed, displace the calipers from the discs (see Section 3) – this makes it easier to push the pistons back into the caliper to allow for the extra thickness of new pads. Otherwise, the pads can be removed with the caliper in situ.

2 Remove the retaining clip from each end of the pad retaining pin **(see illustration)**. Withdraw the pad pin, noting how it fits through the pad spring **(see illustration)**. Remove the pad spring, noting which way round it fits **(see illustration)**. Withdraw the pads from the top of the caliper, noting how they fit **(see illustration)**. If required, remove the anti-chatter shim from the back of each pad, noting how it fits **(see illustration 2.9a)**.

3 Inspect the surface of each pad for contamination and check that the friction material has not worn beyond its service limit (see Chapter 1, Section 8). If either pad is worn down to or beyond the service limit wear indicator, is fouled with oil or grease, or is heavily scored or damaged by dirt and debris, both pads in each caliper must be renewed. Note that it is not possible to degrease the friction material; if the pads are contaminated in any way, new ones must be fitted.

4 If the pads are in good condition clean them carefully, using a fine wire brush which is completely free of oil and grease, to remove all traces of road dirt and corrosion. Using a pointed instrument, clean out the groove in the friction material and dig out any embedded particles of foreign matter. Remove any areas of glazing using emery cloth. Spray the caliper area with a dedicated brake cleaner to remove any dust and remove and traces of corrosion which might cause sticking of the caliper/pad operation.

5 Check the condition of the brake disc (see Section 4).

6 Remove all traces of corrosion from the pad pin. Check it for signs of damage and renew it if necessary.

7 If new pads are being installed, push the pistons as far back into the caliper as possible, using hand pressure or a piece of wood as leverage **(see illustration)**. This will

2.2a Remove the retaining clips . . .

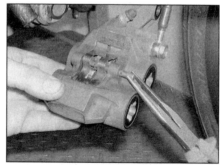

2.2b . . . then withdraw the pad pin

2.2c Remove the pad spring . . .

2.2d . . . and lift the pads out of the caliper

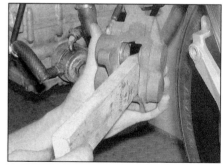

2.7 Push the pistons back into the caliper – using a piece of wood as shown works well

2.9a Fit the shim onto the back of the pad if removed

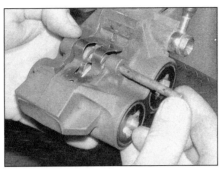

2.9b Insert the pad pin . . .

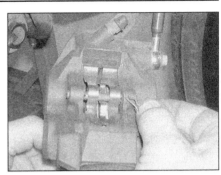

2.9c . . . and secure it with the clips

displace brake fluid back into the reservoir, so it may be necessary to remove the fluid reservoir cap, plate and diaphragm and siphon out some fluid (depending on how much fluid was in there in the first place and how far the pistons have to be pushed in). If the pistons are difficult to push back, attach a length of clear hose to the bleed valve and place the open end in a suitable container, then open the valve and try again. Take great care not to draw any air into the system. If in doubt, bleed the brakes afterwards (see Section 11).

8 Smear the backs of the pads and the shank of the pad pin(s) with copper-based grease, making sure that none gets on the front or sides of the pads.

9 If removed, fit the shim onto the back of each pad **(see illustration)**. Insert the pads into the caliper so that the friction material of each pad faces the disc **(see illustration 2.2d)**. Fit the pad spring onto the pads, making sure the arrow points up, in the direction of normal disc rotation **(see illustration 2.2c)**. Insert the pad retaining pin through the hole in the outer pad, then press down on the pad spring and push the pin through the spring and the hole in the inner pad **(see illustration)**. Install the retaining clips, using new ones if necessary **(see illustration)**.

10 If displaced, install the calipers (see Section 3). Top up the master cylinder reservoir if necessary (see *Daily (pre-ride) checks*), and refit the diaphragm, plate and reservoir cap.

11 Operate the brake lever several times to bring the pads into contact with the disc. Check the operation of the brake before riding the motorcycle.

3 Front brake calipers – removal, overhaul and installation

⚠ **Warning:** *If a caliper indicates the need for an overhaul (usually due to leaking fluid or sticky operation), all old brake fluid should be flushed from the system. Also, the dust created by the brake system may contain asbestos, which is harmful to your health. Never blow it out with compressed air and do not inhale any of it. An approved filtering mask should be worn when working on the brakes. Do not, under any circumstances, use petroleum-based solvents to clean brake parts. Use the specified clean brake fluid, dedicated brake cleaner or denatured alcohol only, as described.*

Removal

1 If the calipers are being overhauled, remove the brake pads (see Section 2). If the calipers are just being displaced or removed, the pads can be left in place.

2 Unscrew the nut and remove the bolt securing the brake hose holder to the mudguard and caliper mounting bracket **(see illustration)**.

3 If the calipers are just being displaced and not completely removed or overhauled, do not disconnect the brake hose. If the calipers are being completely removed or overhauled, remove the brake hose banjo bolt and detach the hose, noting its alignment with the caliper **(see illustration)**. Plug the hose end, or wrap a plastic bag tightly around it, to minimise fluid loss and prevent dirt entering the system. Discard the sealing washers, as new ones must be used on installation. **Note:** *If you are planning to overhaul the caliper and do not have a source of compressed air to blow out the pistons, just loosen the banjo bolt at this stage and retighten it lightly. The bike's hydraulic system can then be used to force the pistons out of the body once the pads have been removed. Disconnect the hose once the pistons have been sufficiently displaced.*

4 Unscrew the caliper mounting bolts and slide the caliper off the disc **(see illustrations)**.

3.2 Unscrew the nut and withdraw the bolt securing the brake hose clamp to the mudguard

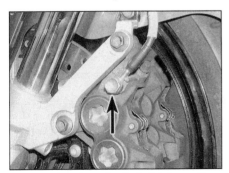

3.3 Brake hose banjo bolt (arrowed)

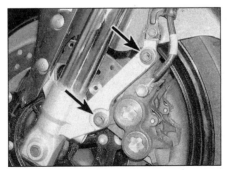

3.4a Unscrew the bolts (arrowed) . . .

3.4b . . . and slide the caliper off the disc

7•4 Brakes, wheels and tyres

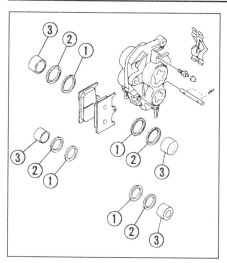

3.5 Front brake caliper components

1 Dust seal 2 Piston seal 3 Piston

Overhaul

5 Clean the exterior of the caliper with denatured alcohol or brake system cleaner **(see illustration)**.

6 Using a flat piece of wood, block the pistons on one side of the caliper in their bores and displace the opposite pistons either by pumping them out by operating the front brake lever, or by forcing them out using compressed air **(see illustration)**. Remove the seals (see Step 7) from the bore of the displaced pistons, then reinstall the pistons and block them, using the wood. Now displace the pistons from the other side using the same method. Remove the wood and all the pistons. Mark each piston head and caliper body with a felt marker to ensure that the pistons can be matched to their original bores on reassembly. If the compressed air method is used, direct the air into the fluid inlet to force the pistons out of the body. Use only low pressure to ease the pistons out and make sure both pistons on the side concerned are displaced at the same time. If the air pressure is too high and the pistons are forced out, the caliper and/or pistons may be damaged.

⚠ **Warning: Never place your fingers in front of the pistons in an attempt to catch or protect them when applying compressed air, as serious injury could result.**

Caution: Do not try to remove the pistons by levering them out, or by using pliers or any other grips. Do not attempt to remove the caliper bore plugs on the outside of the caliper.

7 Using a wooden or plastic tool, remove the dust seals from the caliper bores **(see illustration)**. Discard them, as new ones must be used on installation. If a metal tool is being used, take great care not to damage the caliper bores.

8 Remove and discard the piston seals in the same way.

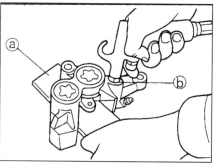

3.6 Using a block of wood (a) and compressed air (B) to remove the pistons

9 Clean the pistons and bores with clean brake fluid of the specified type. If compressed air is available, use it to dry the parts thoroughly (make sure it is filtered and unlubricated).

Caution: Do not, under any circumstances, use a petroleum-based solvent to clean brake parts.

10 Inspect the caliper bores and pistons for signs of corrosion, nicks and burrs and loss of plating. If surface defects are present, the caliper assembly must be renewed. If the necessary measuring equipment is available, compare the dimensions of the caliper bores to those specified at the beginning of the Chapter, and install a new caliper if necessary. If the caliper is in bad shape the master cylinder should also be checked.

11 Lubricate the new piston seals with clean brake fluid and install them in their grooves in the caliper bores. Note that two sizes of bore and piston are used (see Specifications), and care must therefore be taken to ensure that the correct size seals are fitted to the correct bores. The same applies when fitting the new dust seals and pistons.

12 Lubricate the new dust seals with clean brake fluid and install them in their grooves in the caliper bores.

13 Lubricate the pistons with clean brake fluid and install them, closed-end first, into the caliper bores. Using your thumbs, push the pistons all the way in, making sure they enter the bore squarely.

Installation

14 If necessary, push the pistons a little way

3.15a Install the caliper mounting bolts . . .

3.7 Use a plastic or wooden tool (such as a pencil) to remove the seals

back into the caliper using hand pressure or a piece of wood as leverage **(see illustration 2.7)**. Slide the caliper onto the brake disc, making sure the pads sit squarely each side of the disc if they weren't removed **(see illustration 3.4b)**.

15 Install the caliper mounting bolts and tighten them to the torque setting specified at the beginning of the Chapter **(see illustrations)**.

16 If removed, connect the brake hose to the caliper, using new sealing washers on each side of the fitting. Align the hose as noted on removal. Tighten the banjo bolt to the torque setting specified at the beginning of the Chapter **(see illustration 3.3)**. Fit the brake hose holder onto the bracket and tighten the bolt **(see illustration 3.2)**.

17 If removed, install the brake pads (see Section 2).

18 Top up the master cylinder reservoir with DOT 4 brake fluid (see *Daily (pre-ride) checks*) and bleed the hydraulic system as described in Section 11. Check that there are no leaks and thoroughly test the operation of the brake before riding the motorcycle.

4 Front brake discs – inspection, removal and installation

Inspection

1 Visually inspect the surface of the disc for score marks and other damage. Light scratches are normal after use and will not

3.15b . . . and tighten them to the specified torque

Brakes, wheels and tyres 7•5

4.2 Set up a dial indicator with the probe contacting the brake disc, then rotate the wheel to check for runout

4.3 Using a micrometer to measure disc thickness

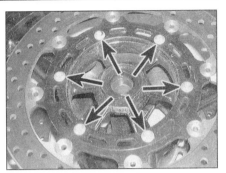

4.5 Unscrew the bolts (arrowed) and remove the disc

affect brake operation, but deep grooves and heavy score marks will reduce braking efficiency and accelerate pad wear. If a disc is badly grooved it must be machined, or a new one fitted.

2 To check disc runout, position the bike on an auxiliary stand and support it so that the front wheel is raised off the ground. Mount a dial gauge to a fork leg, with the plunger on the gauge touching the surface of the disc about 10 mm (1/2 in) from the outer edge (see illustration). Rotate the wheel and watch the gauge needle, comparing the reading with the limit listed in the Specifications at the beginning of the Chapter. If the runout is greater than the service limit, check the wheel bearings for play (see Chapter 1). If the bearings are worn, install new ones (see Section 16) and repeat this check. If the disc runout is still excessive, a new disc will have to be fitted, although machining by an engineer may be possible.

3 The disc must not be machined or allowed to wear down to a thickness less than the service limit as listed in this Chapter's Specifications. The thickness of the disc can be checked with a micrometer or any other measuring tool (see illustration). If the thickness of the disc is less than the service limit, a new one must be fitted.

Removal

4 Remove the wheel (see Section 14).
Caution: Do not lay the wheel down and allow it to rest on the disc – the disc could become warped. Set the wheel on wood blocks so the disc does not support the weight of the wheel.

5 Mark the relationship of the disc to the wheel, so that it can be installed in the same position. Unscrew the disc retaining bolts, loosening them evenly and a little at a time in a criss-cross pattern to avoid distorting the disc, then remove the disc from the wheel (see illustration).

Installation

6 Before installing the disc, make sure there is no dirt or corrosion where the disc seats on the hub, particularly right in the angle of the seat, as this will not allow the disc to sit flat when it is bolted down, and it will appear to be warped when checked or when using the front brake.

7 Mount the disc on the wheel with its marked side facing out, aligning the previously applied matchmarks (if you are reinstalling the original disc).

8 Clean the threads of the disc mounting bolts, then apply a suitable non-permanent thread locking compound. Install the bolts and tighten them evenly and a little at a time in a criss-cross pattern to the torque setting specified at the beginning of the Chapter (see illustration 4.5). Clean off all grease from the brake disc using acetone or brake system cleaner. If a new brake disc has been installed, remove any protective coating from its working surfaces.

9 Install the front wheel (see Section 14).
10 Operate the brake lever several times to bring the pads into contact with the disc.

Check the operation of the brakes carefully before riding the bike.

5 Front brake master cylinder – removal, overhaul and installation

1 If the master cylinder is leaking fluid, or if the lever does not produce a firm feel when the brake is applied, and bleeding the brakes does not help (see Section 11), and the hydraulic hoses and the brake calipers are all in good condition, then master cylinder overhaul is recommended.

2 Before disassembling the master cylinder, read through the entire procedure and make sure that you have the correct rebuild kit. Also, you will need some new DOT 4 brake fluid, some clean rags and internal circlip pliers. **Note:** *To prevent damage to the paint from spilled brake fluid, always cover the fuel tank when working on the master cylinder.*

Caution: Disassembly, overhaul and reassembly of the brake master cylinder must be done in a spotlessly clean work area to avoid contamination and possible failure of the brake hydraulic system components.

Removal

3 Remove the reservoir cap clamp and partially unscrew the cap (see illustration).
4 Disconnect the brake light switch wiring connectors (see illustration).
5 If the master cylinder is being overhauled, remove the front brake lever (see Chapter 6). If it is just being displaced it can remain in situ.
6 If the master cylinder is being completely removed or overhauled, unscrew the brake hose banjo bolt and separate the hoses from the master cylinder, noting their alignment (see illustration). Discard the sealing washers, as they must be replaced with new ones. Wrap the ends of the hoses in a clean rag and suspend them in an upright position or bend them down carefully and place the open ends in a clean container. The objective is to prevent excessive loss of brake fluid, fluid spills and system contamination. If the master cylinder is just being displaced and

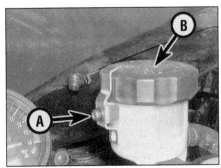

5.3 Undo the screw (A) and remove the clamp, then unscrew the cap (B)

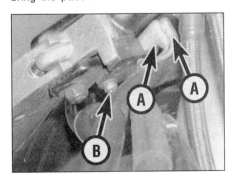

5.4 Brake light switch wiring connectors (A) and mounting screw (B)

7•6 Brakes, wheels and tyres

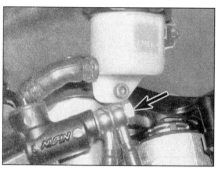

5.6 Brake hose banjo bolt (arrowed)

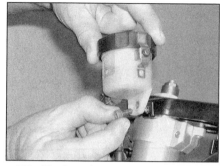

5.7a Displace the reservoir and drain it

5.7b Release the clamp (arrowed) and detach the hose from its union

5.8 Front brake master cylinder clamp bolts (arrowed)

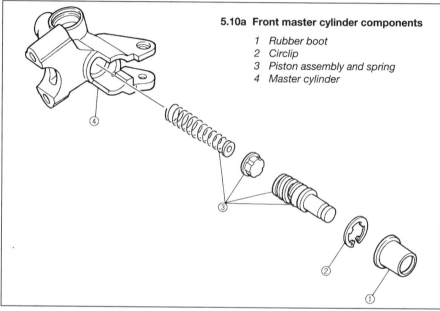

5.10a Front master cylinder components

1. Rubber boot
2. Circlip
3. Piston assembly and spring
4. Master cylinder

not completely removed or overhauled, do not disconnect the brake hoses.

7 Unscrew the bolt securing the reservoir to the top yoke, then remove the reservoir cap and lift off the diaphragm plate and the rubber diaphragm **(see illustration)**. Drain the brake fluid from the reservoir into a suitable container. Release the clamp securing the reservoir hose to the union on the master cylinder, then detach the hose **(see illustration)**. Wipe any remaining fluid out of the reservoir with a clean rag.

8 Unscrew the master cylinder clamp bolts, then lift the master cylinder away from the handlebar **(see illustration)**.

9 If required, remove the screw securing the brake light switch and remove the switch **(see illustration 5.4)**.

Caution: Do not tip the master cylinder upside down, or brake fluid will run out.

Overhaul

10 Carefully remove the dust boot from the master cylinder **(see illustrations)**.

11 Using circlip pliers, remove the circlip and slide out the piston assembly and the spring, noting how they fit **(see illustration)**. If they are difficult to remove, apply low pressure compressed air to the fluid outlet. Lay the parts out in the proper order to prevent confusion during reassembly.

12 Remove the reservoir hose union rubber cap, then remove the circlip and detach the union from the master cylinder. Discard the O-ring, as a new one must be used. Inspect the reservoir hose for cracks or splits and renew if necessary.

13 Clean all parts with clean brake fluid. If compressed air is available, use it to dry the parts thoroughly (make sure it is filtered and unlubricated).

Caution: Do not, under any circumstances, use a petroleum-based solvent to clean brake parts.

14 Check the master cylinder bore for corrosion, scratches, nicks and score marks. If the necessary measuring equipment is available, compare the diameter of the bore to that given in the Specifications Section of this

5.10b Remove the rubber boot from the end of the master cylinder piston . . .

Chapter. If damage or wear is evident, the master cylinder must be replaced with a new one. If the master cylinder is in poor condition, then the calipers should be checked as well. Check that the fluid inlet and outlet ports in the master cylinder are clear.

15 The dust boot, circlip, piston assembly and spring are included in the rebuild kit. Use all of the new parts, regardless of the apparent condition of the old ones. Fit them according to the layout of the old piston assembly **(see illustration 5.10a)**.

5.11 . . . then depress the piston and remove the circlip using a pair of internal circlip pliers

Brakes, wheels and tyres 7•7

16 Fit the spring into the master cylinder with its narrow end facing out.
17 Lubricate the piston assembly with clean brake fluid. Fit the assembly into the master cylinder, making sure it is the correct way round. Make sure the lips on the cup do not turn inside out when they are slipped into the bore. Depress the piston and install the new circlip, making sure that it locates in the groove **(see illustration 5.11)**.
18 Install the rubber dust boot, making sure the lip is seated correctly in the groove **(see illustration 5.10b)**.
19 Fit a new O-ring onto the reservoir hose union, then press the union into the master cylinder and secure it with the circlip. Fit the rubber cap over the circlip.
20 Inspect the reservoir cap rubber diaphragm and renew it if it is damaged or deteriorated.

Installation

21 If removed, install the brake light switch.
22 Attach the master cylinder to the handlebar and fit the clamp with its 'UP' mark facing up, aligning the top mating surfaces of the clamp with the punchmark on the handlebar **(see illustration 5.8)**. Tighten the upper bolt first, then the lower bolt to the torque setting specified at the beginning of the Chapter.
23 If detached, connect the brake hoses to the master cylinder, using new sealing washers on each side of the unions, and aligning the hoses as noted on removal. Tighten the banjo bolt to the torque setting specified at the beginning of this Chapter **(see illustration 5.6)**.
24 If removed, install the brake lever (see Chapter 6).
25 Mount the reservoir and tighten the bolt securely **(see illustration 5.7a)**. Connect the reservoir hose to the union and secure it with the clamp **(see illustration 5.7b)**.
26 Connect the brake light switch wiring connectors **(see illustration 5.4)**.
27 Fill the fluid reservoir with new DOT 4 brake fluid as described in *Daily (pre-ride) checks*. Refer to Section 11 of this Chapter and bleed the air from the system.
28 Fit the rubber diaphragm, making sure it is correctly seated, the diaphragm plate and the cap onto the master cylinder reservoir, then fit the cap clamp **(see illustrations)**.
29 Check the operation of the front brake before riding the motorcycle.

6 Rear brake pads – renewal

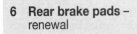

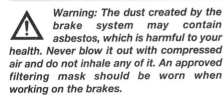

Warning: *The dust created by the brake system may contain asbestos, which is harmful to your health. Never blow it out with compressed air and do not inhale any of it. An approved filtering mask should be worn when working on the brakes.*

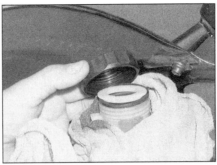

5.28a Fit the diaphragm, plate and cap . . .

1 Displace the rear brake caliper (see Section 7) – there is no need to disconnect the hose.
2 Remove the pad pin retaining clips, then withdraw the pad pins from the caliper using a suitable pair of pliers and noting how they fit through the pad spring **(see illustrations)**. Remove the pad spring, noting how it fits **(see illustration)**. Withdraw the pads from the caliper body **(see illustration)**. If required, remove the anti-chatter shim from the back of each pad, noting how it fits **(see illustration 6.8)**.
3 Inspect the surface of each pad for contamination and check that the friction material has not worn beyond its service limit (see Chapter 1, Section 8). If either pad is worn down to or beyond the service limit wear indicator, is fouled with oil or grease, or is heavily scored or damaged by dirt and debris, both pads must be renewed as a set. Note

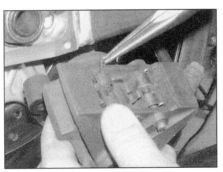

6.2a Remove the retaining clips . . .

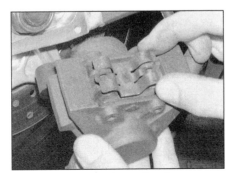

6.2c Remove the pad spring . . .

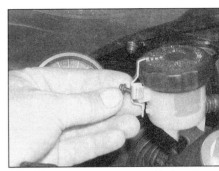

5.28b . . . and secure them with the clamp

that it is not possible to degrease the friction material; if the pads are contaminated in any way, new ones must be fitted.
4 If the pads are in good condition, clean them carefully using a fine wire brush which is completely free of oil and grease, to remove all traces of road dirt and corrosion. Using a pointed instrument, dig out any embedded particles of foreign matter. Any areas of glazing may be removed using emery cloth. Spray with a dedicated brake cleaner to remove any dust.
5 Check the condition of the brake disc (see Section 8).
6 Remove all traces of corrosion from the pad pins. Check them for signs of damage and renew it if necessary.
7 If new pads are being installed, push the pistons as far back into the caliper as possible using hand pressure or a piece of wood as leverage **(see illustration)**. This will displace

6.2b . . . then withdraw the pad pins

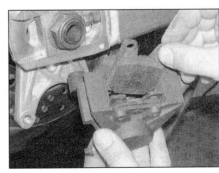

6.2d . . . and lift the pads out of the caliper

7•8 Brakes, wheels and tyres

6.7 Push the pistons back into the caliper – using a piece of wood as shown works well

6.8 Fit the shim onto the back of the pad if removed

6.9a Insert the pad pins . . .

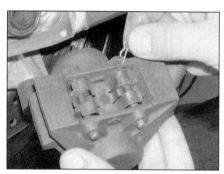

6.9b . . . and secure them with the clips

brake fluid back into the reservoir, so it may be necessary to remove the fluid reservoir cap, plate and diaphragm, and siphon out some fluid (depending on how much fluid was in there in the first place and how far the pistons have to be pushed in). If the pistons are difficult to push back, attach a length of clear hose to the bleed valve and place the open end in a suitable container, then open the valve and try again. Take great care not to draw any air into the system. If in doubt, bleed the brakes afterwards (see Section 11).

8 Smear the backs of the pads and the shank of each pad pin with copper-based grease, making sure that none gets on the front or sides of the pads. Fit the anti-chatter shim onto the back of each pad **(see illustration)**.

9 Insert the pads into the caliper so that the friction material of each pad is facing the disc **(see illustration 6.2d)**. Locate the pad spring, then insert one of the pad pins, making sure it passes correctly through the spring **(see illustrations 6.2c and b)**. Insert the other pad pin, pressing down on the spring end so that the pin fits over it **(see illustration)**. Fit the retaining clips **(see illustration)**.

10 Install the caliper (see Section 7). Top up the master cylinder reservoir if necessary (see *Daily (pre-ride) checks*).

11 Operate the brake pedal several times to bring the pads into contact with the disc. Check the operation of the brake before riding the motorcycle.

7 Rear brake caliper – removal, overhaul and installation

Warning: *If a caliper indicates the need for an overhaul (usually due to leaking fluid or sticky*

operation), all old brake fluid should be flushed from the system. Also, the dust created by the brake system may contain asbestos, which is harmful to your health. Never blow it out with compressed air and do not inhale any of it. An approved filtering mask should be worn when working on the brakes. Do not, under any circumstances, use petroleum-based solvents to clean brake parts. Use the specified clean brake fluid, dedicated brake cleaner or denatured alcohol only, as described.

Removal

1 If the caliper is just being displaced and not completely removed or overhauled, do not disconnect the brake hose. If the caliper is being overhauled, unscrew the brake hose banjo bolt and separate the brake hose from the caliper, noting its alignment **(see illustration)**. Plug the hose end, or wrap a plastic bag tightly around it, to minimise fluid loss and prevent dirt entering the system. Discard the two sealing washers, as they must be replaced with new ones. **Note:** *If you are planning to overhaul the caliper and do not have a source of compressed air to blow out the pistons, just loosen the banjo bolt at this stage and retighten it lightly. The bike's hydraulic system can then be used to force the pistons out of the body once the pads have been removed. Disconnect the hose once the pistons have been sufficiently displaced.*

2 Unscrew the caliper mounting bolts, and slide the caliper off the disc **(see illustrations)**.

3 If the caliper is being overhauled, remove the brake pads (see Section 6). If the caliper is only being displaced or removed, the pads can be left in place.

Overhaul

4 Clean the exterior of the caliper with denatured alcohol or brake system cleaner **(see illustration)**.

5 Using a flat piece of wood, block the piston on one side of the caliper in its bore and displace the opposite piston either by pumping it out by operating the brake pedal, or by forcing it out using compressed air **(see**

7.1 Brake hose banjo bolt (arrowed)

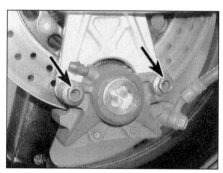

7.2a Unscrew the bolts (arrowed) . . .

7.2b . . . and slide the caliper off the disc

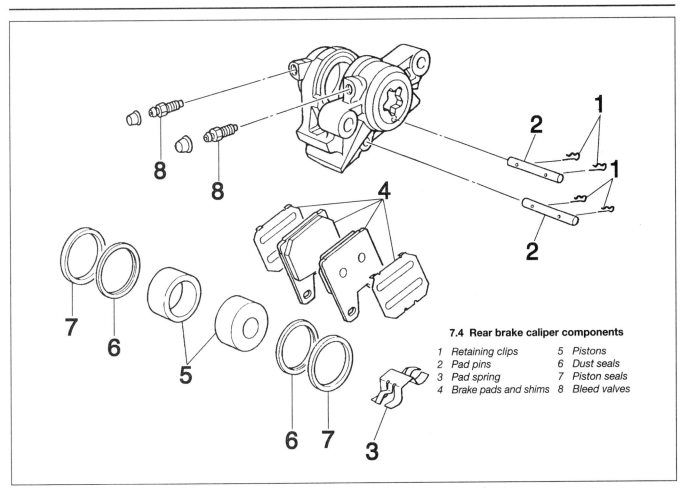

7.4 Rear brake caliper components

1 Retaining clips
2 Pad pins
3 Pad spring
4 Brake pads and shims
5 Pistons
6 Dust seals
7 Piston seals
8 Bleed valves

illustration). Remove the seals (see Step 6) from the bore of the displaced piston, then reinstall the piston and block it, using the wood. Now displace the piston from the other side using the same method. Remove the wood and both pistons. Mark each piston head and caliper body with a felt marker to ensure that the pistons can be matched to their original bores on reassembly. If the compressed air method is used, direct the air into the fluid inlet to force the piston out of the body. Use only low pressure to ease the piston out. If the air pressure is too high and the piston is forced out, the caliper and/or piston may be damaged.

⚠ **Warning: Never place your fingers in front of the piston in an attempt to catch or protect it when applying compressed air, as serious injury could result.**

Caution: Do not try to remove the pistons by levering them out, or by using pliers or any other grips. Do not attempt to remove the caliper bore plugs on the outside of the caliper.

6 Using a wooden or plastic tool, remove the dust seals from the caliper bores **(see illustration 3.7)**. Discard them, as new ones must be used on installation. If a metal tool is being used, take great care not to damage the caliper bores.

7 Remove and discard the piston seals in the same way.

8 Clean the pistons and bores with clean brake fluid of the specified type. If compressed air is available, use it to dry the parts thoroughly (make sure it is filtered and unlubricated).

Caution: Do not, under any circumstances, use a petroleum-based solvent to clean brake parts.

9 Inspect the caliper bores and pistons for signs of corrosion, nicks and burrs and loss of plating. If surface defects are present, the

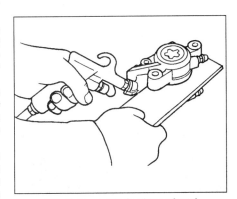

7.5 Using a block of wood and compressed air to remove the pistons

caliper assembly must be renewed. If the necessary measuring equipment is available, compare the dimensions of the caliper bores to those specified at the beginning of the Chapter, and install a new caliper if necessary. If the caliper is in bad shape the master cylinder should also be checked.

10 Lubricate the new piston seals with clean brake fluid and install them in their grooves in the caliper bores.

11 Lubricate the new dust seals with clean brake fluid and install them in their grooves in the caliper bores.

12 Lubricate the pistons with clean brake fluid and install them, closed-end first, into the caliper bores. Using your thumbs, push the pistons all the way in, making sure they enter the bore squarely.

Installation

13 If necessary, push the pistons a little way back into the caliper using hand pressure or a piece of wood as leverage **(see illustration 6.7)**. If removed, install the brake pads (see Section 6).

14 Slide the caliper onto the brake disc, making sure the pads sit squarely each side of the disc **(see illustration 7.2b)**.

15 Install the caliper mounting bolts, and

7•10 Brakes, wheels and tyres

7.15a Install the caliper bolts . . .

7.15b . . . and tighten them to the specified torque

8.3 Rear brake disc bolts (arrowed)

tighten them to the torque setting specified at the beginning of the Chapter **(see illustrations)**.

16 If detached, connect the brake hose to the caliper, making sure it is correctly routed around the torque arm. Use new sealing washers on each side of the union, and align the hose as noted on removal. Tighten the banjo bolt to the torque setting specified at the beginning of the Chapter **(see illustration 7.1)**. Top up the master cylinder reservoir with DOT 4 brake fluid (see *Daily (pre-ride) checks*) and bleed the hydraulic system as described in Section 11.

17 Check that there are no leaks and thoroughly test the operation of the brake before riding the motorcycle.

8 Rear brake disc –
inspection, removal
and installation

Inspection

1 Refer to Section 4 of this Chapter, noting that the dial gauge should be attached to the swingarm.

Removal

2 Remove the rear wheel (see Section 15).
3 Mark the relationship of the disc to the wheel, so that it can be installed in the same position. Unscrew the disc retaining bolts, loosening them evenly and a little at a time in a criss-cross pattern to avoid distorting the disc, and remove the disc **(see illustration)**.

Installation

4 Before installing the disc, make sure there is no dirt or corrosion where the disc seats on the hub, particularly right in the angle of the seat, as this will not allow the disc to sit flat when it is bolted down, and it will appear to be warped when checked or when using the rear brake.
5 Install the disc on the wheel with its marked side facing out, aligning the previously applied matchmarks (if you are reinstalling the original disc).
6 Clean the threads of the disc mounting bolts, then apply a suitable non-permanent thread locking compound. Install the bolts and tighten them evenly and a little at a time in a criss-cross pattern to the torque setting specified at the beginning of the Chapter **(see illustration 8.3)**. Clean off all grease from the brake disc, using acetone or brake system cleaner. If a new brake disc has been installed, remove any protective coating from its working surfaces.
7 Install the rear wheel (see Section 15).
8 Operate the brake pedal several times to bring the pads into contact with the disc. Check the operation of the brake carefully before riding the motorcycle.

9 Rear brake master cylinder –
removal, overhaul
and installation

1 If the master cylinder is leaking fluid, or if the lever does not produce a firm feel when the brake is applied, and bleeding the brakes does not help (see Section 11), and the hydraulic hose and brake caliper are in good condition, then master cylinder overhaul is recommended.

2 Before disassembling the master cylinder, read through the entire procedure and make sure that you have the correct rebuild kit. Also, you will need some new DOT 4 brake fluid, some clean rags and internal circlip pliers. **Note:** *To prevent damage to the paint from spilled brake fluid, always cover the surrounding components when working on the master cylinder.*

Caution: *Disassembly, overhaul and reassembly of the brake master cylinder must be done in a spotlessly clean work area to avoid contamination and possible failure of the brake hydraulic system components.*

Removal

3 Remove the screw securing the master cylinder fluid reservoir to the frame, then remove the reservoir cap clamp (on 1998 and 1999 models), cap, diaphragm plate and diaphragm, and pour the fluid into a container **(see illustration)**. Release the clamp securing the reservoir hose to the union on the master cylinder and detach the hose, being prepared to catch any residual fluid **(see illustration)**.

4 Unscrew the brake hose banjo bolt and separate the brake hose from the master cylinder, noting its alignment **(see illustration)**. Discard the two sealing washers, as they must be replaced with new ones.

9.3a Remove the screw (arrowed) and drain the reservoir . . .

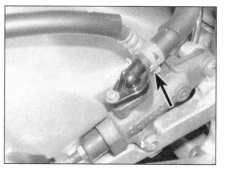

9.3b . . . and detach the hose (arrowed) from the master cylinder

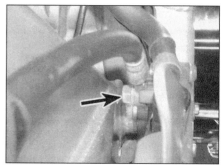

9.4 Brake hose banjo bolt (arrowed)

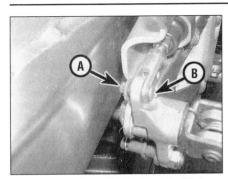

9.5 Remove the split pin (A) and withdraw the clevis pin (B)

9.6 Master cylinder mounting bolts (arrowed)

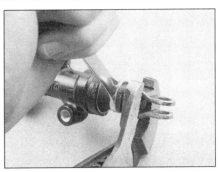

9.7 Hold the clevis and slacken the locknut

Wrap the end of the hose in a clean rag and suspend the hose in an upright position or bend it down carefully and place the open end in a clean container. The objective is to prevent excessive loss of brake fluid, fluid spills and system contamination.

5 Remove the split pin and washer from the clevis pin securing the brake pedal to the master cylinder pushrod **(see illustration)**. Withdraw the clevis pin and separate the pedal from the pushrod. Discard the split pin, as a new one must be used.

6 Unscrew the two bolts securing the master cylinder to the footrest bracket and remove the master cylinder **(see illustration)**.

Overhaul

7 If required, mark the position of the clevis locknut on the pushrod, then slacken the locknut and thread the clevis and nut off the pushrod **(see illustration)**.

8 Dislodge the rubber dust boot from the base of the master cylinder to reveal the pushrod retaining circlip **(see illustration)**.

9 Depress the pushrod and, using circlip pliers, remove the circlip **(see illustration)**. Slide out the pushrod, piston assembly and spring **(see illustration)**. If they are difficult to remove, apply low pressure compressed air to the fluid outlet. Lay the parts out in the proper order to prevent confusion during reassembly.

10 Clean all of the parts with clean brake fluid.

Caution: Do not, under any circumstances, use a petroleum-based solvent to clean brake parts. If compressed air is available, use it to dry the parts thoroughly (make sure it is filtered and unlubricated).

11 Check the master cylinder bore for corrosion, scratches, nicks and score marks. If the necessary measuring equipment is available, compare the diameter of the bore to that given in the Specifications Section of this Chapter. If damage is evident, the master cylinder must be replaced with a new one. If the master cylinder is in poor condition, then the caliper should be checked as well.

12 Inspect the reservoir hose for cracks or splits and replace it with a new one, if necessary. If required, remove the union screw and pull the union from the master cylinder. Discard the O-ring as a new one

9.8 Remove the dust boot from the bottom of the master cylinder

must be used.

13 The dust boot, circlip, piston assembly and spring are included in the rebuild kit. Use all of the new parts, regardless of the

9.9a Depress the pushrod and remove the circlip

apparent condition of the old ones. Fit them according to the layout of the old piston assembly **(see illustration 9.9b)**.

14 Fit the spring in the master cylinder.

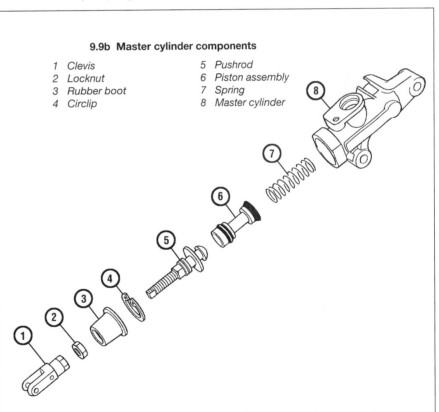

9.9b Master cylinder components

1 Clevis
2 Locknut
3 Rubber boot
4 Circlip
5 Pushrod
6 Piston assembly
7 Spring
8 Master cylinder

7•12 Brakes, wheels and tyres

9.24a Fit the diaphragm, plate and cap...

9.24b ...then mount the reservoir, not forgetting the cap clamp where fitted

15 Lubricate the piston assembly with clean brake fluid. Fit the assembly into the master cylinder, making sure it is the correct way round. Make sure the lips on the cup do not turn inside out when they are slipped into the bore.
16 Install and depress the pushrod, then fit a new circlip, making sure it is properly seated in the groove **(see illustration 9.9a)**.
17 Install the rubber dust boot, making sure the lip is seated properly in the groove **(see illustration 9.8)**.
18 If removed, fit a new reservoir hose union O-ring, then push the union into the master cylinder and secure it with the screw.
19 If removed, thread the clevis locknut and the clevis onto the master cylinder pushrod end. Position the clevis as noted on removal, then tighten the locknut securely **(see illustration 9.7)**.

Installation

20 Fit the master cylinder onto the footrest bracket and tighten its mounting bolts to the torque setting specified at the beginning of the Chapter **(see illustration 9.6)**.
21 Align the brake pedal with the master cylinder pushrod clevis, then slide in the clevis pin and secure it using a new split pin, not forgetting the washer **(see illustration 9.5)**.
22 Connect the brake hose banjo bolt to the master cylinder, using a new sealing washer on each side of the banjo union. Ensure that the hose is positioned so that it butts against the lug and tighten the banjo bolt to the specified torque setting **(see illustration 9.4)**.

23 Locate the reservoir in position but do not fix it to the bracket. Ensure that the hose is correctly routed, then connect it to the union on the master cylinder and secure it with the clamp **(see illustration 9.3b)**. Check that the hose is secure and clamped at the reservoir end as well. If the clamps have weakened, use new ones.
24 Fill the fluid reservoir with new DOT 4 brake fluid (see *Daily (pre-ride) checks*) and bleed the system following the procedure in Section 11. On completion install the reservoir and tighten the screw, not forgetting the cap clamp on 1998 and 1999 models **(see illustrations)**.
25 Check the operation of the brake carefully before riding the motorcycle.

10 Brake hoses and unions – inspection and renewal

Inspection

1 Brake hose condition should be checked regularly and the hoses renewed at the specified interval (see Chapter 1).
2 Twist and flex the rubber hoses while looking for cracks, bulges and seeping fluid **(see illustration)**. Check extra carefully around the areas where the hoses connect with the banjo fittings, as these are common areas for hose failure.
3 Check the banjo union fittings connected to the brake hoses. If the union fittings are rusted, scratched or cracked, fit new hoses.

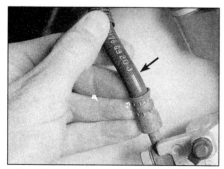

10.2 Flex the brake hoses (arrowed) and check for cracks, bulges and leaking fluid

10.4 Remove the banjo bolt and separate the hose from the caliper; there is a sealing washer on each side of the fitting

Renewal

4 The brake hoses have banjo union fittings on each end. Cover the surrounding area with plenty of rags and unscrew the banjo bolt at each end of the hose or pipe, noting its alignment **(see illustration)**. Free the hose from any clips or guides and remove it, noting its routing. Discard the sealing washers on the hose unions.
5 Position the new hose, making sure it is not twisted or otherwise strained, and abut the tab on the hose union with the lug on the component casting, where present. Otherwise, align the hose as noted on removal. Install the hose banjo bolts, using new sealing washers on both sides of the unions. Tighten the banjo bolts to the torque setting specified at the beginning of this Chapter. Make sure the hoses are correctly aligned and routed clear of all moving components.
6 Flush the old brake fluid from the system, refill with new DOT 4 brake fluid (see *Daily (pre-ride) checks*) and bleed the air from the system (see Section 11). Check the operation of the brakes carefully before riding the motorcycle.

11 Brake system bleeding

1 Bleeding the brakes is simply the process of removing all the air bubbles from the brake fluid reservoirs, the hoses and the brake calipers. Bleeding is necessary whenever a brake system hydraulic connection is loosened, when a component or hose is renewed, or when the master cylinder or caliper is overhauled. Leaks in the system may also allow air to enter, but leaking brake fluid will reveal their presence and warn you of the need for repair.
2 To bleed the brakes, you will need some new DOT 4 brake fluid, a length of clear vinyl or plastic tubing, a small container partially filled with clean brake fluid, some rags and a ring spanner to fit the brake caliper bleed valves.
3 Cover the fuel tank and other painted components to prevent damage in the event that brake fluid is spilled.
4 When bleeding the brakes, look for any 'high spots' in the system where air bubbles could become trapped and eliminate them by displacing the component and positioning it where bubbles are less likely to get trapped. Wiggling or tapping any such spots will help to dislodge any bubbles already trapped.
5 Remove the reservoir cap clamp where fitted, cap, diaphragm plate and diaphragm **(see illustrations 5.28b and a (Front) or 9.24b and a (Rear))**. Slowly pump the brake lever or pedal a few times, until no air bubbles can be seen floating up from the holes in the bottom of the reservoir. This bleeds the air from the master cylinder end of the line. Loosely refit the reservoir cap.

Brakes, wheels and tyres 7•13

11.6a Front brake caliper bleed valve (arrowed)

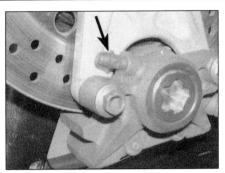

11.6b Rear brake caliper bleed valve (arrowed) – note that there are two, one on each side of the caliper

11.6c To bleed the brakes, you need a spanner, a short section of clear tubing, and a clear container half-filled with brake fluid

6 Pull the dust cap off the bleed valve (see illustrations). Attach one end of the clear vinyl or plastic tubing to the bleed valve and submerge the other end in the brake fluid in the container (see illustration).

7 Remove the reservoir cap and check the fluid level. Do not allow the fluid level to drop below the lower mark during the bleeding process.

8 Carefully pump the brake lever or pedal three or four times and hold it in (Front) or down (Rear) while opening the caliper bleed valve. When the valve is opened, brake fluid will flow out of the caliper into the clear tubing, and the lever will move toward the handlebar, or the pedal will move down.

9 Retighten the bleed valve, then release the brake lever or pedal gradually. Repeat the process until no air bubbles are visible in the brake fluid leaving the caliper, or if the fluid is being changed until new fluid is coming out, and the lever or pedal is firm when applied. On completion, disconnect the bleeding equipment, then tighten the bleed valve to the torque setting specified at the beginning of the Chapter, and install the dust cap. If bleeding the front brake, go on to bleed air from the other caliper. Note that on the rear caliper there is a bleed valve for each side of the caliper, and each must be bled.

 Old brake fluid is invariably much darker in colour than new fluid, making it easy to see when all old fluid has been expelled from the system.

10 Install the diaphragm and cap assembly, wipe up any spilled brake fluid and check the entire system for leaks.

 If it is not possible to produce a firm feel to the lever or pedal, the fluid may be aerated. Let the brake fluid in the system stabilise for a few hours and then repeat the procedure when the tiny bubbles in the system have settled out.

12 Wheels – inspection and repair

1 In order to carry out a proper inspection of the wheels, it is necessary to support the bike upright so that the wheel being inspected is raised off the ground. Position the motorcycle on an auxiliary stand. Clean the wheels thoroughly to remove mud and dirt that may interfere with the inspection procedure or mask defects. Make a general check of the wheels (see Chapter 1) and tyres (see *Daily (pre-ride) checks*).

2 Attach a dial gauge to the fork or the swingarm and position its tip against the side of the rim (see illustration). Spin the wheel slowly and check the axial (side-to-side) runout of the rim. In order to accurately check radial (out of round) runout with the dial gauge, the wheel would have to be removed from the machine, and the tyre from the wheel. With the axle clamped in a vice and the dial gauge positioned on the top of the rim, the wheel can be rotated to check the runout.

3 An easier, though slightly less accurate, method is to attach a stiff wire pointer to the fork or the swingarm and position the end a fraction of an inch from the wheel (where the wheel and tyre join). If the wheel is true, the distance from the pointer to the rim will be

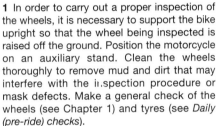

12.2 Check the wheel for radial (out-of-round) runout (A) and axial (side-to-side) runout (B)

constant as the wheel is rotated. **Note:** *If wheel runout is excessive, check the wheel bearings very carefully before renewing the wheel.*

4 The wheels should also be visually inspected for cracks, flat spots on the rim and other damage. Look very closely for dents in the area where the tyre bead contacts the rim. Dents in this area may prevent complete sealing of the tyre against the rim, which leads to deflation of the tyre over a period of time. If damage is evident, or if runout in either direction is excessive, the wheel will have to be replaced with a new one. Never attempt to repair a damaged cast alloy wheel.

13 Wheels – alignment check

1 Misalignment of the wheels, which may be due to a cocked rear wheel or a bent frame or fork yokes, can cause strange and possibly serious handling problems. If the frame or yokes are at fault, repair by a frame specialist or replacement with new parts are the only alternatives.

2 To check the alignment you will need an assistant, a length of string or a perfectly straight piece of wood and a ruler. A plumb bob or other suitable weight will also be required.

3 In order to make a proper check of the wheels it is necessary to support the bike in an upright position, using an auxiliary stand. Measure the width of both tyres at their widest points. Subtract the smaller measurement from the larger measurement, then divide the difference by two. The result is the amount of offset that should exist between the front and rear tyres on both sides.

4 If a string is used, have your assistant hold one end of it about halfway between the floor and the rear axle, touching the rear sidewall of the tyre.

5 Run the other end of the string forward and pull it tight so that it is roughly parallel to the floor (see illustration). Slowly bring the string into contact with the front sidewall of the rear tyre, then turn the front wheel until it is parallel

7•14 Brakes, wheels and tyres

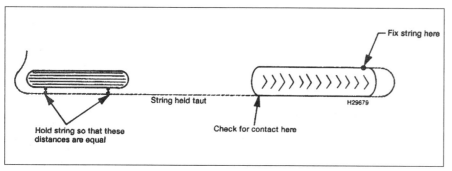

13.5 Wheel alignment check using string

with the string. Measure the distance from the front tyre sidewall to the string.

6 Repeat the procedure on the other side of the motorcycle. The distance from the front tyre sidewall to the string should be equal on both sides.

7 As previously mentioned, a perfectly straight length of wood or metal bar may be substituted for the string **(see illustration)**. The procedure is the same.

8 If the distance between the string and tyre is greater on one side, or if the rear wheel appears to be cocked, refer to Chapter 1, Section 1 and check that the chain adjuster markings coincide on each side of the swingarm.

9 If the front-to-back alignment is correct, the wheels still may be out of alignment vertically.

10 Using a plumb bob, or other suitable weight, and a length of string, check the rear wheel to make sure it is vertical. To do this, hold the string against the tyre upper sidewall and allow the weight to settle just off the floor. When the string touches both the upper and lower tyre sidewalls and is perfectly straight, the wheel is vertical. If it is not, place thin spacers under one leg of the stand until it is.

11 Once the rear wheel is vertical, check the front wheel in the same manner. If both wheels are not perfectly vertical, the frame and/or major suspension components are bent.

14 Front wheel – removal and installation

Removal

1 Remove the lower fairing (see Chapter 8). Put the motorcycle on an auxiliary stand and support it under the crankcase so that the front wheel is off the ground. Always make sure the motorcycle is properly supported.

2 Displace the front brake calipers (see Section 3). Support the calipers with a piece of wire or a bungee cord so that no strain is placed on the hydraulic hoses. There is no need to disconnect the hoses from the calipers. **Note:** *Do not operate the front brake lever with the calipers removed.*

3 Slacken the axle clamp bolt on the bottom of the right-hand fork, then unscrew the axle, using an automotive 19 mm sump plug tool or a bolt and nut arrangement, as shown, if a large hex key is not available **(see illustrations)**.

4 Support the wheel, then withdraw the axle from the right-hand side and remove the wheel **(see illustration)**. Use a drift to drive out the axle if required.

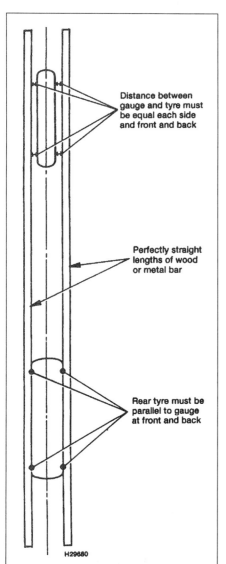

13.7 Wheel alignment check using a straight edge

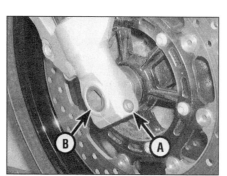

14.3a Slacken the clamp bolt (A) and unscrew the axle (B)

14.3c ... then unscrew the axle using a spanner on the inner nut

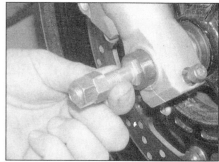

14.3b Use a nut and bolt arrangement as shown if a suitable hex key isn't available ...

14.4 Withdraw the axle and remove the wheel

Brakes, wheels and tyres 7•15

14.5a Remove the spacer and seal cover . . .

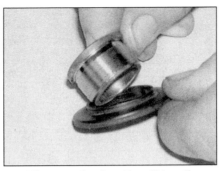

14.5b . . . noting how they fit together

14.10 Tighten the axle to the specified torque, then tighten the clamp bolt

5 Remove the spacer and seal cover from each side of the wheel, noting how they fit together **(see illustrations)**.
Caution: Don't lay the wheel down and allow it to rest on a disc – the disc could become warped. Set the wheel on wood blocks so the disc does not support the weight of the wheel, or keep it upright.
6 Check the axle for straightness by rolling it on a flat surface such as a piece of plate glass (first wipe off all old grease and remove any corrosion using fine emery cloth). If the equipment is available, place the axle in V-blocks and check for runout using a dial gauge. If the axle is bent, replace it with a new one.
7 Check the condition of the wheel bearings (see Section 16).

Installation

8 Apply lithium-based grease to the lips of the bearing seals. Fit the spacers into the seal covers if separated, then fit them into each side of the wheel, with the shouldered end of the spacers facing out **(see illustrations 14.5b and a)**.
9 Manoeuvre the wheel into position. Apply a thin coat of lithium-based grease to the axle.
10 Lift the wheel into place between the forks, making sure the spacers remain in position, and slide the axle in from the right-hand side **(see illustration 14.4)**. Tighten the axle to the torque setting specified at the beginning of the Chapter **(see illustration)**. Now tighten the axle clamp bolt on the bottom of the right-hand fork to the specified torque setting **(see illustration 14.3a)**.
11 Install the brake calipers, making sure the pads sit squarely on each side of the discs (see Section 3).
12 Take the bike off its auxiliary stand and install the lower fairing (see Chapter 8).
13 Apply the front brake a few times to bring the pads back into contact with the discs. Take the bike off the stand, apply the front brake and pump the front forks a few times to settle all components in position.
14 Check for correct operation of the front brake before riding the motorcycle.

15 Rear wheel – removal and installation

Removal

1 Support the motorcycle securely in an upright position using an auxiliary stand.
2 Displace the rear brake caliper (see Section 7). Make sure no strain is placed on the hydraulic hose. There is no need to disconnect the hose from the caliper. **Note:** *Do not operate the brake pedal with the caliper removed.*
3 Slacken the adjuster locknut on each side of the swingarm, then turn the adjusters in to provide some slack in the chain **(see illustration)**. Unscrew the axle nut and remove the washer **(see illustration)**. Remove the adjustment position marker from the right-hand side of the swingarm **(see illustration)**.
4 Support the wheel then withdraw the axle along with the left-hand adjustment position marker and lower the wheel to the ground **(see illustration)**. Note how the caliper bracket locates between the wheel and the swingarm.
5 Disengage the chain from the sprocket and remove the wheel from between the swingarm ends **(see illustration)**. Remove the caliper

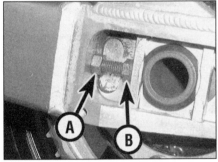

15.3a Slacken the locknut (A) on each side and turn each adjuster (B) in

15.3b Unscrew the axle nut and remove the washer . . .

15.5a Disengage the chain and remove the wheel

15.3c . . . and the position marker

15.4 Withdraw the axle and lower the wheel to the ground

7•16 Brakes, wheels and tyres

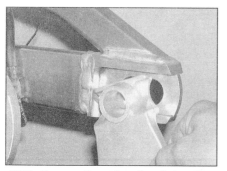

15.5b Remove the caliper bracket, noting how it locates in the swingarm

15.6a Remove the seal cover . . .

15.6b . . . and spacer from each side, noting how they fit

bracket, noting how it fits **(see illustration)**.
Caution: *Do not lay the wheel down and allow it to rest on the disc or the sprocket – they could become warped. Set the wheel on wood blocks so the disc or the sprocket does not support the weight of the wheel. Do not operate the brake pedal with the wheel removed.*

6 Remove the seal cover and spacer from each side of the wheel, noting which fits where **(see illustrations)**.

7 Check the axle for straightness by rolling it on a flat surface such as a piece of plate glass (first wipe off all old grease and remove any corrosion using fine emery cloth). If the equipment is available, place the axle in V-blocks and check for runout using a dial gauge. If the axle is bent, replace it with a new one.

8 Check the condition of the bearing seals and wheel bearings (see Section 16).

Installation

9 Apply a thin coat of lithium-based grease to the lips of each grease seal, and also to the spacers and the axle. Slide the left-hand adjustment position marker onto the axle, making sure it is the correct way around.

10 Fit the seal cover and spacer into each side of the wheel **(see illustrations 15.6b and a)**. Locate the brake caliper bracket on the swingarm **(see illustration 15.5b)**.

11 Manoeuvre the wheel into place between the ends of the swingarm and engage the drive chain with the sprocket **(see illustration 15.5a)**.

12 Lift the wheel into position, making sure the caliper bracket and the spacers remain in place, and slide the axle, with the adjustment marker, through from the left-hand side **(see illustration 15.4)**. Make sure it passes through the caliper bracket. Align the flats on the axle head between the raised sections on the adjustment marker, and make sure the raised sections are vertical, not horizontal **(see illustration)**. Check that everything is correctly aligned, then fit the right-hand adjustment position marker with the side with the tapered edges facing in **(see illustration 15.3c)**. Fit the washer and the axle nut **(see illustration 15.3b)**.

13 Adjust the chain slack as described in Chapter 1, then tighten the axle nut to the torque setting specified at the beginning of the Chapter.

14 Install the brake caliper, making sure the pads sit squarely on each side of the disc (see Section 7).

15 Operate the brake pedal several times to bring the pads into contact with the disc. Check the operation of the rear brake carefully before riding the bike.

16 Wheel bearings – removal, inspection and installation

Front wheel bearings

Note: *Always renew the wheel bearings in pairs, never individually. Avoid using a high pressure cleaner on the wheel bearing area.*

1 Remove the wheel (see Section 14).
2 Set the wheel on blocks so as not to allow

15.12 Locate the flats on the axle head between the raised sections on the position marker

the weight of the wheel to rest on either brake disc.

3 Lever out the grease seal on each side of the wheel using a large flat-bladed screwdriver, taking care not to damage the rim **(see illustration)**. Discard the seals if they are damaged or deteriorated.

> **HAYNES HINT** *Position a piece of wood against the wheel to prevent the screwdriver shaft damaging it when levering the grease seal out.*

4 Using a metal rod (preferably a brass drift punch) inserted through the centre of the one bearing, tap evenly around the inner race of the other bearing to drive it from the hub **(see illustrations)**. The bearing spacer will also come out.

16.3 Lever out the grease seal on each side

16.4a Knock out the bearings using a drift . . .

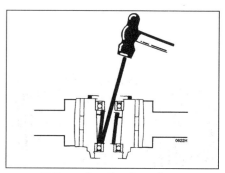

16.4b . . . locating it as shown

Brakes, wheels and tyres

16.9 A socket can be used to drive in the bearing

16.11a Fit the grease seal . . .

16.11b . . . and tap it into place

5 Lay the wheel on its other side so that the remaining bearing faces down. Drive the bearing out of the wheel using the same technique as above.

6 If the bearings are of the unsealed type or are only sealed on one side, clean them with a high flash-point solvent (one which won't leave any residue) and blow them dry with compressed air (do not let the bearings spin as you dry them). Apply a few drops of oil to the bearing. **Note:** *If the bearing is sealed on both sides do not attempt to clean it.*

 Refer to Tools and Workshop Tips (Section 5) in the Reference section for more information about bearings.

7 Hold the outer race of the bearing and rotate the inner race – if the bearing doesn't turn smoothly, has rough spots or is noisy, replace it with a new one.

8 If the bearing is good and can be re-used, wash it in solvent once again and dry it, then pack the bearing with lithium based grease.

9 Thoroughly clean the hub area of the wheel. Install a bearing into its recess in one side of the hub, with the marked or sealed side facing outwards. Using the old bearing (if new ones are being fitted), a bearing driver or a socket large enough to contact the outer race of the bearing, drive it in until it's completely seated **(see illustration)**.

10 Turn the wheel over and install the bearing spacer. Drive the other bearing into place as described above.

11 Apply a smear of lithium-based grease to the lips of one seal, then press it into the wheel, using a seal or bearing driver or a suitable socket **(see illustrations)**. Turn the wheel over and fit the other seal.

12 Clean off all grease from the brake discs using acetone or brake system cleaner, then install the wheel (see Section 14).

Rear wheel bearings

13 Remove the rear wheel (see Section 15). Lift the sprocket coupling out of the wheel, noting how it fits **(see illustration)**. A caged ball bearing is fitted in the right-hand side of the wheel and a needle bearing is fitted in the left-hand side.

14 Set the wheel on blocks with the disc facing up.

15 Lever out the grease seal on the righthand side of the wheel using a large flat-bladed screwdriver, taking care not to damage the rim of the hub **(see Haynes Hint above)** **(see illustration 16.3)**. Discard the seal if it is damaged or deteriorated. Remove the circlip securing the bearing.

16 Turn the wheel over and rest it on the blocks so that the disc is off the ground or bench. Remove the collar from inside the needle bearing **(see illustration)**. Using a metal rod (preferably a brass drift punch) inserted through the centre of the needle bearing, tap evenly around the inner race of the caged ball bearing to drive it from the hub **(see illustration and 16.4b)**. The bearing spacer will also come out.

17 Lay the wheel on its other side so that the needle bearing faces down. Drive the bearing out of the wheel using the same technique as above.

18 Refer to Steps 6 to 8 above and check the bearings. Note that removal of a needle bearing makes it unusable – a new one must be installed, whatever the condition of the old one.

19 Thoroughly clean the hub area of the wheel. First install the new needle bearing into its recess in the left-hand side of the hub. Needle bearings should be pressed or drawn, rather than driven, into position. In the absence of a press, a suitable drawbolt arrangement can be made up as described in *Tools and Workshop Tips* (Section 5) in the *Reference* section.

20 Install the caged ball bearing into its recess in the right-hand side of the hub, with the marked or sealed side facing outwards. Using the old bearing (if new ones are being fitted), a bearing driver or a socket large enough to contact the outer race of the bearing, drive it in squarely until it is completely seated **(see illustration 16.9)**. Fit the circlip, making sure it locates correctly in its groove.

21 Apply a smear of grease to the lips of the new grease seal, and press it into the right-hand side of the wheel, using a seal or bearing driver, a suitable socket or a flat piece of wood **(see illustration)**. Since the seals sit flush with the top surface of its housing, using a piece of wood, as shown, will automatically set it flush without the risk of setting it too deep and having to lever it out again.

16.13 Lift the sprocket coupling out of the wheel

16.16a Remove the collar . . .

16.16b . . . and drive out the ball bearing as described

7•18 Brakes, wheels and tyres

16.21 Using a piece of wood as shown automatically sets the seal flush with the rim

16.25 Lever out the grease seal

16.26 Drive the bearing out from the inside

16.28 A socket can be used to drive in the bearing

16.29a Press or drive the seal into the coupling . . .

16.29b . . . using a suitably-sized socket or bearing driver

22 Turn the wheel over and fit the spacer into the needle bearing **(see illustration 16.16a)**.
23 Clean off all grease from the brake disc using acetone or brake system cleaner. Install the sprocket coupling assembly **(see illustration 16.13)**. Install the wheel (see Section 15).

Sprocket coupling bearing

24 Remove the rear wheel (see Section 15). Lift the sprocket coupling out of the wheel, noting how it fits **(see illustration 16.13)**.
25 Using a large flat-bladed screwdriver lever out the grease seal from the outside of the coupling, taking care not to damage the rim of the coupling (**see Haynes Hint above**) **(see illustration)**. Discard the seal if it is damaged or deteriorated.
26 Support the coupling on blocks of wood and drive the bearing out from the inside using a bearing driver or socket **(see illustration)**.
27 Refer to Steps 6 to 8 above and check the bearings.
28 Thoroughly clean the bearing recess in the coupling then fit the bearing into the outside of the coupling, with the marked or sealed side facing out. Using the old bearing (if a new one is being fitted), a bearing driver or a socket large enough to contact the outer race of the bearing, drive it in until it is completely seated **(see illustration)**.
29 Apply a smear of grease to the lips of the new seal, and press it into the coupling, using a seal or bearing driver or a suitable socket **(see illustrations)**.
30 Check the sprocket coupling/rubber dampers (see Chapter 6, Section 17).
31 Clean off all grease from the brake disc using acetone or brake system cleaner. Fit the sprocket coupling into the wheel **(see illustration 16.13)**, then install the wheel (see Section 15).

17 Tyres – general information and fitting

General information

1 The wheels fitted on all models are designed to take tubeless tyres only. Tyre sizes are given in the Specifications at the beginning of this chapter.

2 Refer to *Daily (pre-ride) checks* at the beginning of this manual for tyre maintenance.

Fitting new tyres

3 When selecting new tyres, refer to the tyre information label on the swingarm and the tyre options listed in the Owner's Handbook. Ensure that front and rear tyre types are compatible, and of the correct size and speed rating; if necessary, seek advice from a Yamaha dealer or tyre fitting specialist **(see illustration)**.
4 It is recommended that tyres are fitted by a motorcycle tyre specialist and that this is not attempted in the home workshop. This is particularly relevant in the case of tubeless tyres because the force required to break the seal between the wheel rim and tyre bead is substantial, and is usually beyond the capabilities of an individual working with normal tyre levers. Additionally, the specialist will be able to balance the wheels after tyre fitting.
5 Note that punctured tubeless tyres can in some cases be repaired. Seek the advice of a Yamaha dealer or a motorcycle tyre fitting specialist concerning tyre repairs.

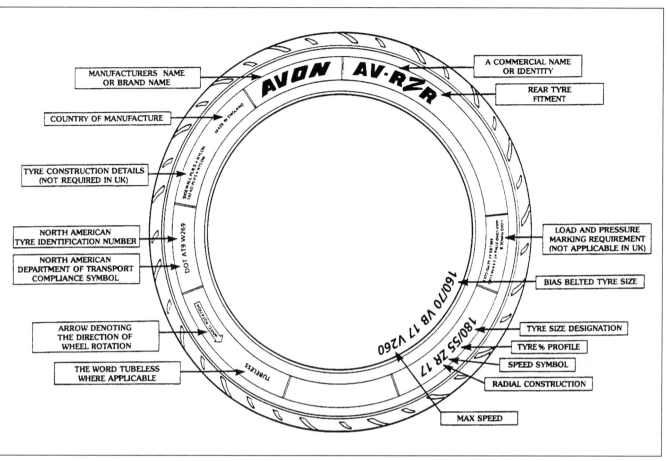

17.3 Common tyre sidewall markings

Notes

Chapter 8
Bodywork

Contents

Fairing and body panels – removal and installation 3
Front mudguard – removal and installation 6
General information 1
Rear view mirrors – removal and installation 4
Seats – removal and installation 2
Windshield – removal and installation 5

Degrees of difficulty

| **Easy,** suitable for novice with little experience | **Fairly easy,** suitable for beginner with some experience | **Fairly difficult,** suitable for competent DIY mechanic | **Difficult,** suitable for experienced DIY mechanic | **Very difficult,** suitable for expert DIY or professional |

1 General information

1 This Chapter covers the procedures necessary to remove and install the body parts. Since many service and repair operations on these motorcycles require the removal of the body parts, the procedures are grouped here and referred to from other Chapters.
2 In the case of damage to the body parts, it is usually necessary to remove the broken component and replace it with a new (or used) one. The material of which the body panels are composed does not lend itself to conventional repair techniques. There are, however, some shops that specialise in 'plastic welding', so it may be worthwhile seeking the advice of one of these specialists before consigning an expensive component to the bin.
3 When attempting to remove any body panel, first study it closely, noting any fasteners and associated fittings, to be sure of returning everything to its correct place on installation. In some cases the aid of an assistant will be required when removing panels, to help avoid the risk of damage to paintwork. Once the evident fasteners have been removed, try to withdraw the panel as described but DO NOT FORCE IT – if it will not release, check that all fasteners have been removed and try again. Where a panel engages another by means of tabs, be careful not to break the tab or its mating slot or to damage the paintwork. Remember that a few moments of patience at this stage will save you a lot of money in replacing broken fairing panels! To release trim clips, push the centre into the body, then draw the body out of the panel **(see illustration)**. To undo quick-release screws, turn them 90° anti-clockwise.
4 When installing a body panel, first study it closely, noting any fasteners and associated fittings removed with it, to be sure of returning

8•2 Bodywork

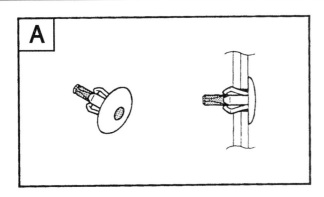
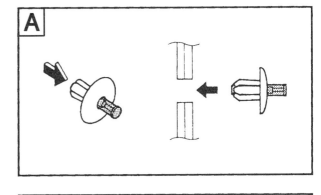
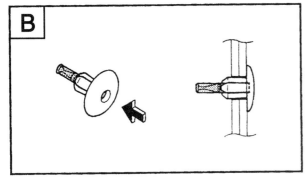

1.3 Trim clip removal

A Clip installed
B Clip with centre pin pushed in and ready for removal from panel

1.4 Trim clip installation

A Clip prepared for installation
B Clip being secured in panel

everything to its correct place. Check that all fasteners are in good condition, including all trim nuts or clips and damping/rubber mounts; any of these must be replaced with new ones, if faulty, before the panel is reassembled. Check also that all mounting brackets are straight, and repair or renew them if necessary before attempting to install the panel. Where assistance was required to remove a panel, make sure your assistant is on hand to install it. To install trim clips, first push the centre back out so that it protrudes from the top of the body **(see illustration)**. Fit the body into its panel, then push the centre in so that it is flush with the top of the body. To install quick-release screws, turn them 90° clockwise.

5 Tighten the fasteners securely, but be careful not to overtighten any of them or the panel may break (not always immediately) due to the uneven stress.

2 Seats – removal and installation

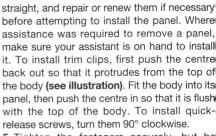

2.1a Turn the key and lift up the front of the seat . . .

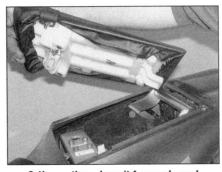

2.1b . . . then draw it forwards and remove it

Removal

1 To remove the passenger seat, insert the ignition key into the seat lock located on the left-hand side of the bike, and turn it anti-clockwise to unlock the seat **(see illustration)**. Lift up the front of the seat and draw it forwards, noting how the tab at the back locates **(see illustration)**.

2 To remove the rider's seat, pull up each rear corner of the seat to access the bolts that hold it, then unscrew the bolts and remove the seat, noting how the tab at the front locates under the tank bracket **(see illustrations)**.

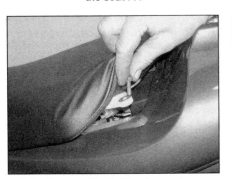

2.2a Remove the bolt under each rear corner . . .

2.2b . . . then lift the rear of the seat and draw it back

Bodywork 8•3

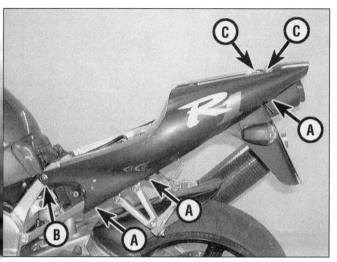

3.2a Release the trim clips (A) on each side, remove the bolt (B) on each side, and the two screws (C) . . .

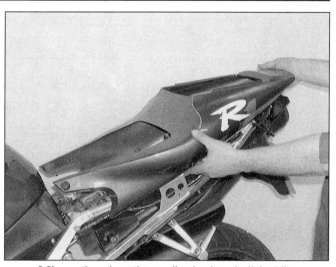

3.2b . . . then draw the cowling back and off the bike

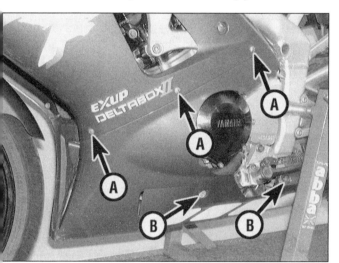

3.5 Undo the quick-screws (A) and the screws (B) . . .

3.6 . . . and remove the lower fairing

Installation

3 Installation is the reverse of removal. Push down on the front of the passenger seat to engage the latch.

3 Fairing and body panels – removal and installation

Seat cowling

1 Remove the seats (see Section 2).
2 On 1998 and 1999 models, release and remove the six trim clips on the underside of the seat cowling (three on each side) **(see illustration)**. Remove the bolt on each side at the front and the screws on the top at the back, then carefully draw the cowling back and off the bike **(see illustration)**.
3 On 2000-on models, release and remove the eight trim clips on the underside of the seat cowling (four on each side). Remove the bolt on each side at the front and the screw on the top at the back, then carefully draw the cowling back and off the bike.
4 Installation is the reverse of removal.

Lower fairing

5 Release the three quick-release screws securing each side of the lower fairing to each side panel, and undo the two screws securing each side of the lower fairing to the frame **(see illustration)**.
6 Carefully lower the panel and remove it, noting how it engages with the fairing side panels along its top edges **(see illustration)**.
7 Installation is the reverse of removal.

Fairing side panels

8 Disconnect the turn signal wiring connectors **(see illustration)**.
9 Remove the cockpit trim panel (see below).
10 Release the three quick-release screws securing the side panel to the lower fairing, the two quick-release screws and trim clip securing the side panel to the fairing, and undo the screw securing the side panel to the frame **(see illustrations)**.

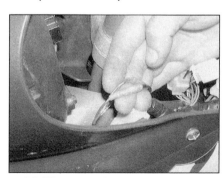

3.8 Disconnect the turn signal wiring connectors

8•4 Bodywork

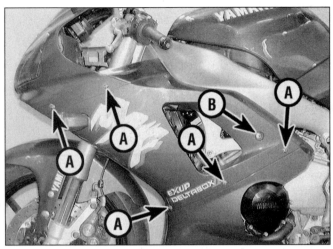

3.10a Undo the quick-screws (A), the screw (B) . . .

3.10b . . . and the trim clip (arrowed) . . .

3.11 . . . and remove the side panel

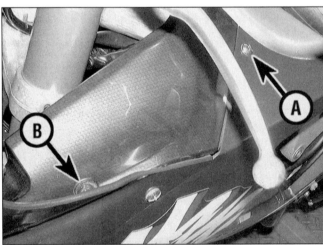

3.13a Undo the quick-release screw (A) and the screw (B) . . .

3.13b . . . and remove the trim panel

3.16 Disconnect the wiring connector

Bodywork 8•5

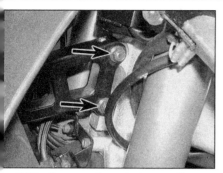

3.18a Unscrew the nuts (arrowed) . . .

3.18b . . . then remove the bolt assembly . . .

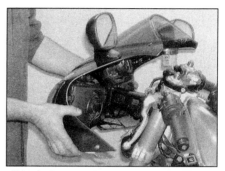

3.19 . . . and remove the fairing

11 Carefully draw the panel away, noting how it engages with the fairing along its top edge and the lower fairing along its bottom edge **(see illustration)**.
12 Installation is the reverse of removal. Make sure the panel locates correctly with the fairing and lower fairing.

Cockpit trim panels

13 Release the quick-release screw and undo the screw securing the trim panel and remove it, noting how it fits **(see illustrations)**.
14 Installation is the reverse of removal. Make sure the panel locates correctly with the fairing and fairing side panels.

Fairing

15 Remove the cockpit trim panels (see above).
16 Disconnect the instrument cluster and lighting wiring connector **(see illustration)**.
17 Release the two quick-release screws and the trim clip securing the fairing to each side panel **(see illustrations 3.10a and b)**.
18 Remove the two nuts securing the fairing stay to the steering head **(see illustration)**. Support the fairing and push the bolt assembly out to the right **(see illustration)**.
19 Carefully lift the fairing assembly up and draw it forwards and remove it – the headlight and instrument cluster come away with the fairing **(see illustration)**. Separate them if required (see Chapter 9).
20 Installation is the reverse of removal. Make sure the wiring connector is correctly and securely connected.

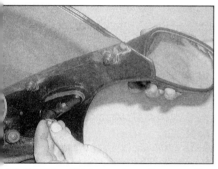

4.1 Unscrew the two nuts and remove the mirror

4 Rear view mirrors – removal and installation

Removal

1 Unscrew the nuts securing each mirror and remove the mirror along with its rubber insulator pad **(see illustration)**.

Installation

2 Installation is the reverse of removal.

5 Windshield – removal and installation

Removal

1 Remove the screws securing the windshield to the fairing and remove the windshield, noting how it fits **(see illustrations)**.

Installation

2 Installation is the reverse of removal. Do not overtighten the screws.

5.1a Undo the screws (arrowed) . . .

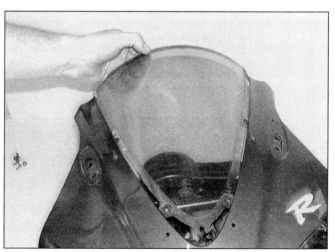

5.1b . . . and remove the windshield

6 Front mudguard – removal and installation

Removal

1 Unscrew the nut and remove the bolt securing the brake hose holder to the mudguard and caliper mounting bracket **(see illustration)**.

2 Remove the four screws securing the mudguard and draw it forward **(see illustration)**.

Installation

3 Installation is the reverse of removal.

6.1 Release the brake hose holder . . .

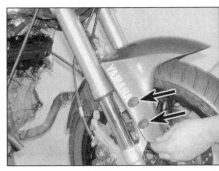

6.2 . . . then undo the two screws on each side (arrowed) and remove the mudguard

Chapter 9
Electrical system

Contents

Alternator rotor and stator – check, removal and installation 32
Battery – charging 4
Battery – removal, installation, inspection and maintenance 3
Brake light switches – check and renewal 14
Brake/tail light bulbs – renewal 9
Charging system – leakage and output test 31
Charging system testing – general information and precautions ... 30
Clutch switch – check and renewal 23
Electrical system – fault finding 2
Fuel pump and relay .... see Chapter 4
Fuses – check and renewal 5
General information 1
Handlebar switches – check 19
Handlebar switches – removal and installation 20
Headlight aim – check and adjustment .... see Chapter 1
Headlight assembly – removal and installation 8
Headlight bulbs and sidelight bulbs – renewal 7
Horn – check and renewal 25
Ignition (main) switch – check, removal and installation 18
Ignition system components .... see Chapter 5
Instrument and warning lights – renewal 17
Instrument cluster – removal and installation 15
Instruments – check and renewal 16
Lighting system – check 6
Neutral switch – check, removal and installation 21
Oil level sensor and relay – check, removal and installation 26
Regulator/rectifier – check and renewal 33
Relay assembly – check and renewal 24
Sidestand switch – check and renewal 22
Starter motor – disassembly, inspection and reassembly 29
Starter motor – removal and installation 28
Starter relay – check and renewal 27
Tail light assembly – removal and installation 10
Turn signal assemblies – removal and installation 13
Turn signal bulbs – renewal 12
Turn signal circuit – check 11

Degrees of difficulty

| Easy, suitable for novice with little experience | Fairly easy, suitable for beginner with some experience | Fairly difficult, suitable for competent DIY mechanic | Difficult, suitable for experienced DIY mechanic | Very difficult, suitable for expert DIY or professional |

Specifications

Battery
Capacity 12V, 10Ah
Type GT12-B4
Charge condition
   Fully charged 12.8V
   Half-charged 12.4V
   Discharged .... 12V or less
Charging time .... Until fully charged (12.8V) (see Section 4)
Current leakage .... 1 mA (max)

Alternator
Nominal output .... 12V, 26A @ 5000 rpm
Stator coil resistance .... 0.45 to 0.55 ohms @ 20°C

Regulator/rectifier
Regulated voltage output (no load) .... 14.1 to 14.9V @ 5000 rpm

Starter relay
Resistance .... 4.18 to 4.62 ohms @ 20°C

Starter motor
1998 and 1999 models
Brush length
  Standard .... 10 mm
  Service limit (min) .... 5 mm
Commutator diameter
  Standard .... 28 mm
  Service limit (min) .... 27 mm
Mica depth .... 0.7 mm
2000-on models
Brush length
  Standard .... 9.8 mm
  Service limit (min) .... 3.65 mm
Commutator diameter
  Standard .... 24.5 mm
  Service limit (min) .... 23.5 mm
Mica depth .... 1.5 mm

Fuses
1998 and 1999 models
Main 30A
Headlight 20A
Signal 20A
Ignition 15A
Cooling fan 7.5A
Backup fuse (odometer) 7.5A
2000-on models
Main 30A
Headlight 20A
Signal 20A
Ignition 15A
Cooling fan 10A
Backup fuse (odometer) 10A

Bulbs
Headlight .... 60/55W halogen x 2
Sidelight .... 5W x 2
Brake/tail light .... 21/5W x 2
Turn signal lights
  UK models .... 21W x 4
  US models .... 27/8W x 2 (front with running light), 27W x 2 (rear)
Instrument illumination lights (1998 and 1999 models only) .... 1.4W x 2

Torque wrench settings
Neutral switch .... 20 Nm
Oil level sensor bolts .... 10 Nm
Starter motor mounting bolts .... 7 Nm
Starter motor long bolts .... 5 Nm
Alternator stator bolts .... 10 Nm
Alternator rotor bolt .... 65 Nm
Alternator cover bolts .... 12 Nm

1 General information

All models have a 12 volt electrical system charged by a three-phase alternator with a separate regulator/rectifier.

The regulator maintains the charging system output within the specified range to prevent overcharging, and the rectifier converts the ac (alternating current) output of the alternator to dc (direct current) to power the lights and other components and to charge the battery. The alternator rotor is mounted on the left-hand end of the crankshaft.

The starter motor is mounted on the top of the crankcase. The starting system includes the motor, the battery, the relay and the various wires and switches. If the engine kill switch is in the 'RUN' position and the ignition (main) switch is ON, the starter relay allows the starter motor to operate only if the transmission is in neutral (neutral switch on) or, if the transmission is in gear, with the clutch lever pulled into the handlebar and the sidestand up.

Note: *Keep in mind that electrical parts, once purchased, cannot be returned. To avoid unnecessary expense, make very sure the faulty component has been positively identified before buying a replacement part.*

2 Electrical system – fault finding

Warning: To prevent the risk of short circuits, the ignition (main) switch must always be OFF and the battery negative (–ve) terminal should be disconnected before any of the bike's other electrical components are disturbed. Don't forget to reconnect the terminal securely once work is finished or if battery power is needed for circuit testing.

1 A typical electrical circuit consists of an electrical component, the switches, relays,

Electrical system 9•3

3.2a Disconnect the negative terminal first, then the positive (arrowed)

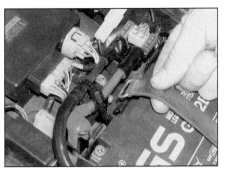

3.2b Unhook the strap . . .

3.2c . . . and remove the battery

etc. related to that component and the wiring and connectors that hook the component to both the battery and the frame. To aid in locating a problem in any electrical circuit, refer to *Wiring Diagrams* at the end of this Chapter.

2 Before tackling any troublesome electrical circuit, first study the wiring diagram (see end of Chapter) thoroughly to get a complete picture of what makes up that individual circuit. Trouble spots, for instance, can often be narrowed down by noting if other components related to that circuit are operating properly or not. If several components or circuits fail at one time, it may be that the fault lies in the fuse or earth (ground) connection, as several circuits are often routed through the same fuse and earth (ground) connections.

3 Electrical problems often stem from simple causes, such as loose or corroded connections or a blown fuse. Prior to any electrical fault finding, always visually check the condition of the fuse, wires and connections in the problem circuit. Intermittent failures can be especially frustrating, since you cannot always duplicate the failure when it is convenient to do a test. In such situations, it is good practice to clean all connections in the affected circuit, whether or not they appear to be good. All the connections and wires should also be wiggled to check for looseness, which can cause intermittent failure.

4 If testing instruments are going to be utilised, use the wiring diagram to plan where you will make the necessary connections in order to pinpoint the trouble spot accurately.

5 The basic tools needed for electrical fault finding include a battery and bulb test circuit, a continuity tester, a test light, and a jumper wire. A multimeter capable of reading volts, ohms and amps is also very useful as an alternative to the above, and is necessary for performing more extensive tests and checks.

 Refer to Fault Finding Equipment in the Reference section for details of how to use electrical test equipment.

3 Battery – removal, installation, inspection and maintenance

Caution: Be extremely careful when handling or working around the battery. The electrolyte is very caustic and an explosive gas (hydrogen) is given off when the battery is charging.

Removal and installation

1 Remove the rider's seat (see Chapter 8).
2 Unscrew the negative (–ve) terminal bolt first and disconnect the lead from the battery **(see illustration)**. Lift up the insulating cover to access the positive (+ve) terminal, then unscrew the bolt and disconnect the lead. Release the battery strap and remove the battery from the bike **(see illustrations)**.
3 On installation, clean the battery terminals and lead ends with a wire brush or knife and emery paper. Reconnect the leads, connecting the positive (+ve) terminal first.

 Battery corrosion can be kept to a minimum by applying a layer of petroleum jelly to the terminals after the cables have been connected.

4 Install the seat (see Chapter 8).

Inspection and maintenance

5 The battery fitted on all models is of the maintenance-free (sealed) type, therefore requiring no specific maintenance. However, the following checks should still be regularly performed.
6 Check the battery terminals and leads for tightness and corrosion. If corrosion is evident, unscrew the terminal bolts and disconnect the leads from the battery, disconnecting the negative (–ve) terminal first, and clean the terminals and lead ends with a wire brush or knife and emery paper. Reconnect the leads, connecting the negative (–ve) terminal last, and apply a thin coat of petroleum jelly to the connections to slow further corrosion.
7 The battery case should be kept clean to prevent current leakage, which can discharge the battery over a period of time (especially when it sits unused). Wash the outside of the case with a solution of baking soda and water. Rinse the battery thoroughly, then dry it.
8 Look for cracks in the case and renew the battery if any are found. If acid has been spilled on the frame or battery box, neutralise it with a baking soda and water solution, then dry it thoroughly.
9 If the motorcycle sits unused for long periods of time, disconnect the leads from the battery terminals, negative (–ve) terminal first. Refer to Section 4 and charge the battery once every month to six weeks.
10 The condition of the battery can be assessed by measuring the voltage present at the battery terminals, and comparing the figure against the chart **(see illustration)**. Connect the voltmeter positive (+ve) probe to the battery positive (+ve) terminal, and the negative (–ve) probe to the battery negative (–ve) terminal. When fully charged, there should be 12.8 volts (or more) present. If the voltage falls below 12.0 volts the battery must be removed, disconnecting the negative (–ve) terminal first, and recharged as described in Section 4.

4 Battery – charging

Caution: Be extremely careful when handling or working around the battery. The electrolyte is very caustic and an explosive gas (hydrogen) is given off when the battery is charging.

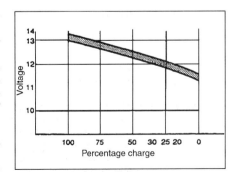

3.10 Measure the voltage to assess the condition of the battery from the chart

9•4 Electrical system

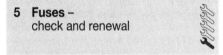

4.1 Measure the voltage to determine the charging time required

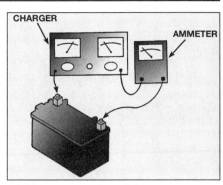

4.2 If the charger doesn't have an ammeter built in, connect one in series as shown. DO NOT connect the ammeter between the battery terminals, or it will be ruined

1 Remove the battery (see Section 3). If not already done, refer to Section 3, Step 10, and check the open circuit voltage of the battery. Refer to the chart **(see illustration)** and read off the charging time required according to the voltage reading taken.

2 Connect the charger to the battery, making sure that the positive (+ve) lead on the charger is connected to the positive (+ve) terminal on the battery, and the negative (–ve) lead is connected to the negative (–ve) terminal. The battery should be charged for the specified time, or until the voltage across the terminals reaches 12.8V (allow the battery to stabilise for 30 minutes after charging, before taking a voltage reading). Exceeding this can cause the battery to overheat, buckling the plates and rendering it useless. Few owners will have access to an expensive current controlled charger, so if a normal domestic charger is used check that after a possible initial peak, the charge rate falls to a safe level **(see illustration)**. If the battery becomes hot during charging **stop**. Further charging will cause damage. **Note:** *In emergencies the battery can be charged at a higher rate of around 3.0 amps for a period of 1 hour. However, this is not recommended and the low amp charge is by far the safer method of charging the battery.*

3 If the recharged battery discharges rapidly when left disconnected it is likely that an internal short caused by physical damage or sulphation has occurred. A new battery will be required. A sound item will tend to lose its charge at about 1% per day.

4 Install the battery (see Section 3).

5 If the motorcycle sits unused for long periods of time, charge the battery once every month to six weeks and leave it disconnected.

5 Fuses – check and renewal

1 The electrical system is protected by fuses of different ratings. All except the main fuse are housed in the fusebox, which is located under the rider's seat on the right-hand side **(see illustration)**. The main fuse is integral with the starter relay, which is located under the rider's seat on the left-hand side.

2 To access the fusebox fuses, remove the rider's seat (see Chapter 8) and unclip the fusebox lid **(see illustration)**. To access the main fuse, remove the rider's seat (see Chapter 8) and remove the fuse cover from the relay **(see illustration)**.

3 The fuses can be removed and checked visually. If you can't pull the fuse out with your fingertips, use a suitable pair of pliers. A blown fuse is easily identified by a break in the element **(see illustration)**, or can be tested for continuity using an ohmmeter or continuity tester – if there is no continuity, it has blown. Each fuse is clearly marked with its rating and must only be replaced by a fuse of the same rating. Spare fuses are housed in the fusebox, and a spare main fuse is housed in the starter relay. If a spare fuse is used, always replace it with a new one so that a spare of each rating is carried on the bike at all times.

⚠ **Warning:** *Never put in a fuse of a higher rating or bridge the terminals with any other substitute, however temporary it may be. Serious damage may be done to the circuit, or a fire may start.*

4 If a fuse blows, be sure to check the wiring circuit very carefully for evidence of a short-circuit. Look for bare wires and chafed, melted or burned insulation. If the fuse is renewed before the cause is located, the new fuse will blow immediately.

5 Occasionally a fuse will blow or cause an open-circuit for no obvious reason. Corrosion of the fuse ends and fusebox terminals may occur and cause poor fuse contact. If this happens, remove the corrosion with a wire brush or emery paper, then spray the fuse end and terminals with electrical contact cleaner.

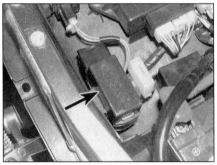

5.1 Fusebox (arrowed)

5.2a Unclip the lid to access the fusebox fuses

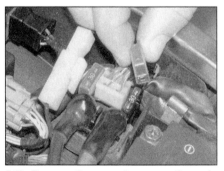

5.2b Remove the cover to access the main fuse

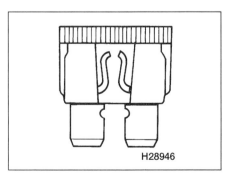

5.3 A blown fuse can be identified by a break in its element

Electrical system

6 Lighting system – check

1 The battery provides power for operation of the headlight, tail light, brake light, turn signals and instrument cluster lights. If none of the lights operate, always check battery voltage before proceeding. Low battery voltage indicates either a faulty battery or a defective charging system. Refer to Section 3 for battery checks and Sections 30 and 31 for charging system tests. Also, check the condition of the fuses (see Section 5). When checking for a blown filament in a bulb, it is advisable to back up a visual check with a continuity test of the filament as it is not always apparent that a bulb has blown. When testing for continuity, remember that on tail light and turn signal bulbs it is often the metal body of the bulb which is the ground or earth.

Headlights and relays

2 If the headlight fails to work, check the bulbs first (see Section 7), then the headlight fuse (see Section 5), and the wiring connectors, then check for battery voltage at the black/yellow (high beam) and/or black/blue (low beam) wire terminal on the supply side of the headlight wiring connector, with the ignition switch and light switch ON. If voltage is present, check for continuity between the black wire terminal and earth (ground). If there is no continuity, check the earth (ground) circuit for an open or poor connection.

3 If no voltage is indicated, check the wiring between the headlight, relays, light switches and the ignition switch, then check the switches themselves.

4 To check the headlight relays, remove the fairing (see Chapter 8), then disconnect the relevant relay wiring connector (see illustration). Make the checks on the relay side of the connector. Use a continuity tester (or a multimeter set to the resistance range) and a 12V battery with insulated jumper wires according to the relevant procedure below to check the relays.

5 On all 1998 and 1999 models two relays are fitted, one for the low beam circuit and another for the high beam circuit. To test, connect the meter between the red/yellow wire and either the black/yellow wire (high beam relay) or black/blue wire (low beam relay); no continuity should be shown. Leaving the meter in place, now use the jumper wires to connect the battery positive (+ve) terminal to either the white/yellow wire (high beam relay) or the white/green wire (low beam relay) and the battery negative (-ve) terminal to the black wire; the relay should now close and continuity (0 ohms) should be shown on the meter.

6 On European 2000-on models the primary relay (4 wires) works in conjuction with the lighting switch and the secondary relay (5 wires) switches the current between the high beam and low beam circuits. Test the primary relay by connecting the meter probes between the red/yellow and black/blue wires; no continuity should be shown. Now connect the battery positive (+ve) terminal to the blue/black wire and the negative (-ve) terminal to the black wire; continuity should be shown on the meter. Test the secondary relay low beam circuit by connecting the meter probes between the black/blue and black/green wires; continuity (0 ohms) should be shown. Test the secondary relay high beam circuit by connecting the meter probes between the black/blue and black/yellow wires; no continuity should be shown. Leaving the meter in place, now connect the battery positive (+ve) terminal to the yellow wire and the negative (-ve) terminal to the black wire; continuity (0 ohms) should be shown on the meter.

7 On US 2000-on models only one relay is fitted. To test the low beam circuit, connect the meter between the red/yellow and black/green wires; continuity (0 ohms) should be shown. To test the high beam circuit, connect the meter between the red/yellow and black/yellow wires; no continuity should be shown. Leaving the meter in place, now connect the positive (+ve) probe of the battery to the yellow wire and the negative (-ve) terminal to the black wire; continuity (0 ohms) should be shown on the meter.

Tail light

8 If the tail light fails to work, check the bulb(s) and the bulb terminals and wiring connector first (see Section 9), then the headlight fuse, then check for battery voltage at the blue/red wire terminal on the supply side of the tail light wiring connector, with the ignition switch and light switch ON. If voltage is present, check for continuity between the wiring connector terminals on the tail light side of the wiring connector and the corresponding terminals in the bulbholder. If voltage and continuity are present, check for continuity between the black wire terminal and earth (ground). If there is no continuity, check the earth (ground) circuit for an open or poor connection.

9 If no voltage is indicated, check the wiring between the tail light, the light switch and the ignition switch, then check the switches themselves.

6.4 Headlight relays (arrowed)

Sidelight (European models)

10 If the sidelight fails to work, check the bulb(s) and the bulb terminals and wiring connectors first (see Section 7), then the headlight fuse, then check for battery voltage at the blue/red wire terminal on the supply side of the sidelight wiring connector, with the ignition switch and light switch ON. If voltage is present, check for continuity between the wiring connector terminals on the sidelight side of the wiring connector and the corresponding terminals in the bulbholder. If voltage and continuity are present, check for continuity between the black wire terminal and earth (ground). If there is no continuity, check the earth (ground) circuit for an open or poor connection.

11 If no voltage is indicated, check the wiring between the sidelight, the light switch and the ignition switch, then check the switches themselves.

Brake light

12 If the brake light fails to work, check the bulb(s) and the bulb terminals and wiring connector first (see Section 14), then the signal fuse, then check for battery voltage at the yellow wire terminal on the supply side of the tail light wiring connector, with the ignition switch ON and the brake lever or pedal applied. If voltage is present, check for continuity between the black wire terminal and earth (ground). If there is no continuity, check the earth (ground) circuit for an open or poor connection.

13 If no voltage is indicated, check the brake light switches (see Section 14), then the wiring between the tail light and the switches.

Instrument and warning lights

14 See Section 17.

Turn signal lights

15 If one light fails to work, check the bulb and the bulb terminals first, then the wiring connectors. If none of the turn signals work, first check the signal fuse.

16 If the fuse is good, see Section 11 for the turn signal circuit check.

7 Headlight bulbs and sidelight bulbs – renewal

Note: *The headlight bulb is of the quartz-halogen type. Do not touch the bulb glass as skin acids will shorten the bulb's service life. If the bulb is accidentally touched, it should be wiped carefully when cold with a rag soaked in methylated spirit and dried before fitting.*

 Warning: Allow the bulb time to cool before removing it if the headlight has just been on.

Headlights

1 Disconnect the relevant wiring connector from the back of the headlight assembly and

7.1a Disconnect the wiring connector . . .

7.1b . . . and remove the dust cover

7.2a Release the clip . . .

7.2b . . . and remove the bulb

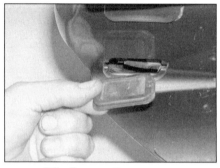

7.6a Remove the cover . . .

7.6b . . . then pull the bulbholder out of the headlight . . .

remove the rubber dust cover, noting how it fits **(see illustrations)**.
2 Release the bulb retaining clip, noting how it fits, then remove the bulb **(see illustrations)**.
3 Fit the new bulb, bearing in mind the information in the **Note** above. Make sure the tabs on the bulb fit correctly in the slots in the bulb housing, and secure it in position with the retaining clip.
4 Install the dust cover, making sure it is correctly seated and with the 'TOP' mark at the top, and connect the wiring connector **(see illustrations 7.1b and a)**.
5 Check the operation of the headlight.

> **HAYNES HiNT** *Always use a paper towel or dry cloth when handling new bulbs to prevent injury if the bulb should break, and to increase bulb life.*

Sidelights (European models)

6 The bulbholder is accessed via the underside of the fairing. Remove the rubber cover, then pull the bulbholder out of the base of the headlight **(see illustrations)**. Carefully pull the bulb out of the holder **(see illustration)**. Install the new bulb, then install the bulbholder and the rubber cover.
7 Check the operation of the sidelight.

8 Headlight assembly – removal and installation

Removal

1 Remove the fairing (see Chapter 8), then remove the instrument cluster (see Section 15).
2 Remove the four screws securing the headlight assembly and lift it out of the fairing **(see illustrations)**. If required, remove the headlight and sidelight bulbs and the relays, and the wiring sub-loom, noting how it fits.

Installation

3 Installation is the reverse of removal. Make sure all the wiring is correctly connected and secured. Check the operation of the headlight and sidelight. Check the headlight aim (see Chapter 1).

9 Brake/tail light bulbs – renewal

1 Remove the passenger seat (see Chapter 8). Remove the rubber bulbholder cover, noting how it fits **(see illustration)**.

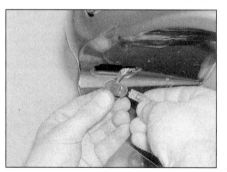

7.6c . . . and the bulb out of the holder

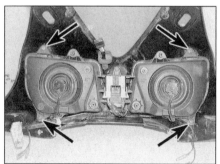

8.2a Undo the screws (arrowed) . . .

8.2b . . . and remove the headlight assembly

Electrical system

9.1 Remove the cover

9.2a Release the bulbholder from the tail light . . .

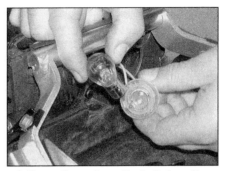

9.2b . . . then release the bulb from the holder

2 Turn the bulbholder anti-clockwise and withdraw it from the tail light **(see illustration)**. Push the bulb into the holder and twist it anti-clockwise to remove it **(see illustration)**.

3 Check the socket terminals for corrosion and clean them if necessary. Line up the pins of the new bulb with the slots in the socket, then push the bulb in and turn it clockwise until it locks into place. **Note:** *The pins on the bulb are offset so it can only be installed one way. It is a good idea to use a paper towel or dry cloth when handling the new bulb to prevent injury if the bulb should break, and to increase bulb life.*

4 Fit the bulbholder into the tail light and turn it clockwise to secure it, then fit the rubber cover, with the 'UP' mark and arrow facing up **(see illustration 9.1)**. Install the seat (see Chapter 8).

10 Tail light assembly – removal and installation

Removal

1 Remove the seat cowling (see Chapter 8).
2 Release the turn signal wiring from the clip on each side of the tail light assembly backing plate **(see illustration)**. Disconnect the tail light assembly wiring connector **(see illustration)**.
3 Unscrew the two bolts securing the mudguard to the rear sub-frame **(see illustration)**. Slacken the bolt securing each bungee hook **(see illustration 10.4a)**.
4 Undo the two screws and release the three trim clips securing the tail light assembly and remove it, bending the mudguard down to provide clearance **(see illustrations)**. Refer to Chapter 8, Section 1 for details on how to release the trim clips, if required.

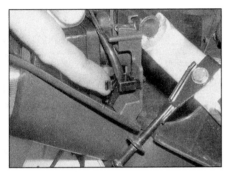

10.2a Release the turn signal wiring . . .

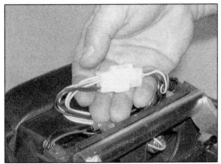

10.2b . . . and disconnect the tail light wiring connector

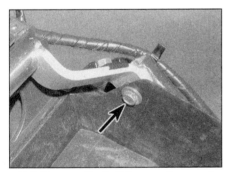

10.3 Remove the bolt (arrowed) on each side

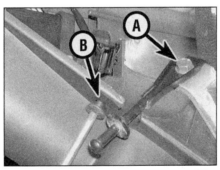

10.4a Slacken the bungee hook bolt (A) and remove the screw (B) on each side . . .

10.4b . . . then release the front trim clips . . .

10.4c . . . and the top trim clips (arrowed) . . .

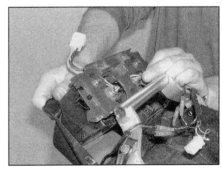

10.4d . . . and manoeuvre the assembly out

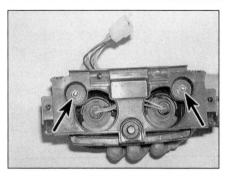

10.5 Unscrew the nuts (arrowed) and separate the light from the plate if required

11.3 Turn signal relay (arrowed)

12.1 Remove the screw and detach the lens . . .

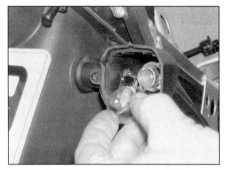

12.2 . . . and remove the bulb

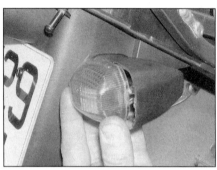

12.3 Make sure the tab locates correctly when fitting the lens

either LEFT or RIGHT. If no voltage is present, replace the relay with a new one. If voltage was present, check the wiring between the relay, turn signal switch and turn signal lights for continuity. Turn the ignition OFF when the check is complete.

12 Turn signal bulbs – renewal

1 Remove the screw securing the turn signal lens and remove the lens, noting how it fits **(see illustration)**.
2 Push the bulb into the holder and twist it anti-clockwise to remove it **(see illustration)**. Check the socket terminals for corrosion and clean them, if necessary. Line up the pins of the new bulb with the slots in the socket, then push the bulb in and turn it clockwise until it locks into place.
3 Fit the lens onto the holder, making sure the tab locates correctly **(see illustration)**. Do not overtighten the screw as the lens or threads could be damaged.

13 Turn signal assemblies – removal and installation

Front

Removal

1 Remove the fairing side panel (see Chapter 8).
2 Remove the screw securing the turn signal to the inside of the fairing, accessing it via the hole in the inner panel, and remove the mounting plate **(see illustrations)**. Remove the turn signal from the panel, noting how it fits. Take care not to snag the wiring as you pull it through.

Installation

3 Installation is the reverse of removal. Make sure the wiring is correctly routed and securely connected. Check the operation of the turn signals.

5 If required, unscrew the two nuts and separate the tail light unit from the backing plate **(see illustration)**.

Installation

6 Installation is the reverse of removal. Check the operation of the tail light and the brake light.

11 Turn signal circuit – check

1 Most turn signal problems are the result of a burned-out bulb or corroded socket. This is especially true when the turn signals function properly in one direction, but fail to flash in the other direction. Check the bulbs and the sockets (see Section 12) and the wiring connectors. Also, check the signal fuse (see Section 5) and the switch (see Section 20).
2 The battery provides power for operation of the turn signal lights, so if they do not operate, also check the battery voltage. Low battery voltage indicates either a faulty battery or a defective charging system. Refer to Section 3 for battery checks and Sections 30 and 31 for charging system tests.
3 If the bulbs, sockets, connectors, fuse, switch and battery are good, check the turn signal relay, which is mounted under the rider's seat **(see illustration)**. Remove the rider's seat for access (see Chapter 8).
4 Check for voltage at the brown wire in the relay wiring connector with the ignition ON. If no voltage is present, using the appropriate wiring diagram at the end of this Chapter check the wiring between the relay and the ignition (main) switch. If voltage was present, check for voltage at the brown/white wire with the ignition ON, and with the switch turned to

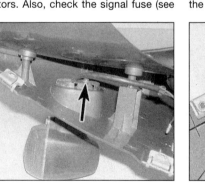

13.2a Undo the screw (arrowed) . . .

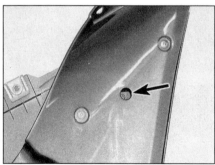

13.2b . . . accessing it via the hole in the inner panel (arrowed)

Electrical system 9•9

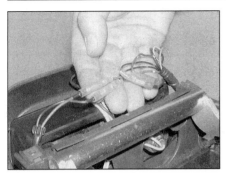

13.5 Rear turn signal wiring connectors

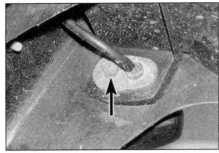

13.6 The turn signal is secured by a nut (arrowed) or screw on the inside of the mudguard

Rear

Removal

4 Remove the passenger seat (see Chapter 8).
5 Trace the wiring back from the turn signal and disconnect it at the connectors **(see illustration)**. Release the wiring from its clip and feed it through to the turn signal **(see illustration 10.2a)**.
6 Unscrew the nut (1998 and 1999 models) or remove the screw (2000-on models) securing the turn signal to the inside of the mudguard and remove the mounting plate **(see illustration)**. Remove the turn signal from the mudguard, taking care not to snag the wiring as you pull it through.

Installation

7 Installation is the reverse of removal. Make sure the wiring is correctly routed and securely connected. Check the operation of the turn signals.

14 Brake light switches – check and renewal

Circuit check

1 Before checking the switches, check the brake light circuit (see Section 6, Step 12).
2 The front brake light switch is mounted on the underside of the brake master cylinder. Disconnect the wiring connectors from the switch **(see illustration)**. Using a continuity tester, connect the probes to the terminals of the switch. With the brake lever at rest, there should be no continuity. With the brake lever applied, there should be continuity. If the switch does not behave as described, replace it with a new one.
3 The rear brake light switch is mounted on the right-hand side, above the brake pedal **(see illustration)**. Remove the fuel tank (see Chapter 4) to access the wiring connector. Trace the wiring from the switch and disconnect it at the connector **(see illustration)**. Using a continuity tester, connect the probes to the terminals on the switch side of the wiring connector. With the brake pedal at rest, there should be no continuity. With the brake pedal applied, there should be continuity. If the switch does not behave as described, replace it with a new one.

4 If the switches are good, rejoin the wire connector and check for voltage at the brown wire terminal on the connector with the ignition switch ON – there should be battery voltage. If there is no voltage present, check the wiring between the switch and the ignition switch (see *Wiring Diagrams* at the end of this Chapter).

Switch renewal

Front brake light switch

5 The switch is mounted on the underside of the brake master cylinder. Disconnect the wiring connectors from the switch **(see illustration 14.2)**.
6 Remove the single screw and washers securing the switch to the bottom of the master cylinder and remove the switch.
7 Installation is the reverse of removal. The switch is not adjustable.

Rear brake light switch

8 The switch is mounted on the right-hand side, above the brake pedal **(see illustration 14.3a)**. Remove the fuel tank (see Chapter 4) to access the wiring connector. Trace the wiring from the switch and disconnect it at the connector **(see illustration 14.3b)**. Free the wiring from any clips or ties and feed it through to the switch.
9 Detach the lower end of the switch spring from the brake pedal **(see illustration)**. Unscrew and remove the switch from the adjuster nut in the bracket **(see illustration 14.3a)**.
10 Installation is the reverse of removal. Make sure the brake light is activated just before the rear brake pedal takes effect. If adjustment is necessary, hold the switch and turn the adjuster nut on the switch body until the brake light is activated when required.

15 Instrument cluster – removal and installation

Removal

1 Remove the fairing, then remove the windshield and the rear view mirrors (see Chapter 8).

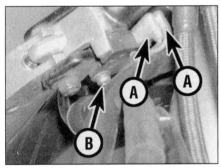

14.2 Front brake switch wiring connectors (A) and mounting screw (B)

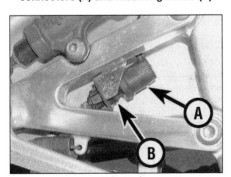

14.3a Rear brake light switch (A) and adjuster nut (B)

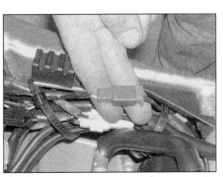

14.3b Rear brake light switch wiring connector

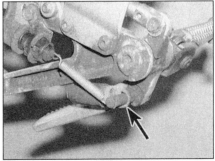

14.9 Unhook the spring from the lug (arrowed)

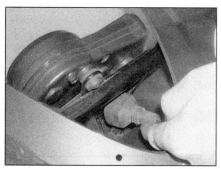

15.2a Pull back the rubber cover . . .

15.2b . . . then disconnect the wiring connector (arrowed)

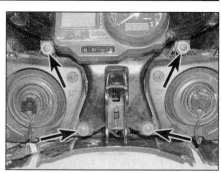

15.3a Undo the screws (arrowed) . . .

2 Pull back the rubber cover and disconnect the wiring connector from the instrument cluster **(see illustrations)**.

3 Remove the screws securing the fairing stay and remove it along with the instrument cluster **(see illustrations)**.

4 Remove the screws securing the cluster and lift it off the stay **(see illustration)**.

Installation

5 Installation is the reverse of removal. Check all the rubber mounting grommets in the fairing stay for cracks and deterioration and replace them with new ones, if necessary.

16 Instruments – check and renewal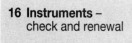

15.3b . . . and remove the fairing stay with the instruments attached

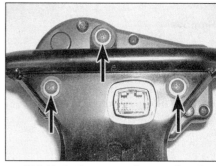

15.4 Undo the screws (arrowed) and remove the instruments

Check

1 Yamaha provide no specific test data for the instruments. If any of them are suspected of being faulty, take the instrument cluster to a Yamaha dealer for further assessment. If there are any faults, even if just a warning light LED is faulty, a new cluster will have to be fitted, as no individual components are available. However, a few checks can be made to determine whether it is the instruments themselves or the components which supply the information to them, that are faulty.

2 The speedometer is controlled by the speed sensor, which is mounted in the back of the crankcase. On 2000-on models, when the ignition is first switched ON, or while the engine is running, the speed sensor performs its own self-diagnosis in the event of failure or faulty wiring. When a fault occurs, the tachometer will be seen to display zero rpm for 3 seconds, then 4000 rpm for 2.5 seconds, then the actual engine speed for 3 seconds, whereupon it will repeat the cycle until the engine is switched off. If any problems occur on any models, check the speed sensor wiring circuit for shorts or breaks, and the connectors for loose, damaged or corroded terminals, referring to *Wiring Diagrams* at the end of the Chapter **(see illustrations)**. To test the output from the sensor, place the motorcycle on an auxiliary stand so the rear wheel is off the ground. Make sure the transmission is in neutral. Disconnect the sensor wiring connector **(see illustration 16.2b)**, then connect the positive (+ve) probe of a voltmeter set to the DC20V scale to the white wire terminal on the sensor side of the connector, and connect the negative (–ve) probe to earth (ground). Turn the ignition (main) switch ON. Turn the rear wheel in its normal direction of rotation and check the readings on the meter – it should be seen to fluctuate between zero and 5 volts as the wheel is turned. If not, and the wiring to the connector is good, replace the sensor with a new one. Otherwise, assuming all the wiring and connectors are good, the ignition control unit could be faulty (see Chapter 5). To fit a new sensor, unscrew the bolt securing the old one and withdraw it from the crankcase **(see illustration 16.2a)**. Install the sensor using a new O-ring.

3 The tachometer is controlled by the ignition control unit – refer to Chapter 5 for further information.

4 The engine temperature display is controlled by the temperature sender (1998 and 1999 models) or thermo unit (2000-on models) – refer to Chapter 3 for test details.

5 The fuel warning display is controlled by the fuel level sensor – refer to Chapter 4 for test details.

6 The oil warning display is controlled by the oil level sensor – refer to Section 26 for test details.

7 If any of the warning lights or instrument lights fail, refer to Section 17.

Renewal

8 Remove the instrument cluster and replace it with a new one (see Section 15). No individual components are available.

16.2a Speed sensor

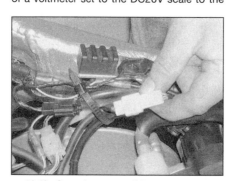

16.2b Speed sensor wiring connector

Electrical system 9•11

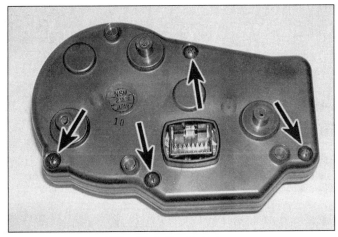

17.2a Undo the screws...

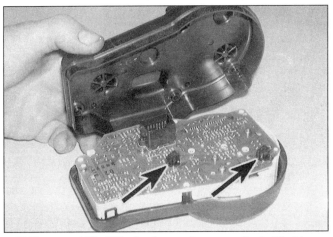

17.2b ...and lift off the back cover to access the bulbholders (arrowed)

17 Instrument and warning lights – renewal

1 On 1998 and 1999 models, the instrument illumination bulbs are conventional and replaceable. On 2000-on models, the instruments are illuminated by LED's which are not available individually – if they fail, a new instrument cluster must be fitted. On all models, the warning and indicator functions (neutral, high beam, turn signals, oil level, fuel level and coolant temperature) are all illuminated by LED's which are not available individually – if they fail, a new instrument cluster must be fitted.

2 To renew an instrument illumination bulb on 1998 and 1999 models, first remove the instrument cluster (see Section 15). Undo the screws securing the rear cover and lift it off **(see illustrations)**. Release the bulbholder by twisting it and pulling it out **(see illustration)**. Pull the bulb out of the holder and install the new one **(see illustration)**.

3 On all models, to test whether an LED has failed, remove the instrument cluster (see Section 15). Using a fully charged 12V battery and two suitable jump leads, and referring to Wiring Diagrams at the end of the Chapter, connect the jumper leads directly to the relevant terminals on the instrument cluster for the LED being tested. If the LED comes on, the fault lies elsewhere in the circuit. If it doesn't, the LED has failed and so a new instrument cluster must be installed.

18 Ignition (main) switch – check, removal and installation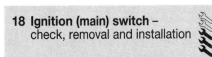

⚠ **Warning:** To prevent the risk of short circuits, remove the seat and disconnect the battery negative (–ve) lead before making any ignition (main) switch checks.

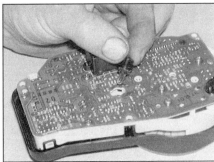

17.2c Release the bulbholder...

Check

1 Remove the left-hand cockpit trim panel (see Chapter 8). Trace the ignition (main) switch wiring back from the base of the switch and disconnect it at the connectors **(see illustration)**. Make the checks on the switch side of the connector.

2 Using an ohmmeter or a continuity tester, check the continuity of the connector terminal pairs (see Wiring Diagrams at the end of this Chapter). Continuity should exist between the terminals connected by a solid line on the diagram when the switch key is turned to the indicated position.

3 If the switch fails any of the tests, replace it with a new one.

18.1 Disconnect the ignition switch wiring connectors

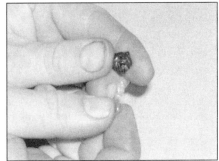

17.2d ...and install a new bulb

Removal

4 Remove the left-hand cockpit trim panel (see Chapter 8). Trace the ignition (main) switch wiring back from the base of the switch and disconnect it at the connectors **(see illustration 18.1)**.

5 Remove the fuel tank (see Chapter 4) and the fairing (see Chapter 8). **Note:** *Although it is not strictly necessary to remove the fuel tank and fairing, doing so will prevent the possibility of damage should a tool slip.*

6 Unscrew the bolt securing the front brake master cylinder to the top yoke and displace it **(see illustration)**. Keep the master cylinder upright to prevent fluid leakage.

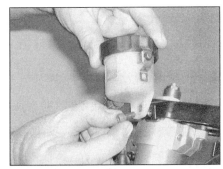

18.6 Unscrew the bolt and displace the reservoir

9•12 Electrical system

18.7a Remove the blanking caps...

18.7b ... then unscrew the handlebar positioning bolts

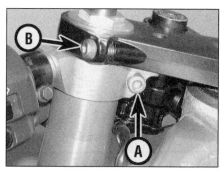
18.7c Slacken the handlebar clamp bolts (A) and the fork clamp bolts (B)

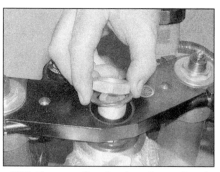
18.8 Unscrew the steering stem nut and remove the washer

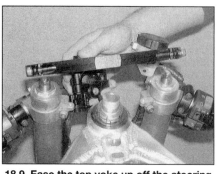
18.9 Ease the top yoke up off the steering stem and forks

18.10 Ignition switch bolts (arrowed)

7 Remove the blanking caps from the handlebar positioning bolts using a small flat-bladed screwdriver, then unscrew the bolts **(see illustrations)**. Slacken the handlebar clamp bolts, and the fork clamp bolts in the top yoke **(see illustration)**.

8 Unscrew the steering stem nut and remove it along with its washer **(see illustration)**.

9 Gently ease the top yoke upwards off the fork tubes and position it clear, using a rag to protect the tank or other components **(see illustration)**.

10 Two shear-head bolts mount the ignition switch to the underside of the top yoke **(see illustration)**. The heads of the bolts must be tapped around using a suitable punch or drift, or drilled off, before the switch can be removed. To do this, mount the yoke in a vice equipped with soft jaws and padded out with rags. Remove the bolts and withdraw the switch from the top yoke.

Installation

11 Installation is the reverse of removal. Obtain new special shear-head bolts and tighten them until their heads shear off. Make sure wiring connectors are securely connected and correctly routed. Tighten the steering stem nut, the fork clamp bolts, the handlebar positioning bolts and the handlebar clamp bolts, in that order, to the torque settings specified at the beginning of Chapter 6.

19 Handlebar switches – check

1 Generally speaking, the switches are reliable and trouble-free. Most troubles, when they do occur, are caused by dirty or corroded contacts, but wear and breakage of internal parts is a possibility that should not be overlooked. If breakage does occur, the entire switch and related wiring harness will have to be replaced with a new one, as individual parts are not available.

2 The switches can be checked for continuity using an ohmmeter or a continuity test light.

3 Remove the left-hand cockpit trim panel to access the left-hand switch wiring connectors and/or the ignition coils to access the right-hand switch wiring connectors (see Chapters 8 and/or 5). Trace the wiring harness of the switch in question back to its connectors and disconnect them **(see illustrations)**.

4 Check for continuity between the terminals of the switch harness with the switch in the various positions (i.e. switch OFF – no continuity, switch ON – continuity) – see *Wiring Diagrams* at the end of this Chapter.

5 If the continuity check indicates a problem exists, refer to Section 20, remove the switch and spray the switch contacts with electrical contact cleaner. If they are accessible, the contacts can be scraped clean with a knife or polished with wire wool. If switch components are damaged or broken, it should be obvious when the switch is disassembled.

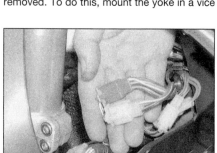
19.3a Left-hand switch wiring connectors

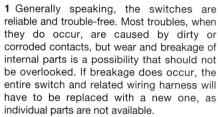

19.3b Right-hand switch connectors

Electrical system 9•13

20.3a Undo the screws (arrowed) . . .

20.3b . . . and separate the halves

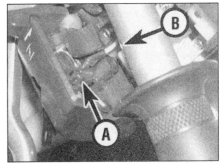

20.4 Locate the pin (A) in the hole (B)

20 Handlebar switches – removal and installation

Removal

1 If the switch is to be removed from the bike, rather than just displaced from the handlebar, trace the wiring harness of the switch in question back to its connectors and disconnect them. Remove the left-hand cockpit trim panel to access the left-hand switch wiring connectors and/or the ignition coils to access the right-hand switch wiring connectors (see Chapters 8 and/or 5) **(see illustrations 19.3a and/or b)**. Work back along the harness, freeing it from all the relevant clips and ties, noting its correct routing.

2 Disconnect the wiring connectors from the brake light switch (if removing the right-hand switch) or the clutch switch (if removing the left-hand switch) **(see illustration 14.2 or 23.2)**.

3 Unscrew the handlebar switch screws and free the switch from the handlebar by separating the halves **(see illustrations)**. When removing the left-hand switch, remove the choke cable lever, noting how it fits.

Installation

4 Installation is the reverse of removal. Refer to Chapter 4 for installation of the choke cable, if required. Make sure the locating pin in the switch housing locates in the hole in the handlebar **(see illustration)**. Make sure the wiring connectors are correctly routed and securely connected.

21 Neutral switch – check, removal and installation

Check

1 Before checking the electrical circuit, check the signal fuse (see Section 5).

2 The switch is located on the back of the engine on the left-hand side. Make sure the transmission is in neutral. To access the neutral switch wiring connector, remove the fuel tank (see Chapter 4). Trace the wiring from the switch and disconnect it at the connector **(see illustration)**.

3 With the connector disconnected and the ignition switched ON, the neutral light should be out. If not, the wire between the connector and instrument cluster must be earthed (grounded) at some point.

4 Check for continuity between the light blue wire terminal on the switch side of the wiring connector, and the crankcase. With the transmission in neutral, there should be continuity. With the transmission in gear, there should be no continuity. If there is continuity when in gear, check that the wire is not earthed (grounded). If there is no continuity when in neutral, check for a break in the wire, then remove the switch (see below), and check that the contact plunger is not damaged or seized in the switch body **(see illustration)**.

5 If the continuity tests prove the switch and wiring is good, check for voltage (ignition ON) at the brown terminal on the wiring loom side of the instrument cluster wiring connector – remove the windshield to access the connector (see Chapter 8). If there is no voltage present, check the wire between the connector and signal fuse (see *Wiring Diagrams* at the end of this Chapter). If the voltage is good, check the LED in the instrument cluster (see Section 17), then check the starter circuit cut-off relay (Section 24) and other components in the starter circuit as described in the relevant Sections of this Chapter. If all components are good, check the wiring between the various components (see *Wiring Diagrams* at the end of this Chapter).

Removal

6 The switch is on the back of the engine on the left-hand side. Access to it is quite restricted, and could be impossible without the correct tools. For best access, remove the swingarm (see Chapter 6).

7 Remove the screw securing the wire terminal to the switch, and detach the wire **(see illustration)**.

8 Unscrew the switch and withdraw it from the casing. Check the condition of the sealing washer and replace it with a new one if it is damaged or deformed.

Installation

9 Install the switch using a new sealing washer if necessary, and tighten it to the torque setting specified at the beginning of the Chapter.

10 Connect the wire and secure it with the screw. Check the operation of the neutral light.

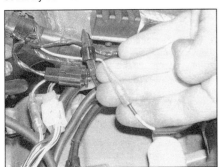

21.2 Disconnect the neutral switch wiring connector

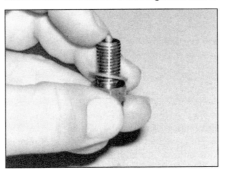

21.4 Check the plunger in the end of the switch

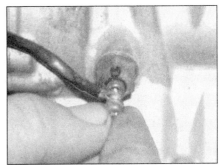

21.7 Undo the screw and detach the wire

9•14 Electrical system

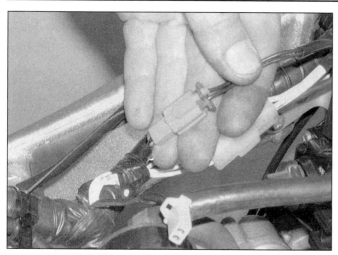

22.2 Sidestand switch wiring connector

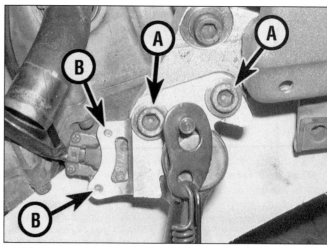

22.7 Sidestand bracket bolts (A), sidestand switch screws (B)

22 Sidestand switch – check and renewal

Check

1 The sidestand switch is mounted on the back of the sidestand bracket. The switch is part of the safety circuit which prevents or stops the engine running if the transmission is in gear whilst the sidestand is down, and prevents the engine from starting if the transmission is in gear unless the sidestand is up and the clutch lever is pulled in. Before checking the electrical circuit, check the fuse (see Section 5).

2 To access the wiring connector, remove the fuel tank (see Chapter 4). Trace the wiring from the switch and disconnect it at the connector **(see illustration)**.

3 Check the operation of the switch using an ohmmeter or continuity test light. Connect the meter probes to the terminals on the switch side of the connector. With the sidestand up there should be continuity (zero resistance) between the terminals, and with the stand down there should be no continuity (infinite resistance).

4 If the switch does not perform as expected, it is defective and must be renewed.

5 If the switch is good, check the starter circuit cut-off relay (Section 24) and other components in the starter circuit as described in the relevant Sections of this Chapter. If all components are good, check the wiring between the various components (see *Wiring Diagrams* at the end of this Chapter).

Renewal

6 The sidestand switch is mounted on the back of the sidestand bracket. Remove the fuel tank (see Chapter 4) to access the wiring connector. Trace the wiring from the switch and disconnect it at the connector **(see illustration 22.2)**. Work back along the switch wiring, freeing it from any relevant retaining clips and ties, noting its correct routing.

7 Unscrew the bolts securing the sidestand bracket to the frame and displace it **(see illustration)**.

8 Undo the screws securing the switch and remove the switch, noting how it fits.

9 Fit the new switch, making sure the plunger locates correctly, and tighten the screws securely. Install the sidestand assembly and tighten the bolts securely.

10 Make sure the wiring is correctly routed up to the connector and retained by all the necessary clips and ties. Reconnect the wiring connector and check the operation of the sidestand switch.

11 Install the fuel tank (see Chapter 4).

23 Clutch switch – check and renewal

Check

1 The clutch switch is mounted on the underside of the clutch lever bracket. The switch is part of the safety circuit which prevents or stops the engine running if the transmission is in gear whilst the sidestand is down, and prevents the engine from starting if the transmission is in gear unless the sidestand is up and the clutch lever is pulled in. The switch is not adjustable.

2 To check the switch, disconnect the wiring connectors **(see illustration)**. Connect the probes of an ohmmeter or a continuity tester to the two switch terminals. With the clutch lever pulled in, continuity should be indicated. With the clutch lever out, no continuity (infinite resistance) should be indicated.

3 If the switch is good, check the starter circuit cut-off relay (Section 24) and other components in the starter circuit as described in the relevant Sections of this Chapter. If all components are good, check the wiring between the various components (see *Wiring Diagrams* at the end of this Chapter).

Renewal

4 The clutch switch is mounted on the underside of the clutch lever bracket.

5 Disconnect the wiring connectors **(see illustration 23.2)**, then remove the screw and detach the switch.

6 Installation is the reverse of removal.

24 Relay assembly – check and renewal

Starter circuit cut-off relay

Check

1 The starter circuit cut-off relay is part of the safety circuit which prevents or stops the engine running if the transmission is in gear whilst the sidestand is down, and prevents the engine from starting if the transmission is in gear unless the sidestand is up and the clutch lever is pulled in.

2 The starter circuit cut-off relay and its diodes are contained within the relay assembly, which is mounted behind the battery below the rider's seat **(see illustration)**. Remove the rider's seat for access (see Chapter 8).

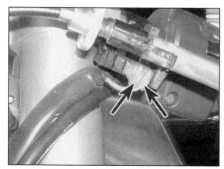

23.2 Clutch switch wiring connectors (arrowed)

Electrical system 9•15

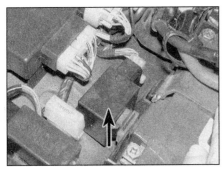

24.2 Relay assembly (arrowed)

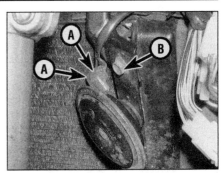

25.3 Horn wiring connectors (A) and mounting bolt (B)

3 Disconnect the battery terminals (see Section 3) and move the leads aside to access the relay. Displace the relay and disconnect the wiring connector. Refer to the wiring diagram for your model (see end of chapter) and the following procedures:

4 To check the operation of the relay, connect a meter set to the ohms x 100 range, or a continuity tester, between the blue/white and black wire terminals of the relay. No continuity should be shown. Now, using jumper wires, connect a fully charged 12V battery to the relay's red/black and black/yellow wire terminals (positive (+ve) lead to red/black and negative (–ve) lead to black/yellow). The meter should now show continuity with battery voltage applied. If it doesn't, replace the relay assembly with a new one.

5 The diodes contained within the relay assembly can be checked by performing a continuity test. Refer to the appropriate wiring diagram at the end of this Chapter and connect the meter (set to the ohms function) or continuity tester across the wire terminals for the diode being tested. The diodes should show continuity in one direction and no continuity when the meter or tester probes are reversed. If any diode shows the same condition in both directions it should be considered faulty, and the relay assembly must be replaced with a new one.

6 If the cut-out relay and diodes are proven good, but the starting system fault still exists, check all other components in the starting circuit (i.e. the neutral switch, side stand switch, clutch switch, starter switch and starter relay) as described in the relevant Sections of this Chapter. If all components are good, check the wiring between the various components (see *Wiring Diagrams* at the end of this Chapter).

Renewal

7 The relay assembly is mounted behind the battery below the rider's seat **(see illustration 24.2)**. Remove the rider's seat for access (see Chapter 8). Disconnect the battery terminals (see Section 3) and move the leads aside to access the relay. Displace the relay and disconnect the wiring connector.

8 Installation is the reverse of removal.

Fuel pump relay

9 The fuel pump relay is housed within the relay assembly. Refer to Chapter 4, Section 15 for test details.

25 Horn – check and renewal

Check

1 If the horn, doesn't work, first check the fuse (see Section 5) and the battery (see Section 3).

2 The horn is mounted on the left-hand end of the radiator. Remove the left-hand fairing side panel for best access to it, though it is accessible by just removing the left-hand cockpit trim panel (see Chapter 8).

3 Unplug the wiring connectors from the horn **(see illustration)**. Using two jumper wires, apply battery voltage directly to the terminals on the horn. If the horn sounds, check the switch (see Section 19) and the wiring between the switch and the horn (see *Wiring Diagrams* at the end of this Chapter).

4 If the horn doesn't sound, replace it with a new one.

Renewal

5 The horn is mounted on the left-hand end of the radiator. Remove the left-hand fairing side panel for best access to it, though it is accessible by just removing the left-hand cockpit trim panel (see Chapter 8).

6 Unplug the wiring connectors from the horn **(see illustration 25.3)**. Unscrew the bolt securing the horn and remove it from the bike.

7 Install the horn and securely tighten the bolt. Connect the wiring connectors to the horn. Install the fairing (see Chapter 8).

26 Oil level sensor and relay – check, removal and installation

Check

1 On 1998 and 1999 models the combined oil level /coolant temperature warning light will come on for a few seconds when the ignition is turned on as a check of the warning light LED. It should then go out and the motorcycle can be started. If the warning light remains on and/or the oil level symbol in the digital display is illuminated, check the oil level as described in Daily (pre-ride) checks. If the oil level is correct, check the sensor and relay as described below. Equally if the warning light comes on and the oil level symbol flashes whilst the motorcycle is being ridden, check the oil level immediately.

2 For 2000-on models, the oil level warning light will come on for a few seconds when the ignition is turned on as a check of the warning light LED. It should then go out and the motorcycle can be started. If the warning light remains on, check the oil level as described in Daily (pre-ride) checks. If the oil level is correct, check the sensor as described below. Equally if the warning light comes on whilst the motorcycle is being ridden, check the oil level immediately.

> **HAYNES HiNT** *The warning light may flicker during sudden acceleration or deceleration or when riding up or down hill. Note that this is a characteristic of the system and provided the oil level is correct, does not indicate a fault*

3 On all models, if the warning light does not come on during the self-checking procedure described above, check the LED as described in Section 17.

4 To check the sensor on all models, remove it from the sump (see Steps 5 and 6 below). Connect one probe of an ohmmeter or continuity tester to the sensor wire and the other probe to its base. With the sensor in its normal installed position (wiring at the bottom), there should be continuity. Turn the sensor upside down. There should be no continuity. If either condition does not occur, replace the sensor with a new one.

5 To check the relay (1998 and 1999 models only), remove the rider's seat (see Chapter 8), then disconnect the relay wiring connector and release it from its holder **(see illustration)**. Set a multimeter to the ohms x 1 scale and connect the positive (+ve) probe to the red/blue wire terminal, and the negative (–ve) probe to the black terminal. There should be no continuity.

26.5 Oil level sensor relay (arrowed)

9•16 Electrical system

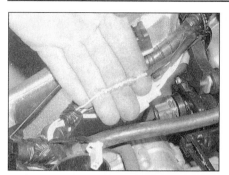

26.7a Oil level sensor wiring connector

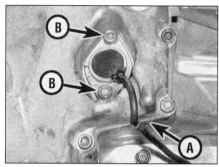

26.7b Free the wiring from its clamp (A) then unscrew the bolts (B) and remove the sensor

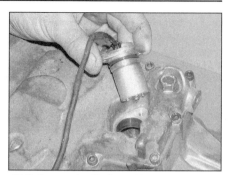

26.9 Smear the O-ring with grease before installing the sensor

Leaving the meter connected, and using a fully-charged 12 volt battery and two insulated jumper wires, connect the battery positive (+ve) lead to the brown wire terminal, and the negative (–ve) lead to the white wire terminal. There should now be continuity shown on the meter. If the relay doesn't behave as stated, replace it with a new one.

Removal

6 Drain the engine oil (see Chapter 1). Remove the fuel tank (see Chapter 4).
7 Trace the wiring back from the sensor and disconnect it at the single bullet connector **(see illustration)**. Free it from its clamp. Unscrew the two bolts securing the sensor to the bottom of the sump and withdraw it from the sump, being prepared to catch any residue oil **(see illustration)**. Check the condition of the O-ring and replace it with a new one if it is damaged, deformed or deteriorated (but note that Yamaha do not actually list it as being available separately from the sensor, so you will have to have it matched for size by them).
8 To remove the relay (1998 and 1999 models), remove the rider's seat (see Chapter 8), then disconnect the relay wiring connector and release it from its holder **(see illustration 26.5)**.

Installation

9 Fit a new O-ring onto the oil level sensor if required. Smear the O-ring (old or new) with lithium grease, then fit the sensor into the sump **(see illustration)**. Tighten its bolts to the torque setting specified at the beginning of the Chapter.

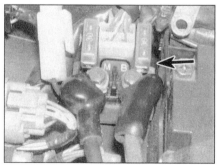

27.2 Starter relay (arrowed)

10 Connect the wiring at the connector **(see illustration 26.7a)** and check the operation of the sensor (see Steps 1 to 4 above). Secure the wiring in its clamp **(see illustration 26.7b)**.
11 Fill the engine with oil (see Chapter 1).
12 Fit the relay into its holder and connect the wiring connector **(see illustration 26.5)**.

27 Starter relay – check and renewal

Check

1 If the starter circuit is faulty, first check the main fuse and ignition fuse (see Section 5).
2 Remove the rider's seat (see Chapter 8). The starter relay is located behind the battery **(see illustration)**. Lift the rubber terminal cover and unscrew the bolt securing the thick starter motor lead to the terminal marked 'M'; position the lead away from the relay terminal. With the ignition switch ON, the engine kill switch in the 'RUN' position, and the transmission in neutral, press the starter switch. The relay should be heard to click.
3 If the relay doesn't click, switch off the ignition and remove the relay as described below; test it as follows.
4 Set a multimeter to the ohms x 1 scale and connect it across the relay's starter motor and battery lead terminals (marked 'M' and 'B' respectively) **(see illustration 27.2)**. Using a fully-charged 12 volt battery and two insulated jumper wires, connect the positive (+ve) terminal of the battery to the red/white wire terminal on the relay, and the negative (–ve) terminal to the blue/white wire terminal on the relay. At this point the relay should be heard to click and the multimeter read 0 ohms (continuity). If this is the case, the relay is proved good. If the relay does not click when battery voltage is applied and indicates no continuity (infinite resistance) across its terminals, it is faulty and must be replaced with a new one.
5 If the relay is good, check for battery voltage at the red/white wire terminal on the loom side of the relay wiring connector when the starter button is pressed. If voltage is present, check the other components in the starter circuit as described in the relevant Sections of this Chapter. If no voltage was present, check the wiring between the various components (see Wiring Diagrams at the end of this Chapter).

Renewal

6 Remove the rider's seat (see Chapter 8). The starter relay is located behind the battery **(see illustration 27.2)**. Disconnect the battery negative (–ve) lead before removing the relay.
7 Disconnect the relay wiring connector **(see illustration)**. Unscrew the two bolts securing the starter motor and battery leads to the relay and detach the leads. Remove the relay with its rubber sleeve from its mounting lug on the frame.
8 Installation is the reverse of removal. The starter motor lead connects to the terminal marked 'M', and the battery lead to the terminal marked 'B' **(see illustration 27.2)**. Make sure the terminal nuts are securely tightened. Connect the negative (–ve) lead last when reconnecting the battery.

28 Starter motor – removal and installation

Removal

1 Remove the rider's seat and the left-hand fairing side panel (see Chapter 8), and the fuel tank (see Chapter 4). Disconnect the battery negative (–ve) lead. The starter motor is mounted on the crankcase, behind the

27.7 Disconnect the relay wiring connector

Electrical system 9•17

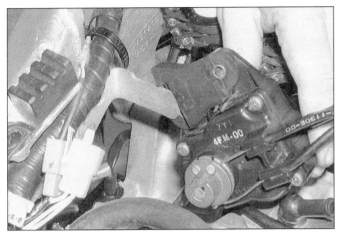

28.1a Displace the EXUP servo . . .   28.1b . . . and the idle speed adjuster

cylinder block. Displace the EXUP servo from its mount and the idle speed adjuster from its holder **(see illustrations)**.

2 Peel back the rubber terminal cover and unscrew the nut (1998 and 1999 models) or bolt (2000-on models) securing the lead to the starter motor **(see illustration)**. Detach the lead.

3 Unscrew the two bolts securing the starter motor, noting the earth lead secured by the rear bolt **(see illustration)**. Draw the starter motor out of the crankcase and remove it from the machine **(see illustration)**.

4 Remove the O-ring on the end of the starter motor and discard it, as a new one must be used.

Installation

5 Fit a new O-ring onto the end of the starter motor, making sure it is seated in its groove, and smear it with grease **(see illustration)**.

6 Manoeuvre the motor into position and slide it into the crankcase **(see illustration 28.3b)**. Ensure that the starter motor teeth mesh correctly with those of the starter idle/reduction gear. Install the

mounting bolts and tighten them to the torque setting specified at the beginning of the Chapter, not forgetting to secure the earth lead with the rear bolt **(see illustration)**.

7 Connect the lead to the starter motor and secure it with the nut or bolt **(see illustration 28.2)**. Make sure the rubber cover is correctly seated over the terminal.

8 Fit the EXUP servo onto its mount and the idle speed adjuster into its holder **(see illustrations 28.1a and b)**. Connect the battery negative (–ve) lead and install the seat (see Chapter 8).

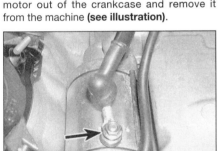
28.2 Pull back the rubber cover and unscrew the terminal nut (arrowed)

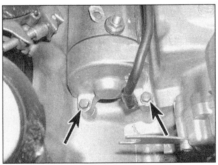
28.3a Unscrew the bolts (arrowed) . . .

28.3b . . . and remove the starter motor

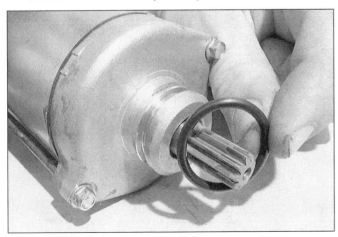
28.5 Fit a new O-ring and smear it with grease

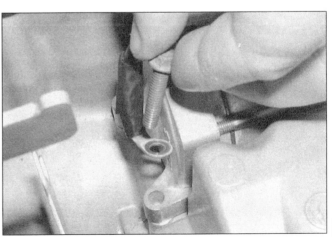
28.6 Secure the earth lead with the rear bolt

9•18 Electrical system

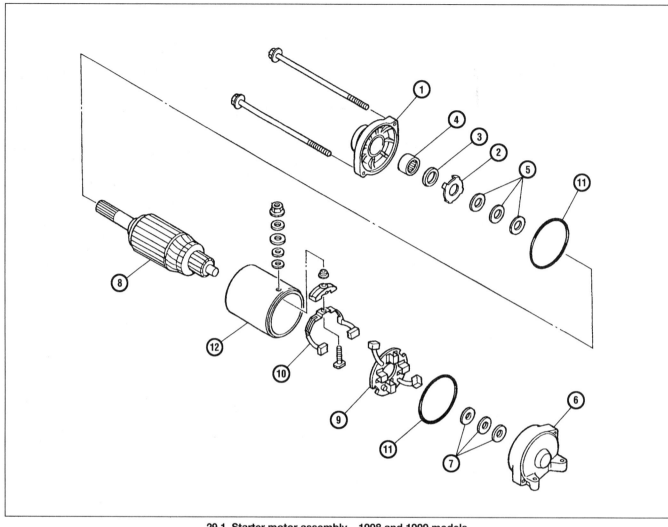

29.1 Starter motor assembly – 1998 and 1999 models

1. Front cover
2. Tabbed washer
3. Oil seal
4. Needle bearing
5. Washer and shims
6. Rear cover
7. Shims
8. Armature
9. Brushplate assembly
10. Brush holder, terminal bolt and insulator piece
11. Sealing ring
12. Main housing

29.2 Note the alignment marks between the housing and the covers

29.3a Unscrew and remove the two bolts (arrowed) . . .

Electrical system

29.3b ... then remove the front cover and sealing ring (arrowed)

29.3c Remove the tabbed washer ...

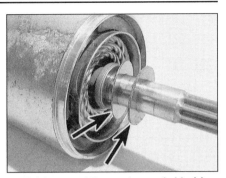

29.3d ... and the washer and shim(s) (arrowed)

29.4a Remove the rear cover and its sealing ring (arrowed) ...

29.4b ... and remove the shims (arrowed)

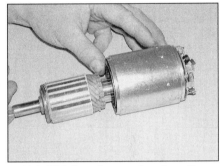

29.5 Withdraw the armature from the housing

29 Starter motor – disassembly, inspection and reassembly

1998 and 1999 models

Disassembly

1 Remove the starter motor (see Section 28) **(see illustration)**.
2 Note the alignment marks between the main housing and the front and rear covers, or make your own if they are unclear **(see illustration)**.
3 Unscrew the two long bolts, then remove the front cover from the motor along with its sealing ring **(see illustrations)**. Discard the sealing ring, as a new one must be used. Remove the tabbed washer from the cover and slide the washer and shim(s) from the front end of the armature, noting their correct fitted order and the number of shims **(see illustrations)**.
4 Remove the rear cover from the motor, along with its sealing ring **(see illustration)**. Discard the sealing ring, as a new one must be used. Remove the shim(s) from the rear end of the armature, noting how many and their correct fitted positions **(see illustration)**.
5 Withdraw the armature from the main housing **(see illustration)**.
6 Lift each brush spring end onto the top of each brush housing and slide the brushes out **(see illustrations)**.

7 At this stage check for continuity between the terminal bolt and the two brushes with yellow insulation. There should be continuity (zero resistance). Check for continuity between the terminal bolt and the housing. There should be no continuity (infinite resistance). Also check for continuity between the brushes with uninsulated wires and the brushplate. There should be continuity (zero resistance). If there is no continuity when there should be or vice versa, identify the faulty component and replace it with a new one.
8 Remove the brushplate assembly, noting how it locates **(see illustration)**. Noting the correct fitted location of each component, unscrew the nut from the terminal bolt and remove the plain washer, the various

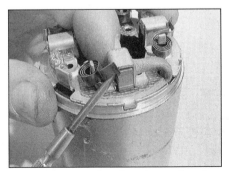

29.6a Place the brush spring ends onto the top of the brush housings ...

29.6b ... and slide the brushes out

29.8a Remove the brushplate assembly, noting how it locates (arrowed)

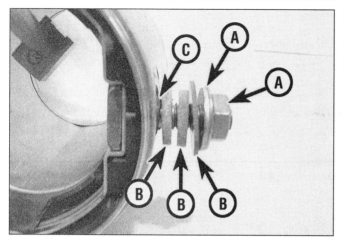

29.8b Remove the nut and washer (A), the insulating washers (B) and O-ring (C)

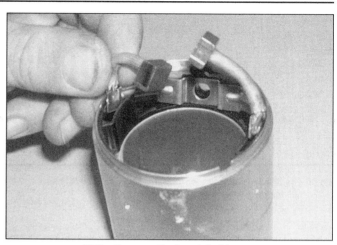

29.8c Remove the brush holder . . .

29.8d . . . and the insulator . . .

29.8e . . . and, if required, the brushplate seat

29.9 Check the length of each brush

insulating washers and the rubber O-ring **(see illustration)**. Remove the terminal bolt and the brush holder from the main housing, noting how they locate, then remove the insulator, and if required, the brushplate seat **(see illustrations)**.

Inspection

9 The parts of the starter motor that are most likely to require attention are the brushes. Measure the length of each brush and compare the results to the Specifications at the beginning of the Chapter **(see illustration)**. If any of the brushes are worn beyond the service limit, replace them with new ones. If the brushes are not worn excessively, nor cracked, chipped, or otherwise damaged, they may be re-used.

10 Inspect the commutator bars on the armature for scoring, scratches and discoloration. The commutator can be cleaned and polished with wire wool, but do not use sandpaper or emery paper. After cleaning, wipe away any residue with a cloth soaked in electrical system cleaner or denatured alcohol.

11 Using an ohmmeter or a continuity test light, check for continuity between the commutator bars **(see illustration)**. Continuity should exist between each bar and all of the others. Also, check for continuity between the commutator bars and the armature shaft **(see illustration)**. There should be no continuity

29.11a Continuity should exist between the commutator bars

29.11b There should be no continuity between the commutator bars and the armature shaft

29.13 On 1998 and 1999 models, check the oil seal and needle bearing in the front cover

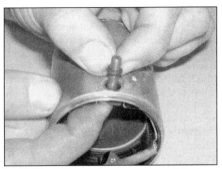

29.15 Insert the terminal bolt and fit the O-ring onto it

29.18 Locate the spring ends onto the brushes

(infinite resistance) between the commutator and the shaft. If the checks indicate otherwise, the armature is defective. Also check the depth of the insulating Mica between the commutator bars – if it is less than the amount specified at the beginning of the Chapter, scrape it away using a pointed tool or suitable hacksaw blade until its is correct. Measure the diameter of the commutator and replace the starter motor with a new one if it has worn below the minimum diameter specified.

12 Check the starter pinion gear for worn, cracked, chipped and broken teeth. If the gear is damaged or worn, replace the starter motor with a new one.

13 Inspect the end covers for signs of cracks or wear. On 1998 and 1999 models, check the oil seal and needle bearing in the front cover and the bush in the rear cover for wear and damage **(see illustration)**. On 2000-on models, check the bearing in each cover. Inspect the magnets in the main housing and the housing itself, for cracks.

14 On 1998 and 1999 models, inspect the insulating washers, the O-ring and insulator piece for signs of damage, and renew if necessary.

Reassembly

15 If removed, fit the brushplate seat **(see illustration 29.8e)**. Fit the insulator into the main housing **(see illustration 29.8d)**, then install the brush holder **(see illustration 29.8c)**. Insert the terminal bolt through the brush holder, insulator and the housing and fit the O-ring onto it **(see illustration)**. Slide the small insulating washer(s) onto the terminal bolt, followed by the large insulating washer(s) and the plain washer **(see illustration 29.8b)**. Fit the nut onto the terminal bolt and tighten it securely.

16 Install the brushplate assembly, making sure its tab is correctly located in the housing slot, and that each of the insulated brush wires sits in its cutout in the brushplate **(see illustration 29.8a)**.

17 Slide all the brushes back into position in their housings **(see illustration 29.6b)**. Make sure that each brush spring is retained against the top of its housing so that it will not exert any pressure on the brush **(see illustration 29.6a)**.

> **HAYNES HiNT** Lifting the end of the brush spring so that it is against the top of the brush holder and not pressing the brush inwards, makes it much easier to install the armature on reassembly.

18 Insert the armature into the front of the housing and locate the brushes on the commutator bars **(see illustration 29.5)**. Slip each brush spring end off the top of the brush housing and onto the brush end **(see illustration)**. Check that each brush is securely pressed against the commutator by its spring and is free to move easily in its housing.

19 Fit the shim(s) onto the rear of the armature shaft **(see illustration 29.4b)**. Apply a smear of grease to the end of the shaft.

20 Fit a new sealing ring onto the rear of the housing **(see illustration)**. Align the rear cover groove with the brushplate outer tab and install the cover – aligning the marks (See Step 2) between the cover and housing will help **(see illustrations 29.4a and 29.2)**.

21 Apply a smear of grease to the front cover oil seal lip. Fit the tabbed washer into the cover so that its teeth are correctly located with the cover ribs **(see illustration 29.3c)**.

22 Fit a new sealing ring onto the front of the housing **(see illustration)**. Slide the shim(s) onto the front end of the armature shaft then fit the washer **(see illustration 29.3d)**. Slide the front cover into position, aligning the marks made on removal **(see illustrations 29.3b and 29.2)**.

23 Check the marks made on removal are correctly aligned, then fit the long bolts and tighten them to the torque setting specified at the beginning of the Chapter **(see illustration)**.

24 Install the starter motor (see Section 28).

29.20 Fit a new sealing ring onto the rear of the housing

29.22 Fit a new sealing ring onto the front of the housing

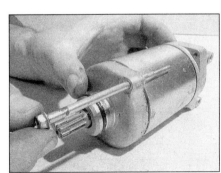

29.23 Fit the long bolts and tighten them to the specified torque

2000-on models

Disassembly

25 Remove the starter motor (see Section 28) **(see illustration)**.
26 Note the alignment marks between the main housing and the front and rear covers, or make your own if they are unclear.
27 Unscrew the two long bolts, then remove the rear cover from the motor along with its sealing ring. Discard the sealing ring, as a new one must be used.
28 Remove the front cover from the motor along with the brushplate assembly. Remove the sealing ring and discard it, as a new one must be used.
29 Withdraw the armature from the main housing.
30 Lift each brush spring end onto the top of each brush housing and slide the brushes out.
31 At this stage check for continuity between the terminal bolt and the two brushes with insulation. There should be continuity (zero resistance). Check for continuity between the terminal and the housing. There should be no continuity (infinite resistance). Also check for continuity between the brushes with uninsulated wires and the brushplate. There should be continuity (zero resistance). If there is no continuity when there should be or vice versa, identify the faulty component and replace it with a new one.
32 Remove the brushplate assembly from the front cover, noting how it locates. Remove the insulated brushes if required, noting how they fit. The uninsulated brushes are secured to the brushplate by a bolt each.

Inspection

33 Refer to Steps 9 to 14 above.

Reassembly

34 Install the brushes if removed, then fit the brushplate assembly, making sure the tab locates correctly in the slot in the cover.
35 Slide all the brushes back into position in their housings. Make sure that each brush spring is retained against the top of its housing so that it will not exert any pressure on the brush.

 HAYNES HiNT *Lifting the end of the brush spring so that it is against the top of the brush holder and not pressing the brush inwards makes it much easier to install the armature on reassembly.*

36 Insert the armature into the front cover and locate the brushes on the commutator bars. Slip each brush spring end off the top of the brush housing and onto the brush end. Check that each brush is securely pressed against the commutator by its spring and is free to move easily in its housing.
37 Fit a new sealing ring onto the front cover. Fit the main housing over the armature, aligning the marks between the cover and housing (Step 26).

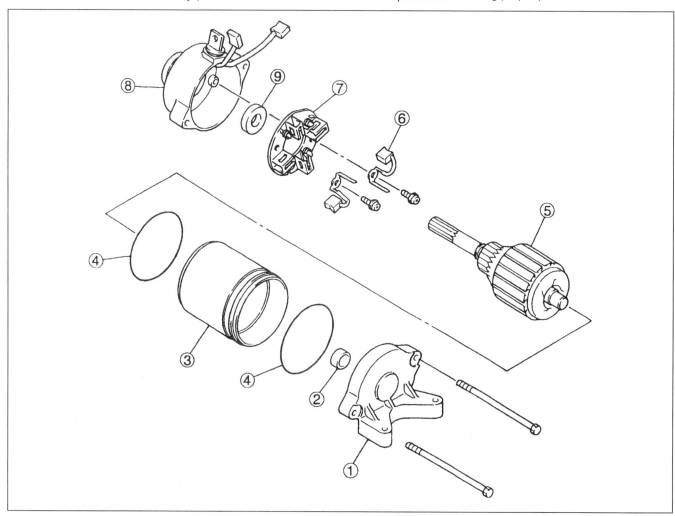

29.25 Starter motor assembly – 2000 models

1 Rear cover	4 Sealing ring	7 Brushplate assembly
2 Bearing	5 Armature	8 Front cover
3 Main housing	6 Brushes	9 Bearing

38 Fit a new sealing ring onto the rear of the housing. Slide the rear cover into position, aligning the marks made on removal.
39 Check the marks made on removal are correctly aligned, then fit the long bolts and tighten them to the torque setting specified at the beginning of the Chapter.
40 Install the starter motor (see Section 28).

30 Charging system testing – general information and precautions

1 If the performance of the charging system is suspect, the system as a whole should be checked first, followed by testing of the individual components. **Note:** *Before beginning the checks, make sure the battery is fully charged and that all system connections are clean and tight.*
2 Checking the output of the charging system and the performance of the various components within the charging system requires the use of a multimeter (with voltage, current and resistance checking facilities).
3 When making the checks, follow the procedures carefully to prevent incorrect connections or short circuits, as irreparable damage to electrical system components may result if short circuits occur.
4 If a multimeter is not available, the job of checking the charging system should be left to a Yamaha dealer or automotive electrician.

31 Charging system – leakage and output test

1 If the charging system of the machine is thought to be faulty, remove the rider's seat (see Chapter 8) and perform the following checks.

Leakage test

Caution: *Always connect an ammeter in series, never in parallel with the battery, otherwise it will be damaged. Do not turn the ignition ON or operate the starter motor when the ammeter is connected – a sudden surge in current will blow the meter's fuse.*

2 Turn the ignition switch OFF and disconnect the lead from the battery negative (–ve) terminal.
3 Set the multimeter to the Amps function and connect its negative (–ve) probe to the battery negative (–ve) terminal, and positive (+ve) probe to the disconnected negative (–ve) lead **(see illustration)**. Always set the meter to a high amps range initially and then bring it down to the mA (milli Amps) range; if there is a high current flow in the circuit it may blow the meter's fuse.
4 No current flow should be indicated. If current leakage is indicated (generally greater than 0.1 mA), there is a short circuit in the wiring. Using the wiring diagrams at the end of this Chapter, systematically disconnect individual electrical components, checking the meter each time until the source is identified.
5 If no leakage is indicated, disconnect the meter and connect the negative (–ve) lead to the battery, tightening it securely.

Output test

6 Start the engine and warm it up to normal operating temperature.
7 To check the regulated voltage output, allow the engine to idle and connect a multimeter set to the 0 to 20 volts DC scale (voltmeter) across the terminals of the battery (positive (+ve) lead to battery positive (+ve) terminal, negative (–ve) lead to battery negative (–ve) terminal). Slowly increase the engine speed briefly to 5000 rpm and note the reading obtained. The regulated voltage should be as specified at the beginning of the Chapter. If the voltage is outside these limits, check the alternator, then the regulator/rectifier (see Sections 32 and 33).

> **HAYNES HiNT** *Clues to a faulty regulator are constantly blowing bulbs, with brightness varying considerably with engine speed, and battery overheating.*

32 Alternator rotor and stator – check, removal and installation

Check

1 To access the wiring connector, remove the fuel tank (see Chapter 4).
2 Trace the wiring back from the alternator cover on the left-hand side of the engine and disconnect it at the white connector containing the three white wires **(see illustration)**.
3 Using a multimeter set to the ohms x 1 (ohmmeter) scale, measure the resistance between each of the white wires on the alternator side of the connector, taking a total of three readings, then check for continuity between each terminal and ground (earth). If

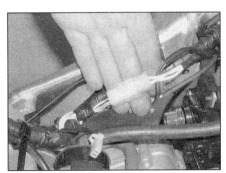

32.2 Alternator wiring connector

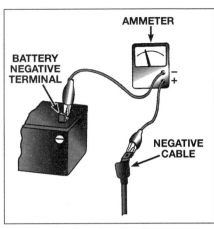

31.3 Checking the charging system leakage rate – connect the meter as shown

the stator coil windings are in good condition the three readings should be within the range shown in the Specifications at the start of this Chapter and there should be no continuity (infinite resistance) between any of the terminals and ground (earth). If not, check the fault is not due to damaged wiring between the connector and coils. If the wiring is good, the alternator stator coil assembly is at fault and should be replaced with a new one.

Removal

4 Drain the engine oil (see Chapter 1). To access the wiring connector, remove the fuel tank (see Chapter 4).
5 Trace the wiring back from the alternator cover on the left-hand side of the engine and disconnect it at the white connector containing the three white wires **(see illustration 32.2)**. Free the wiring from any clips or guides and feed it through to the alternator cover.
6 Remove the lower fairing and the left-hand fairing side panel (see Chapter 8).
7 Working in a criss-cross pattern, unscrew the bolts securing the alternator cover and remove the cover **(see illustration)**. Note the position of the dowels and remove them if loose. Discard the gasket, as a new one must be used.
8 To remove the rotor bolt it is necessary to stop the rotor from turning. If a rotor holding

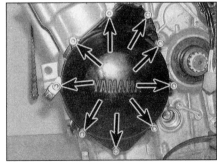

32.7 Unscrew the bolts (arrowed) and remove the cover

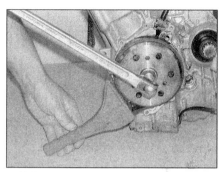

32.8 Unscrewing the rotor bolt using a rotor holding strap

32.9 Removing the rotor using a commercially available rotor puller

32.10a Unscrew the bolt (arrowed) and remove the wiring clamp . . .

32.10b . . . then unscrew the stator bolts (arrowed)

32.11a Install the stator . . .

strap or tool is not available, and if the engine is still in the fame, place the transmission in gear and have an assistant apply the rear brake, then unscrew the bolt **(see illustration)**. Discard the bolt and washer, as new ones must be used.

Caution: If a rotor holding strap is used, make sure it does not contact the raised sections on the outside of the rotor.

9 To remove the rotor from the shaft it is necessary to use a rotor puller. Yamaha provide a special tool (Part No. 90890-01080), or alternatively a similar tool can be obtained commercially **(see illustration)**.

10 To remove the stator, remove the bolt securing the wiring clamp and the three bolts securing the stator, then remove the assembly, noting how the rubber wiring grommet fits **(see illustrations)**.

Installation

11 Install the stator, aligning the rubber wiring grommet with the groove **(see illustration)**. Apply a suitable non-permanent thread locking compound to the stator bolt threads, then install the bolts and tighten them to the torque setting specified at the beginning of the Chapter **(see illustration)**.

Apply a suitable sealant to the wiring grommet, then press it into the cut-out in the cover **(see illustration)**. Secure the wiring with the clamp **(see illustration 32.10a)**.

12 Clean the tapered end of the crankshaft and the corresponding mating surface on the inside of the rotor with a suitable solvent. Make sure that no metal objects have attached themselves to the magnet on the inside of the rotor, then slide the rotor onto the shaft **(see illustration)**.

13 Apply some clean engine oil to the new rotor bolt threads and to the underside of the head, and to both sides of the new washer

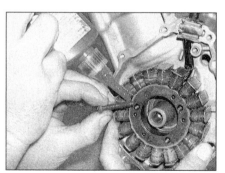

32.11b . . . using a threadlock on the bolts

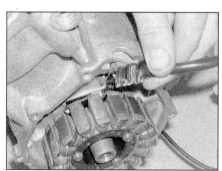

32.11c Fit the wiring grommet into the cutout

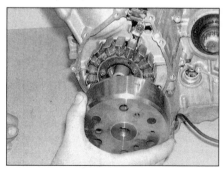

32.12 Slide the rotor onto the end of the crankshaft

Electrical system 9•25

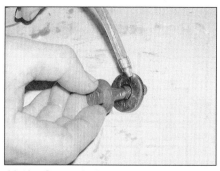

32.13a Smear the bolt and washer with oil as described...

32.13b ... then install them ...

32.13c ... and tighten the bolt to the specified torque

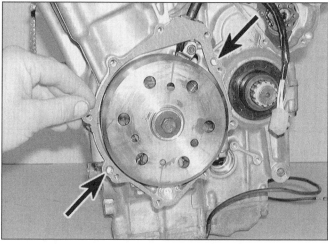

32.14a Locate the gasket onto the dowels (arrowed) ...

32.14b ... then install the cover

(see illustration). Tighten the bolt to the torque setting specified at the beginning of the Chapter, using the method employed on removal to prevent the rotor from turning (see illustration).

14 Fit the new gasket, making sure it locates onto the dowels (see illustration). Install the cover and tighten the bolts evenly in a criss-cross pattern to the specified torque setting (see illustration). Connect the alternator coil wiring connector, making sure it is correctly routed and secured by any clips or ties (see illustration 32.2).

15 Install the left-hand fairing side panel and lower fairing, and the fuel tank (see Chapters 8 and 4). Fill the engine with oil (see Chapter 1).

33 Regulator/rectifier – check and renewal

Check

1 Yamaha provide no test specifications for the regulator/rectifier. If it is suspected of being faulty, first check all other components and the wiring and connectors in the charging circuit, referring to the relevant Sections in this Chapter and to the wiring diagrams at the end.

2 If all other components and the wiring are good, then the regulator/rectifier could be faulty. Remove the unit (see below) and take it to a Yamaha dealer for testing. Alternatively, substitute the suspect unit with a known good one and see if the fault is cured.

Renewal

3 The regulator/rectifier is mounted underneath the rear mounting for the fuel tank – remove the tank for access (see Chapter 4). Disconnect the wiring connector (see illustration).

4 Unscrew the two bolts securing the regulator/rectifier and remove it.

5 Install the new unit and tighten its bolts securely. Connect the wiring connector. Install the fuel tank (see Chapter 4).

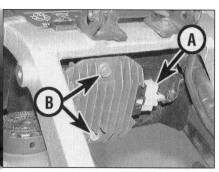

33.3 Regulator/rectifier wiring connector (A) and mounting bolts (B)

9•26 Wiring diagrams

1998 and 1999 models (Europe)

Wiring diagrams 9•27

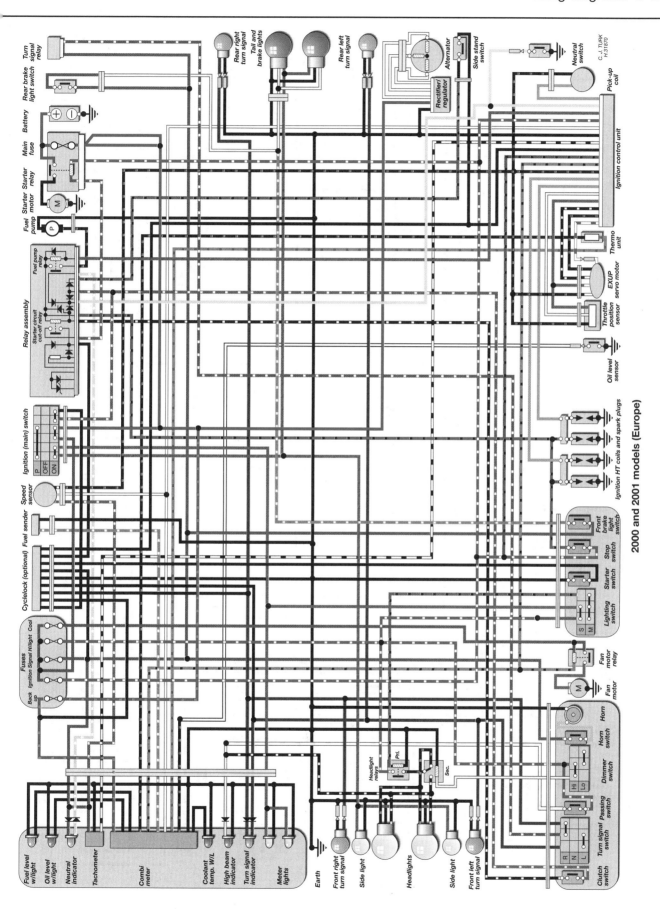

2000 and 2001 models (Europe)

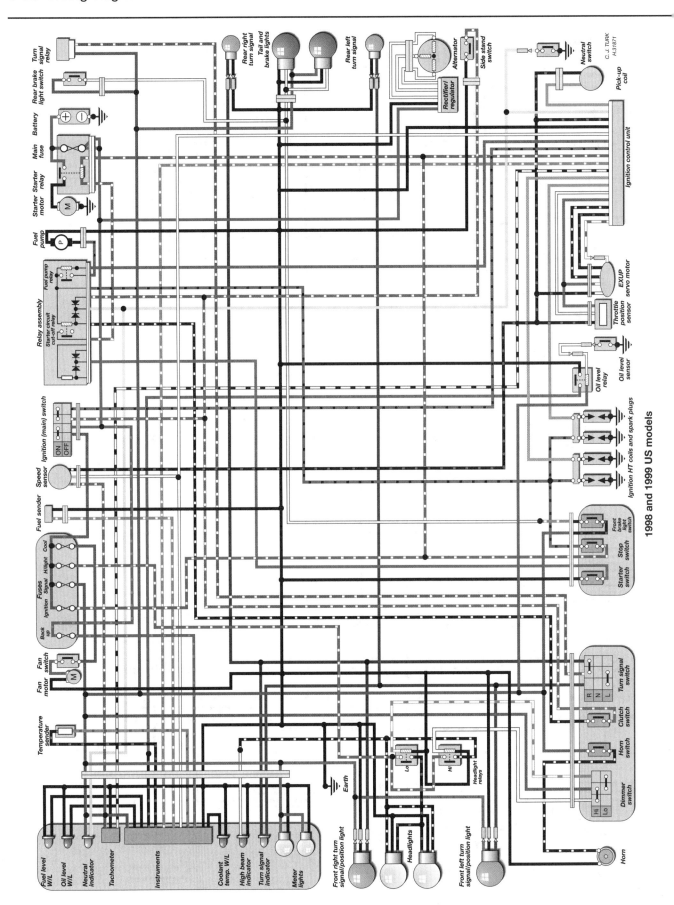

Wiring diagrams

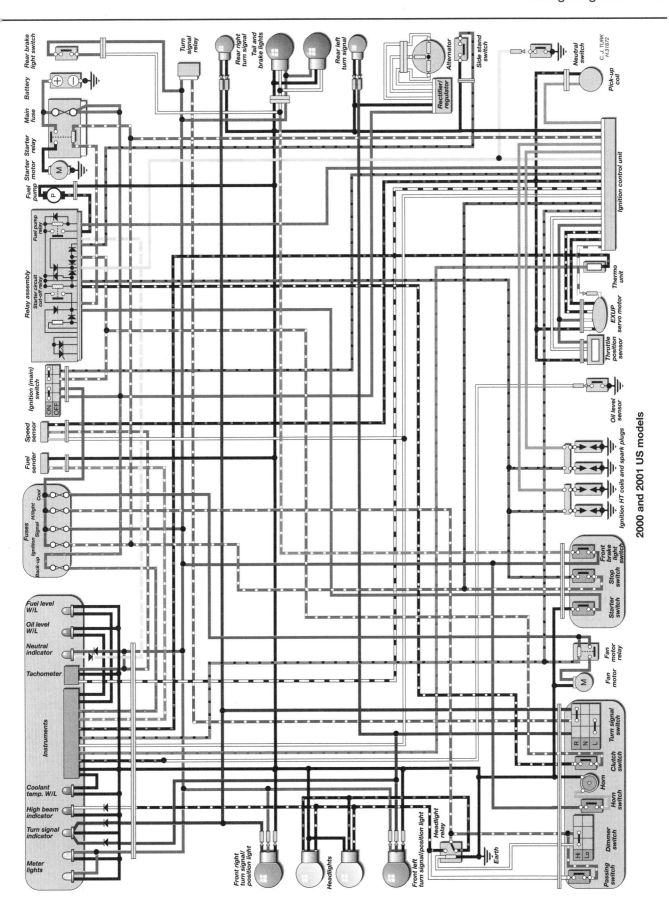

2000 and 2001 US models

Notes

Reference

Tools and Workshop Tips REF•2

- Building up a tool kit and equipping your workshop ● Using tools ● Understanding bearing, seal, fastener and chain sizes and markings ● Repair techniques

Security REF•20

- Locks and chains
- U-locks ● Disc locks
- Alarms and immobilisers
- Security marking systems ● Tips on how to prevent bike theft

Lubricants and fluids 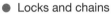 REF•23

- Engine oils
- Transmission (gear) oils
- Coolant/anti-freeze
- Fork oils and suspension fluids ● Brake/clutch fluids
- Spray lubes, degreasers and solvents

Conversion Factors REF•26

- Formulae for conversion of the metric (SI) units used throughout the manual into Imperial measures

MOT Test Checks REF•27

- A guide to the UK MOT test ● Which items are tested ● How to prepare your motorcycle for the test and perform a pre-test check

Storage REF•32

- How to prepare your motorcycle for going into storage and protect essential systems ● How to get the motorcycle back on the road

Fault Finding REF•35

- Common faults and their likely causes ● How to check engine cylinder compression ● How to make electrical tests and use test meters

Technical Terms Explained REF•49

- Component names, technical terms and common abbreviations explained

Index REF•53

REF•2 Tools and Workshop Tips

Buying tools

A toolkit is a fundamental requirement for servicing and repairing a motorcycle. Although there will be an initial expense in building up enough tools for servicing, this will soon be offset by the savings made by doing the job yourself. As experience and confidence grow, additional tools can be added to enable the repair and overhaul of the motorcycle. Many of the specialist tools are expensive and not often used so it may be preferable to hire them, or for a group of friends or motorcycle club to join in the purchase.

As a rule, it is better to buy more expensive, good quality tools. Cheaper tools are likely to wear out faster and need to be renewed more often, nullifying the original saving.

> **Warning: To avoid the risk of a poor quality tool breaking in use, causing injury or damage to the component being worked on, always aim to purchase tools which meet the relevant national safety standards.**

The following lists of tools do not represent the manufacturer's service tools, but serve as a guide to help the owner decide which tools are needed for this level of work. In addition, items such as an electric drill, hacksaw, files, soldering iron and a workbench equipped with a vice, may be needed. Although not classed as tools, a selection of bolts, screws, nuts, washers and pieces of tubing always come in useful.

For more information about tools, refer to the Haynes *Motorcycle Workshop Practice TechBook* (Bk. No. 3470).

Manufacturer's service tools

Inevitably certain tasks require the use of a service tool. Where possible an alternative tool or method of approach is recommended, but sometimes there is no option if personal injury or damage to the component is to be avoided. Where required, service tools are referred to in the relevant procedure.

Service tools can usually only be purchased from a motorcycle dealer and are identified by a part number. Some of the commonly-used tools, such as rotor pullers, are available in aftermarket form from mail-order motorcycle tool and accessory suppliers.

Maintenance and minor repair tools

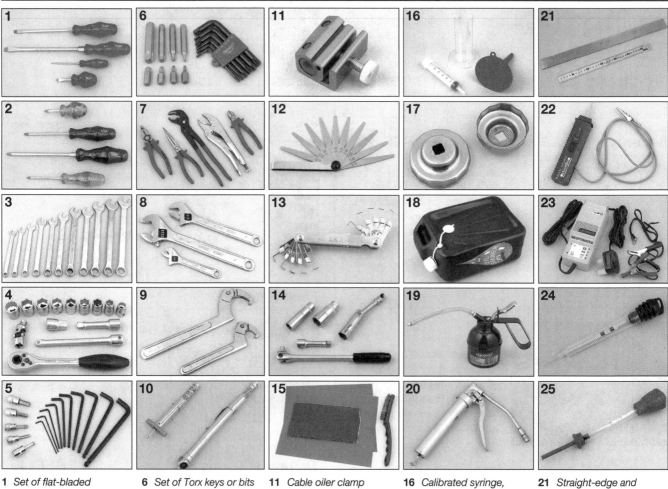

1 Set of flat-bladed screwdrivers
2 Set of Phillips head screwdrivers
3 Combination open-end and ring spanners
4 Socket set (3/8 inch or 1/2 inch drive)
5 Set of Allen keys or bits
6 Set of Torx keys or bits
7 Pliers, cutters and self-locking grips (Mole grips)
8 Adjustable spanners
9 C-spanners
10 Tread depth gauge and tyre pressure gauge
11 Cable oiler clamp
12 Feeler gauges
13 Spark plug gap measuring tool
14 Spark plug spanner or deep plug sockets
15 Wire brush and emery paper
16 Calibrated syringe, measuring vessel and funnel
17 Oil filter adapters
18 Oil drainer can or tray
19 Pump type oil can
20 Grease gun
21 Straight-edge and steel rule
22 Continuity tester
23 Battery charger
24 Hydrometer (for battery specific gravity check)
25 Anti-freeze tester (for liquid-cooled engines)

Tools and Workshop Tips REF•3

Repair and overhaul tools

 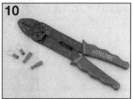   
    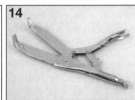
  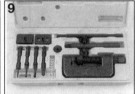  

1 Torque wrench (small and mid-ranges)
2 Conventional, plastic or soft-faced hammers
3 Impact driver set
4 Vernier gauge
5 Circlip pliers (internal and external, or combination)
6 Set of cold chisels and punches
7 Selection of pullers
8 Breaker bars
9 Chain breaking/riveting tool set
10 Wire stripper and crimper tool
11 Multimeter (measures amps, volts and ohms)
12 Stroboscope (for dynamic timing checks)
13 Hose clamp (wingnut type shown)
14 Clutch holding tool
15 One-man brake/clutch bleeder kit

Specialist tools

 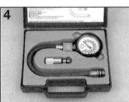   

    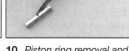 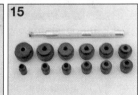

1 Micrometers (external type)
2 Telescoping gauges
3 Dial gauge
4 Cylinder compression gauge
5 Vacuum gauges (left) or manometer (right)
6 Oil pressure gauge
7 Plastigauge kit
8 Valve spring compressor (4-stroke engines)
9 Piston pin drawbolt tool
10 Piston ring removal and installation tool
11 Piston ring clamp
12 Cylinder bore hone (stone type shown)
13 Stud extractor
14 Screw extractor set
15 Bearing driver set

REF•4 Tools and Workshop Tips

1 Workshop equipment and facilities

The workbench

● Work is made much easier by raising the bike up on a ramp - components are much more accessible if raised to waist level. The hydraulic or pneumatic types seen in the dealer's workshop are a sound investment if you undertake a lot of repairs or overhauls (see illustration 1.1).

1.1 Hydraulic motorcycle ramp

● If raised off ground level, the bike must be supported on the ramp to avoid it falling. Most ramps incorporate a front wheel locating clamp which can be adjusted to suit different diameter wheels. When tightening the clamp, take care not to mark the wheel rim or damage the tyre - use wood blocks on each side to prevent this.
● Secure the bike to the ramp using tie-downs (see illustration 1.2). If the bike has only a sidestand, and hence leans at a dangerous angle when raised, support the bike on an auxiliary stand.

1.2 Tie-downs are used around the passenger footrests to secure the bike

● Auxiliary (paddock) stands are widely available from mail order companies or motorcycle dealers and attach either to the wheel axle or swingarm pivot (see illustration 1.3). If the motorcycle has a centrestand, you can support it under the crankcase to prevent it toppling whilst either wheel is removed (see illustration 1.4).

1.3 This auxiliary stand attaches to the swingarm pivot

1.4 Always use a block of wood between the engine and jack head when supporting the engine in this way

Fumes and fire

● Refer to the Safety first! page at the beginning of the manual for full details. Make sure your workshop is equipped with a fire extinguisher suitable for fuel-related fires (Class B fire - flammable liquids) - it is not sufficient to have a water-filled extinguisher.
● Always ensure adequate ventilation is available. Unless an exhaust gas extraction system is available for use, ensure that the engine is run outside of the workshop.
● If working on the fuel system, make sure the workshop is ventilated to avoid a build-up of fumes. This applies equally to fume build-up when charging a battery. Do not smoke or allow anyone else to smoke in the workshop.

Fluids

● If you need to drain fuel from the tank, store it in an approved container marked as suitable for the storage of petrol (gasoline) (see illustration 1.5). Do not store fuel in glass jars or bottles.

1.5 Use an approved can only for storing petrol (gasoline)

● Use proprietary engine degreasers or solvents which have a high flash-point, such as paraffin (kerosene), for cleaning off oil, grease and dirt - never use petrol (gasoline) for cleaning. Wear rubber gloves when handling solvent and engine degreaser. The fumes from certain solvents can be dangerous - always work in a well-ventilated area.

Dust, eye and hand protection

● Protect your lungs from inhalation of dust particles by wearing a filtering mask over the nose and mouth. Many frictional materials still contain asbestos which is dangerous to your health. Protect your eyes from spouts of liquid and sprung components by wearing a pair of protective goggles (see illustration 1.6).

1.6 A fire extinguisher, goggles, mask and protective gloves should be at hand in the workshop

● Protect your hands from contact with solvents, fuel and oils by wearing rubber gloves. Alternatively apply a barrier cream to your hands before starting work. If handling hot components or fluids, wear suitable gloves to protect your hands from scalding and burns.

What to do with old fluids

● Old cleaning solvent, fuel, coolant and oils should not be poured down domestic drains or onto the ground. Package the fluid up in old oil containers, label it accordingly, and take it to a garage or disposal facility. Contact your local authority for location of such sites or ring the oil care hotline.

Note: It is antisocial and illegal to dump oil down the drain. To find the location of your local oil recycling bank, call this number free.

In the USA, note that any oil supplier must accept used oil for recycling.

Tools and Workshop Tips

2 Fasteners - screws, bolts and nuts

Fastener types and applications

Bolts and screws

● Fastener head types are either of hexagonal, Torx or splined design, with internal and external versions of each type **(see illustrations 2.1 and 2.2)**; splined head fasteners are not in common use on motorcycles. The conventional slotted or Phillips head design is used for certain screws. Bolt or screw length is always measured from the underside of the head to the end of the item **(see illustration 2.11)**.

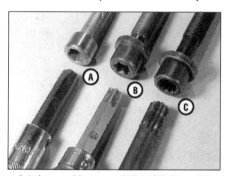

2.1 Internal hexagon/Allen (A), Torx (B) and splined (C) fasteners, with corresponding bits

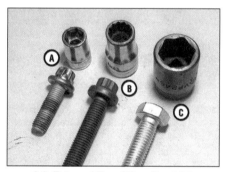

2.2 External Torx (A), splined (B) and hexagon (C) fasteners, with corresponding sockets

● Certain fasteners on the motorcycle have a tensile marking on their heads, the higher the marking the stronger the fastener. High tensile fasteners generally carry a 10 or higher marking. Never replace a high tensile fastener with one of a lower tensile strength.

Washers (see illustration 2.3)

● Plain washers are used between a fastener head and a component to prevent damage to the component or to spread the load when torque is applied. Plain washers can also be used as spacers or shims in certain assemblies. Copper or aluminium plain washers are often used as sealing washers on drain plugs.

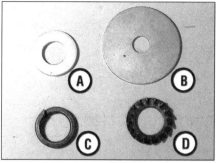

2.3 Plain washer (A), penny washer (B), spring washer (C) and serrated washer (D)

● The split-ring spring washer works by applying axial tension between the fastener head and component. If flattened, it is fatigued and must be renewed. If a plain (flat) washer is used on the fastener, position the spring washer between the fastener and the plain washer.

● Serrated star type washers dig into the fastener and component faces, preventing loosening. They are often used on electrical earth (ground) connections to the frame.

● Cone type washers (sometimes called Belleville) are conical and when tightened apply axial tension between the fastener head and component. They must be installed with the dished side against the component and often carry an OUTSIDE marking on their outer face. If flattened, they are fatigued and must be renewed.

● Tab washers are used to lock plain nuts or bolts on a shaft. A portion of the tab washer is bent up hard against one flat of the nut or bolt to prevent it loosening. Due to the tab washer being deformed in use, a new tab washer should be used every time it is disturbed.

● Wave washers are used to take up endfloat on a shaft. They provide light springing and prevent excessive side-to-side play of a component. Can be found on rocker arm shafts.

Nuts and split pins

● Conventional plain nuts are usually six-sided **(see illustration 2.4)**. They are sized by thread diameter and pitch. High tensile nuts carry a number on one end to denote their tensile strength.

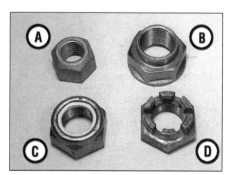

2.4 Plain nut (A), shouldered locknut (B), nylon insert nut (C) and castellated nut (D)

● Self-locking nuts either have a nylon insert, or two spring metal tabs, or a shoulder which is staked into a groove in the shaft - their advantage over conventional plain nuts is a resistance to loosening due to vibration. The nylon insert type can be used a number of times, but must be renewed when the friction of the nylon insert is reduced, ie when the nut spins freely on the shaft. The spring tab type can be reused unless the tabs are damaged. The shouldered type must be renewed every time it is disturbed.

● Split pins (cotter pins) are used to lock a castellated nut to a shaft or to prevent slackening of a plain nut. Common applications are wheel axles and brake torque arms. Because the split pin arms are deformed to lock around the nut a new split pin must always be used on installation - always fit the correct size split pin which will fit snugly in the shaft hole. Make sure the split pin arms are correctly located around the nut **(see illustrations 2.5 and 2.6)**.

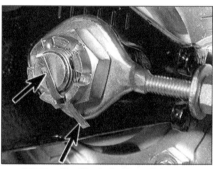

2.5 Bend split pin (cotter pin) arms as shown (arrows) to secure a castellated nut

2.6 Bend split pin (cotter pin) arms as shown to secure a plain nut

Caution: If the castellated nut slots do not align with the shaft hole after tightening to the torque setting, tighten the nut until the next slot aligns with the hole - never slacken the nut to align its slot.

● R-pins (shaped like the letter R), or slip pins as they are sometimes called, are sprung and can be reused if they are otherwise in good condition. Always install R-pins with their closed end facing forwards **(see illustration 2.7)**.

REF•6 Tools and Workshop Tips

2.7 Correct fitting of R-pin. Arrow indicates forward direction

Circlips (see illustration 2.8)

● Circlips (sometimes called snap-rings) are used to retain components on a shaft or in a housing and have corresponding external or internal ears to permit removal. Parallel-sided (machined) circlips can be installed either way round in their groove, whereas stamped circlips (which have a chamfered edge on one face) must be installed with the chamfer facing away from the direction of thrust load **(see illustration 2.9)**.

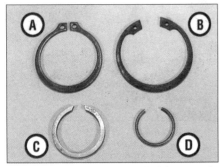

2.8 External stamped circlip (A), internal stamped circlip (B), machined circlip (C) and wire circlip (D)

● Always use circlip pliers to remove and install circlips; expand or compress them just enough to remove them. After installation, rotate the circlip in its groove to ensure it is securely seated. If installing a circlip on a splined shaft, always align its opening with a shaft channel to ensure the circlip ends are well supported and unlikely to catch **(see illustration 2.10)**.

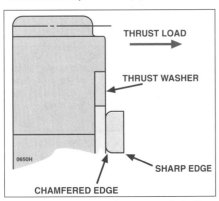

2.9 Correct fitting of a stamped circlip

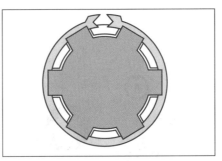

2.10 Align circlip opening with shaft channel

● Circlips can wear due to the thrust of components and become loose in their grooves, with the subsequent danger of becoming dislodged in operation. For this reason, renewal is advised every time a circlip is disturbed.

● Wire circlips are commonly used as piston pin retaining clips. If a removal tang is provided, long-nosed pliers can be used to dislodge them, otherwise careful use of a small flat-bladed screwdriver is necessary. Wire circlips should be renewed every time they are disturbed.

Thread diameter and pitch

● Diameter of a male thread (screw, bolt or stud) is the outside diameter of the threaded portion **(see illustration 2.11)**. Most motorcycle manufacturers use the ISO (International Standards Organisation) metric system expressed in millimetres, eg M6 refers to a 6 mm diameter thread. Sizing is the same for nuts, except that the thread diameter is measured across the valleys of the nut.

● Pitch is the distance between the peaks of the thread **(see illustration 2.11)**. It is expressed in millimetres, thus a common bolt size may be expressed as 6.0 x 1.0 mm (6 mm thread diameter and 1 mm pitch). Generally pitch increases in proportion to thread diameter, although there are always exceptions.

● Thread diameter and pitch are related for conventional fastener applications and the accompanying table can be used as a guide. Additionally, the AF (Across Flats), spanner or socket size dimension of the bolt or nut **(see illustration 2.11)** is linked to thread and pitch specification. Thread pitch can be measured with a thread gauge **(see illustration 2.12)**.

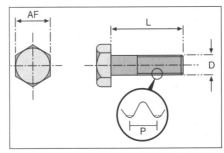

2.11 Fastener length (L), thread diameter (D), thread pitch (P) and head size (AF)

2.12 Using a thread gauge to measure pitch

AF size	Thread diameter x pitch (mm)
8 mm	M5 x 0.8
8 mm	M6 x 1.0
10 mm	M6 x 1.0
12 mm	M8 x 1.25
14 mm	M10 x 1.25
17 mm	M12 x 1.25

● The threads of most fasteners are of the right-hand type, ie they are turned clockwise to tighten and anti-clockwise to loosen. The reverse situation applies to left-hand thread fasteners, which are turned anti-clockwise to tighten and clockwise to loosen. Left-hand threads are used where rotation of a component might loosen a conventional right-hand thread fastener.

Seized fasteners

● Corrosion of external fasteners due to water or reaction between two dissimilar metals can occur over a period of time. It will build up sooner in wet conditions or in countries where salt is used on the roads during the winter. If a fastener is severely corroded it is likely that normal methods of removal will fail and result in its head being ruined. When you attempt removal, the fastener thread should be heard to crack free and unscrew easily - if it doesn't, stop there before damaging something.

● A smart tap on the head of the fastener will often succeed in breaking free corrosion which has occurred in the threads **(see illustration 2.13)**.

● An aerosol penetrating fluid (such as WD-40) applied the night beforehand may work its way down into the thread and ease removal. Depending on the location, you may be able to make up a Plasticine well around the fastener head and fill it with penetrating fluid.

2.13 A sharp tap on the head of a fastener will often break free a corroded thread

Tools and Workshop Tips

- If you are working on an engine internal component, corrosion will most likely not be a problem due to the well lubricated environment. However, components can be very tight and an impact driver is a useful tool in freeing them **(see illustration 2.14)**.

2.14 Using an impact driver to free a fastener

- Where corrosion has occurred between dissimilar metals (eg steel and aluminium alloy), the application of heat to the fastener head will create a disproportionate expansion rate between the two metals and break the seizure caused by the corrosion. Whether heat can be applied depends on the location of the fastener - any surrounding components likely to be damaged must first be removed **(see illustration 2.15)**. Heat can be applied using a paint stripper heat gun or clothes iron, or by immersing the component in boiling water - wear protective gloves to prevent scalding or burns to the hands.

2.15 Using heat to free a seized fastener

- As a last resort, it is possible to use a hammer and cold chisel to work the fastener head unscrewed **(see illustration 2.16)**. This will damage the fastener, but more importantly extreme care must be taken not to damage the surrounding component.

Caution: Remember that the component being secured is generally of more value than the bolt, nut or screw - when the fastener is freed, do not unscrew it with force, instead work the fastener back and forth when resistance is felt to prevent thread damage.

2.16 Using a hammer and chisel to free a seized fastener

Broken fasteners and damaged heads

- If the shank of a broken bolt or screw is accessible you can grip it with self-locking grips. The knurled wheel type stud extractor tool or self-gripping stud puller tool is particularly useful for removing the long studs which screw into the cylinder mouth surface of the crankcase or bolts and screws from which the head has broken off **(see illustration 2.17)**. Studs can also be removed by locking two nuts together on the threaded end of the stud and using a spanner on the lower nut **(see illustration 2.18)**.

2.17 Using a stud extractor tool to remove a broken crankcase stud

2.18 Two nuts can be locked together to unscrew a stud from a component

- A bolt or screw which has broken off below or level with the casing must be extracted using a screw extractor set. Centre punch the fastener to centralise the drill bit, then drill a hole in the fastener **(see illustration 2.19)**. Select a drill bit which is approximately half to three-quarters the diameter of the fastener and drill to a depth which will accommodate the extractor. Use the largest size extractor possible, but avoid leaving too small a wall thickness otherwise the extractor will merely force the fastener walls outwards wedging it in the casing thread.
- If a spiral type extractor is used, thread it anti-clockwise into the fastener. As it is screwed in, it will grip the fastener and unscrew it from the casing **(see illustration 2.20)**.

2.19 When using a screw extractor, first drill a hole in the fastener . . .

2.20 . . . then thread the extractor anti-clockwise into the fastener

- If a taper type extractor is used, tap it into the fastener so that it is firmly wedged in place. Unscrew the extractor (anti-clockwise) to draw the fastener out.

 Warning: Stud extractors are very hard and may break off in the fastener if care is not taken - ask an engineer about spark erosion if this happens.

- Alternatively, the broken bolt/screw can be drilled out and the hole retapped for an oversize bolt/screw or a diamond-section thread insert. It is essential that the drilling is carried out squarely and to the correct depth, otherwise the casing may be ruined - if in doubt, entrust the work to an engineer.
- Bolts and nuts with rounded corners cause the correct size spanner or socket to slip when force is applied. Of the types of spanner/socket available always use a six-point type rather than an eight or twelve-point type - better grip

REF•8 Tools and Workshop Tips

2.21 Comparison of surface drive ring spanner (left) with 12-point type (right)

is obtained. Surface drive spanners grip the middle of the hex flats, rather than the corners, and are thus good in cases of damaged heads **(see illustration 2.21)**.

● Slotted-head or Phillips-head screws are often damaged by the use of the wrong size screwdriver. Allen-head and Torx-head screws are much less likely to sustain damage. If enough of the screw head is exposed you can use a hacksaw to cut a slot in its head and then use a conventional flat-bladed screwdriver to remove it. Alternatively use a hammer and cold chisel to tap the head of the fastener around to slacken it. Always replace damaged fasteners with new ones, preferably Torx or Allen-head type.

HAYNES HiNT

A dab of valve grinding compound between the screw head and screwdriver tip will often give a good grip.

Thread repair

● Threads (particularly those in aluminium alloy components) can be damaged by overtightening, being assembled with dirt in the threads, or from a component working loose and vibrating. Eventually the thread will fail completely, and it will be impossible to tighten the fastener.

● If a thread is damaged or clogged with old locking compound it can be renovated with a thread repair tool (thread chaser) **(see illustrations 2.22 and 2.23)**; special thread

2.22 A thread repair tool being used to correct an internal thread

2.23 A thread repair tool being used to correct an external thread

chasers are available for spark plug hole threads. The tool will not cut a new thread, but clean and true the original thread. Make sure that you use the correct diameter and pitch tool. Similarly, external threads can be cleaned up with a die or a thread restorer file **(see illustration 2.24)**.

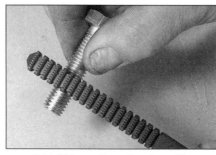

2.24 Using a thread restorer file

● It is possible to drill out the old thread and retap the component to the next thread size. This will work where there is enough surrounding material and a new bolt or screw can be obtained. Sometimes, however, this is not possible - such as where the bolt/screw passes through another component which must also be suitably modified, also in cases where a spark plug or oil drain plug cannot be obtained in a larger diameter thread size.

● The diamond-section thread insert (often known by its popular trade name of Heli-Coil) is a simple and effective method of renewing the thread and retaining the original size. A kit can be purchased which contains the tap, insert and installing tool **(see illustration 2.25)**. Drill out the damaged thread with the size drill specified **(see illustration 2.26)**. Carefully retap the thread **(see illustration 2.27)**. Install the

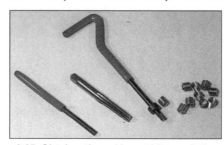

2.25 Obtain a thread insert kit to suit the thread diameter and pitch required

2.26 To install a thread insert, first drill out the original thread . . .

2.27 . . . tap a new thread . . .

2.28 . . . fit insert on the installing tool . . .

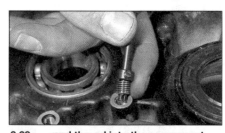

2.29 . . . and thread into the component . . .

2.30 . . . break off the tang when complete

insert on the installing tool and thread it slowly into place using a light downward pressure **(see illustrations 2.28 and 2.29)**. When positioned between a 1/4 and 1/2 turn below the surface withdraw the installing tool and use the break-off tool to press down on the tang, breaking it off **(see illustration 2.30)**.

● There are epoxy thread repair kits on the market which can rebuild stripped internal threads, although this repair should not be used on high load-bearing components.

Tools and Workshop Tips

Thread locking and sealing compounds

● Locking compounds are used in locations where the fastener is prone to loosening due to vibration or on important safety-related items which might cause loss of control of the motorcycle if they fail. It is also used where important fasteners cannot be secured by other means such as lockwashers or split pins.

● Before applying locking compound, make sure that the threads (internal and external) are clean and dry with all old compound removed. Select a compound to suit the component being secured - a non-permanent general locking and sealing type is suitable for most applications, but a high strength type is needed for permanent fixing of studs in castings. Apply a drop or two of the compound to the first few threads of the fastener, then thread it into place and tighten to the specified torque. Do not apply excessive thread locking compound otherwise the thread may be damaged on subsequent removal.

● Certain fasteners are impregnated with a dry film type coating of locking compound on their threads. Always renew this type of fastener if disturbed.

● Anti-seize compounds, such as copper-based greases, can be applied to protect threads from seizure due to extreme heat and corrosion. A common instance is spark plug threads and exhaust system fasteners.

3 Measuring tools and gauges

Feeler gauges

● Feeler gauges (or blades) are used for measuring small gaps and clearances **(see illustration 3.1)**. They can also be used to measure endfloat (sideplay) of a component on a shaft where access is not possible with a dial gauge.

● Feeler gauge sets should be treated with care and not bent or damaged. They are etched with their size on one face. Keep them clean and very lightly oiled to prevent corrosion build-up.

3.1 Feeler gauges are used for measuring small gaps and clearances - thickness is marked on one face of gauge

● When measuring a clearance, select a gauge which is a light sliding fit between the two components. You may need to use two gauges together to measure the clearance accurately.

Micrometers

● A micrometer is a precision tool capable of measuring to 0.01 or 0.001 of a millimetre. It should always be stored in its case and not in the general toolbox. It must be kept clean and never dropped, otherwise its frame or measuring anvils could be distorted resulting in inaccurate readings.

● External micrometers are used for measuring outside diameters of components and have many more applications than internal micrometers. Micrometers are available in different size ranges, eg 0 to 25 mm, 25 to 50 mm, and upwards in 25 mm steps; some large micrometers have interchangeable anvils to allow a range of measurements to be taken. Generally the largest precision measurement you are likely to take on a motorcycle is the piston diameter.

● Internal micrometers (or bore micrometers) are used for measuring inside diameters, such as valve guides and cylinder bores. Telescoping gauges and small hole gauges are used in conjunction with an external micrometer, whereas the more expensive internal micrometers have their own measuring device.

External micrometer

Note: *The conventional analogue type instrument is described. Although much easier to read, digital micrometers are considerably more expensive.*

● Always check the calibration of the micrometer before use. With the anvils closed (0 to 25 mm type) or set over a test gauge (for

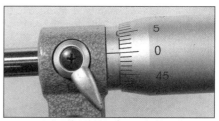

3.2 Check micrometer calibration before use

the larger types) the scale should read zero **(see illustration 3.2)**; make sure that the anvils (and test piece) are clean first. Any discrepancy can be adjusted by referring to the instructions supplied with the tool. Remember that the micrometer is a precision measuring tool - don't force the anvils closed, use the ratchet (4) on the end of the micrometer to close it. In this way, a measured force is always applied.

● To use, first make sure that the item being measured is clean. Place the anvil of the micrometer (1) against the item and use the thimble (2) to bring the spindle (3) lightly into contact with the other side of the item **(see illustration 3.3)**. Don't tighten the thimble down because this will damage the micrometer - instead use the ratchet (4) on the end of the micrometer. The ratchet mechanism applies a measured force preventing damage to the instrument.

● The micrometer is read by referring to the linear scale on the sleeve and the annular scale on the thimble. Read off the sleeve first to obtain the base measurement, then add the fine measurement from the thimble to obtain the overall reading. The linear scale on the sleeve represents the measuring range of the micrometer (eg 0 to 25 mm). The annular scale

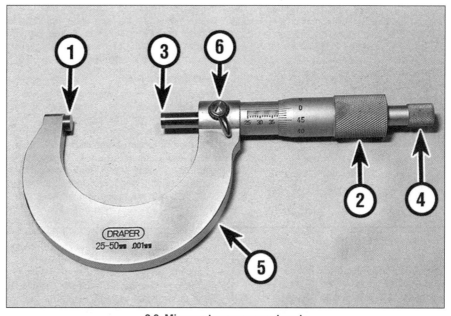

3.3 Micrometer component parts

1 Anvil	3 Spindle	5 Frame
2 Thimble	4 Ratchet	6 Locking lever

REF•10 Tools and Workshop Tips

on the thimble will be in graduations of 0.01 mm (or as marked on the frame) - one full revolution of the thimble will move 0.5 mm on the linear scale. Take the reading where the datum line on the sleeve intersects the thimble's scale. Always position the eye directly above the scale otherwise an inaccurate reading will result.

In the example shown the item measures 2.95 mm **(see illustration 3.4)**:

Linear scale	2.00 mm
Linear scale	0.50 mm
Annular scale	0.45 mm
Total figure	**2.95 mm**

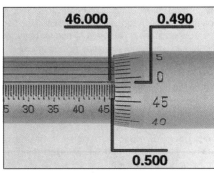

3.5 Micrometer reading of 46.99 mm on linear and annular scales . . .

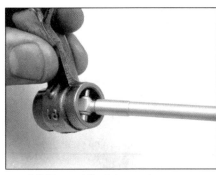

3.7 Expand the telescoping gauge in the bore, lock its position . . .

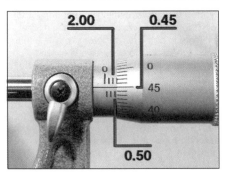

3.4 Micrometer reading of 2.95 mm

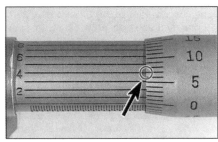

3.6 . . . and 0.004 mm on vernier scale

3.8 . . . then measure the gauge with a micrometer

Most micrometers have a locking lever (6) on the frame to hold the setting in place, allowing the item to be removed from the micrometer.

● Some micrometers have a vernier scale on their sleeve, providing an even finer measurement to be taken, in 0.001 increments of a millimetre. Take the sleeve and thimble measurement as described above, then check which graduation on the vernier scale aligns with that of the annular scale on the thimble **Note:** *The eye must be perpendicular to the scale when taking the vernier reading - if necessary rotate the body of the micrometer to ensure this.* Multiply the vernier scale figure by 0.001 and add it to the base and fine measurement figures.

In the example shown the item measures 46.994 mm **(see illustrations 3.5 and 3.6)**:

Linear scale (base)	46.000 mm
Linear scale (base)	00.500 mm
Annular scale (fine)	00.490 mm
Vernier scale	00.004 mm
Total figure	**46.994 mm**

Internal micrometer

● Internal micrometers are available for measuring bore diameters, but are expensive and unlikely to be available for home use. It is suggested that a set of telescoping gauges and small hole gauges, both of which must be used with an external micrometer, will suffice for taking internal measurements on a motorcycle.

● Telescoping gauges can be used to measure internal diameters of components. Select a gauge with the correct size range, make sure its ends are clean and insert it into the bore. Expand the gauge, then lock its position and withdraw it from the bore **(see illustration 3.7)**. Measure across the gauge ends with a micrometer **(see illustration 3.8)**.

● Very small diameter bores (such as valve guides) are measured with a small hole gauge. Once adjusted to a slip-fit inside the component, its position is locked and the gauge withdrawn for measurement with a micrometer **(see illustrations 3.9 and 3.10)**.

Vernier caliper

Note: *The conventional linear and dial gauge type instruments are described. Digital types are easier to read, but are far more expensive.*

● The vernier caliper does not provide the precision of a micrometer, but is versatile in being able to measure internal and external diameters. Some types also incorporate a depth gauge. It is ideal for measuring clutch plate friction material and spring free lengths.

● To use the conventional linear scale vernier, slacken off the vernier clamp screws (1) and set its jaws over (2), or inside (3), the item to be measured **(see illustration 3.11)**. Slide the jaw into contact, using the thumbwheel (4) for fine movement of the sliding scale (5) then tighten the clamp screws (1). Read off the main scale (6) where the zero on the sliding scale (5) intersects it, taking the whole number to the left of the zero; this provides the base measurement. View along the sliding scale and select the division which

3.9 Expand the small hole gauge in the bore, lock its position . . .

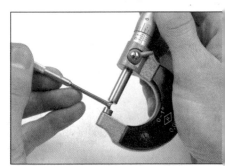

3.10 . . . then measure the gauge with a micrometer

lines up exactly with any of the divisions on the main scale, noting that the divisions usually represents 0.02 of a millimetre. Add this fine measurement to the base measurement to obtain the total reading.

Tools and Workshop Tips REF•11

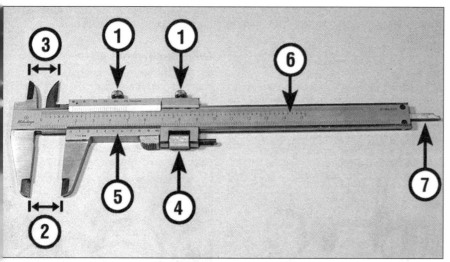

3.11 Vernier component parts (linear gauge)

1 Clamp screws
2 External jaws
3 Internal jaws
4 Thumbwheel
5 Sliding scale
6 Main scale
7 Depth gauge

In the example shown the item measures 55.92 mm **(see illustration 3.12)**:

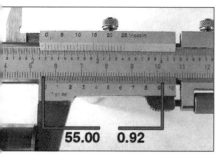

3.12 Vernier gauge reading of 55.92 mm

Base measurement	55.00 mm
Fine measurement	00.92 mm
Total figure	**55.92 mm**

● Some vernier calipers are equipped with a dial gauge for fine measurement. Before use, check that the jaws are clean, then close them fully and check that the dial gauge reads zero. If necessary adjust the gauge ring accordingly. Slacken the vernier clamp screw (1) and set its jaws over (2), or inside (3), the item to be measured **(see illustration 3.13)**. Slide the jaws into contact, using the thumbwheel (4) for fine movement. Read off the main scale (5) where the edge of the sliding scale (6) intersects it, taking the whole number to the left of the zero; this provides the base measurement. Read off the needle position on the dial gauge (7) scale to provide the fine measurement; each division represents 0.05 of a millimetre. Add this fine measurement to the base measurement to obtain the total reading.

In the example shown the item measures 55.95 mm **(see illustration 3.14)**:

Base measurement	55.00 mm
Fine measurement	00.95 mm
Total figure	**55.95 mm**

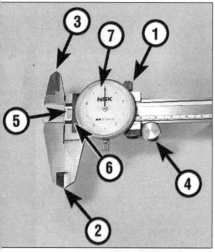

3.13 Vernier component parts (dial gauge)

1 Clamp screw
2 External jaws
3 Internal jaws
4 Thumbwheel
5 Main scale
6 Sliding scale
7 Dial gauge

3.14 Vernier gauge reading of 55.95 mm

Plastigauge

● Plastigauge is a plastic material which can be compressed between two surfaces to measure the oil clearance between them. The width of the compressed Plastigauge is measured against a calibrated scale to determine the clearance.

● Common uses of Plastigauge are for measuring the clearance between crankshaft journal and main bearing inserts, between crankshaft journal and big-end bearing inserts, and between camshaft and bearing surfaces. The following example describes big-end oil clearance measurement.

● Handle the Plastigauge material carefully to prevent distortion. Using a sharp knife, cut a length which corresponds with the width of the bearing being measured and place it carefully across the journal so that it is parallel with the shaft **(see illustration 3.15)**. Carefully install both bearing shells and the connecting rod. Without rotating the rod on the journal tighten its bolts or nuts (as applicable) to the specified torque. The connecting rod and bearings are then disassembled and the crushed Plastigauge examined.

3.15 Plastigauge placed across shaft journal

● Using the scale provided in the Plastigauge kit, measure the width of the material to determine the oil clearance **(see illustration 3.16)**. Always remove all traces of Plastigauge after use using your fingernails.

Caution: Arriving at the correct clearance demands that the assembly is torqued correctly, according to the settings and sequence (where applicable) provided by the motorcycle manufacturer.

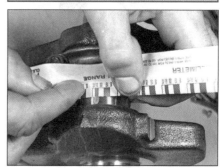

3.16 Measuring the width of the crushed Plastigauge

REF•12 Tools and Workshop Tips

Dial gauge or DTI (Dial Test Indicator)

● A dial gauge can be used to accurately measure small amounts of movement. Typical uses are measuring shaft runout or shaft endfloat (sideplay) and setting piston position for ignition timing on two-strokes. A dial gauge set usually comes with a range of different probes and adapters and mounting equipment.

● The gauge needle must point to zero when at rest. Rotate the ring around its periphery to zero the gauge.

● Check that the gauge is capable of reading the extent of movement in the work. Most gauges have a small dial set in the face which records whole millimetres of movement as well as the fine scale around the face periphery which is calibrated in 0.01 mm divisions. Read off the small dial first to obtain the base measurement, then add the measurement from the fine scale to obtain the total reading.

In the example shown the gauge reads 1.48 mm (see illustration 3.17):

Base measurement	1.00 mm
Fine measurement	0.48 mm
Total figure	**1.48 mm**

3.17 Dial gauge reading of 1.48 mm

● If measuring shaft runout, the shaft must be supported in vee-blocks and the gauge mounted on a stand perpendicular to the shaft. Rest the tip of the gauge against the centre of the shaft and rotate the shaft slowly whilst watching the gauge reading **(see illustration 3.18)**. Take several measurements along the length of the shaft and record the maximum gauge reading as the amount of runout in the shaft. **Note:** *The reading obtained will be total runout at that point - some manufacturers specify that the runout figure is halved to compare with their specified runout limit.*

● Endfloat (sideplay) measurement requires that the gauge is mounted securely to the surrounding component with its probe touching the end of the shaft. Using hand pressure, push and pull on the shaft noting the maximum endfloat recorded on the gauge **(see illustration 3.19)**.

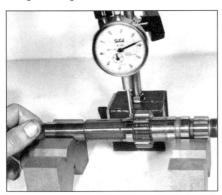

3.18 Using a dial gauge to measure shaft runout

3.19 Using a dial gauge to measure shaft endfloat

● A dial gauge with suitable adapters can be used to determine piston position BTDC on two-stroke engines for the purposes of ignition timing. The gauge, adapter and suitable length probe are installed in the place of the spark plug and the gauge zeroed at TDC. If the piston position is specified as 1.14 mm BTDC, rotate the engine back to 2.00 mm BTDC, then slowly forwards to 1.14 mm BTDC.

Cylinder compression gauges

● A compression gauge is used for measuring cylinder compression. Either the rubber-cone type or the threaded adapter type can be used. The latter is preferred to ensure a perfect seal against the cylinder head. A 0 to 300 psi (0 to 20 Bar) type gauge (for petrol/gasoline engines) will be suitable for motorcycles.

● The spark plug is removed and the gauge either held hard against the cylinder head (cone type) or the gauge adapter screwed into the cylinder head (threaded type) **(see illustration 3.20)**. Cylinder compression is measured with the engine turning over, but not running - carry out the compression test as described in *Fault Finding Equipment*. The gauge will hold the reading until manually released.

Oil pressure gauge

● An oil pressure gauge is used for measuring engine oil pressure. Most gauges come with a set of adapters to fit the thread of the take-off point **(see illustration 3.21)**. If the take-off point specified by the motorcycle manufacturer is an external oil pipe union, make sure that the specified replacement union is used to prevent oil starvation.

3.21 Oil pressure gauge and take-off point adapter (arrow)

● Oil pressure is measured with the engine running (at a specific rpm) and often the manufacturer will specify pressure limits for a cold and hot engine.

Straight-edge and surface plate

● If checking the gasket face of a component for warpage, place a steel rule or precision straight-edge across the gasket face and measure any gap between the straight edge and component with feeler gauges **(see illustration 3.22)**. Check diagonally across the component and between mounting holes **(see illustration 3.23)**.

3.22 Use a straight-edge and feeler gauges to check for warpage

3.20 Using a rubber-cone type cylinder compression gauge

3.23 Check for warpage in these directions

Tools and Workshop Tips

• Checking individual components for warpage, such as clutch plain (metal) plates, requires a perfectly flat plate or piece or plate glass and feeler gauges.

4 Torque and leverage

What is torque?

• Torque describes the twisting force about a shaft. The amount of torque applied is determined by the distance from the centre of the shaft to the end of the lever and the amount of force being applied to the end of the lever; distance multiplied by force equals torque.
• The manufacturer applies a measured torque to a bolt or nut to ensure that it will not slacken in use and to hold two components securely together without movement in the joint. The actual torque setting depends on the thread size, bolt or nut material and the composition of the components being held.
• Too little torque may cause the fastener to loosen due to vibration, whereas too much torque will distort the joint faces of the component or cause the fastener to shear off. Always stick to the specified torque setting.

Using a torque wrench

• Check the calibration of the torque wrench and make sure it has a suitable range for the job. Torque wrenches are available in Nm (Newton-metres), kgf m (kilograms-force metre), lbf ft (pounds-feet), lbf in (inch-pounds). Do not confuse lbf ft with lbf in.
• Adjust the tool to the desired torque on the scale (see illustration 4.1). If your torque wrench is not calibrated in the units specified, carefully convert the figure (see *Conversion Factors*). A manufacturer sometimes gives a torque setting as a range (8 to 10 Nm) rather than a single figure - in this case set the tool midway between the two settings. The same torque may be expressed as 9 Nm ± 1 Nm. Some torque wrenches have a method of locking the setting so that it isn't inadvertently altered during use.

• Install the bolts/nuts in their correct location and secure them lightly. Their threads must be clean and free of any old locking compound. Unless specified the threads and flange should be dry - oiled threads are necessary in certain circumstances and the manufacturer will take this into account in the specified torque figure. Similarly, the manufacturer may also specify the application of thread-locking compound.
• Tighten the fasteners in the specified sequence until the torque wrench clicks, indicating that the torque setting has been reached. Apply the torque again to double-check the setting. Where different thread diameter fasteners secure the component, as a rule tighten the larger diameter ones first.
• When the torque wrench has been finished with, release the lock (where applicable) and fully back off its setting to zero - do not leave the torque wrench tensioned. Also, do not use a torque wrench for slackening a fastener.

Angle-tightening

• Manufacturers often specify a figure in degrees for final tightening of a fastener. This usually follows tightening to a specific torque setting.
• A degree disc can be set and attached to the socket (see illustration 4.2) or a protractor can be used to mark the angle of movement on the bolt/nut head and the surrounding casting (see illustration 4.3).

4.2 Angle tightening can be accomplished with a torque-angle gauge ...

4.3 ... or by marking the angle on the surrounding component

Loosening sequences

• Where more than one bolt/nut secures a component, loosen each fastener evenly a little at a time. In this way, not all the stress of the joint is held by one fastener and the components are not likely to distort.
• If a tightening sequence is provided, work in the REVERSE of this, but if not, work from the outside in, in a criss-cross sequence (see illustration 4.4).

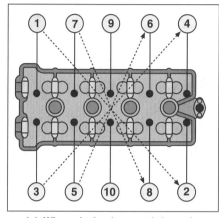

4.4 When slackening, work from the outside inwards

Tightening sequences

• If a component is held by more than one fastener it is important that the retaining bolts/nuts are tightened evenly to prevent uneven stress build-up and distortion of sealing faces. This is especially important on high-compression joints such as the cylinder head.
• A sequence is usually provided by the manufacturer, either in a diagram or actually marked in the casting. If not, always start in the centre and work outwards in a criss-cross pattern (see illustration 4.5). Start off by securing all bolts/nuts finger-tight, then set the torque wrench and tighten each fastener by a small amount in sequence until the final torque is reached. By following this practice,

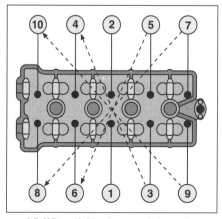

4.5 When tightening, work from the inside outwards

4.1 Set the torque wrench index mark to the setting required, in this case 12 Nm

REF•14 Tools and Workshop Tips

the joint will be held evenly and will not be distorted. Important joints, such as the cylinder head and big-end fasteners often have two- or three-stage torque settings.

Applying leverage

● Use tools at the correct angle. Position a socket wrench or spanner on the bolt/nut so that you pull it towards you when loosening. If this can't be done, push the spanner without curling your fingers around it (see illustration 4.6) - the spanner may slip or the fastener loosen suddenly, resulting in your fingers being crushed against a component.

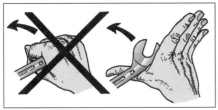

4.6 If you can't pull on the spanner to loosen a fastener, push with your hand open

● Additional leverage is gained by extending the length of the lever. The best way to do this is to use a breaker bar instead of the regular length tool, or to slip a length of tubing over the end of the spanner or socket wrench.
● If additional leverage will not work, the fastener head is either damaged or firmly corroded in place (see *Fasteners*).

5 Bearings

Bearing removal and installation

Drivers and sockets

● Before removing a bearing, always inspect the casing to see which way it must be driven out - some casings will have retaining plates or a cast step. Also check for any identifying markings on the bearing and if installed to a certain depth, measure this at this stage. Some roller bearings are sealed on one side - take note of the original fitted position.
● Bearings can be driven out of a casing using a bearing driver tool (with the correct size head) or a socket of the correct diameter. Select the driver head or socket so that it contacts the outer race of the bearing, not the balls/rollers or inner race. Always support the casing around the bearing housing with wood blocks, otherwise there is a risk of fracture. The bearing is driven out with a few blows on the driver or socket from a heavy mallet. Unless access is severely restricted (as with wheel bearings), a pin-punch is not recommended unless it is moved around the bearing to keep it square in its housing.

● The same equipment can be used to install bearings. Make sure the bearing housing is supported on wood blocks and line up the bearing in its housing. Fit the bearing as noted on removal - generally they are installed with their marked side facing outwards. Tap the bearing squarely into its housing using a driver or socket which bears only on the bearing's outer race - contact with the bearing balls/rollers or inner race will destroy it (see illustrations 5.1 and 5.2).
● Check that the bearing inner race and balls/rollers rotate freely.

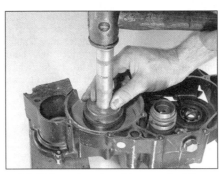

5.1 Using a bearing driver against the bearing's outer race

5.2 Using a large socket against the bearing's outer race

Pullers and slide-hammers

● Where a bearing is pressed on a shaft a puller will be required to extract it (see illustration 5.3). Make sure that the puller clamp or legs fit securely behind the bearing and are unlikely to slip out. If pulling a bearing

5.3 This bearing puller clamps behind the bearing and pressure is applied to the shaft end to draw the bearing off

off a gear shaft for example, you may have to locate the puller behind a gear pinion if there is no access to the race and draw the gear pinion off the shaft as well (see illustration 5.4).

Caution: Ensure that the puller's centre bolt locates securely against the end of the shaft and will not slip when pressure is applied. Also ensure that puller does not damage the shaft end.

5.4 Where no access is available to the rear of the bearing, it is sometimes possible to draw off the adjacent component

● Operate the puller so that its centre bolt exerts pressure on the shaft end and draw the bearing off the shaft.
● When installing the bearing on the shaft tap only on the bearing's inner race - contact with the balls/rollers or outer race will destroy the bearing. Use a socket or length of tubing as a drift which fits over the shaft end (see illustration 5.5).

5.5 When installing a bearing on a shaft use a piece of tubing which bears only on the bearing's inner race

● Where a bearing locates in a blind hole in a casing, it cannot be driven or pulled out as described above. A slide-hammer with knife edged bearing puller attachment will be required. The puller attachment passes through the bearing and when tightened expands to fit firmly behind the bearing (see illustration 5.6). By operating the slide hammer part of the tool the bearing is jarred out of its housing (see illustration 5.7).
● It is possible, if the bearing is of reasonable weight, for it to drop out of its housing if the casing is heated as described opposite. If th

Tools and Workshop Tips

5.6 Expand the bearing puller so that it locks behind the bearing . . .

5.7 . . . attach the slide hammer to the bearing puller

method is attempted, first prepare a work surface which will enable the casing to be tapped face down to help dislodge the bearing - a wood surface is ideal since it will not damage the casing's gasket surface. Wearing protective gloves, tap the heated casing several times against the work surface to dislodge the bearing under its own weight **(see illustration 5.8)**.

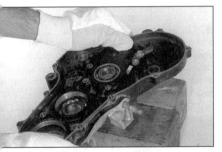

5.8 Tapping a casing face down on wood blocks can often dislodge a bearing

● Bearings can be installed in blind holes using the driver or socket method described above.

Drawbolts

● Where a bearing or bush is set in the eye of a component, such as a suspension linkage arm or connecting rod small-end, removal by drift may damage the component. Furthermore, a rubber bushing in a shock absorber eye cannot successfully be driven out of position. If access is available to a engineering press, the task is straightforward. If not, a drawbolt can be fabricated to extract the bearing or bush.

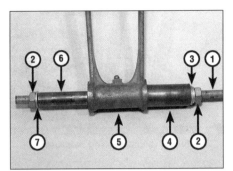

5.9 Drawbolt component parts assembled on a suspension arm

1 Bolt or length of threaded bar
2 Nuts
3 Washer (external diameter greater than tubing internal diameter)
4 Tubing (internal diameter sufficient to accommodate bearing)
5 Suspension arm with bearing
6 Tubing (external diameter slightly smaller than bearing)
7 Washer (external diameter slightly smaller than bearing)

5.10 Drawing the bearing out of the suspension arm

● To extract the bearing/bush you will need a long bolt with nut (or piece of threaded bar with two nuts), a piece of tubing which has an internal diameter larger than the bearing/bush, another piece of tubing which has an external diameter slightly smaller than the bearing/bush, and a selection of washers **(see illustrations 5.9 and 5.10)**. Note that the pieces of tubing must be of the same length, or longer, than the bearing/bush.

● The same kit (without the pieces of tubing) can be used to draw the new bearing/bush back into place **(see illustration 5.11)**.

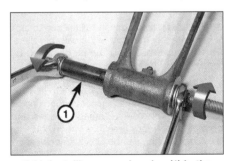

5.11 Installing a new bearing (1) in the suspension arm

Temperature change

● If the bearing's outer race is a tight fit in the casing, the aluminium casing can be heated to release its grip on the bearing. Aluminium will expand at a greater rate than the steel bearing outer race. There are several ways to do this, but avoid any localised extreme heat (such as a blow torch) - aluminium alloy has a low melting point.

● Approved methods of heating a casing are using a domestic oven (heated to 100°C) or immersing the casing in boiling water **(see illustration 5.12)**. Low temperature range localised heat sources such as a paint stripper heat gun or clothes iron can also be used **(see illustration 5.13)**. Alternatively, soak a rag in boiling water, wring it out and wrap it around the bearing housing.

> ⚠ **Warning: All of these methods require care in use to prevent scalding and burns to the hands. Wear protective gloves when handling hot components.**

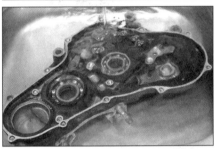

5.12 A casing can be immersed in a sink of boiling water to aid bearing removal

5.13 Using a localised heat source to aid bearing removal

● If heating the whole casing note that plastic components, such as the neutral switch, may suffer - remove them beforehand.

● After heating, remove the bearing as described above. You may find that the expansion is sufficient for the bearing to fall out of the casing under its own weight or with a light tap on the driver or socket.

● If necessary, the casing can be heated to aid bearing installation, and this is sometimes the recommended procedure if the motorcycle manufacturer has designed the housing and bearing fit with this intention.

REF•16 Tools and Workshop Tips

- Installation of bearings can be eased by placing them in a freezer the night before installation. The steel bearing will contract slightly, allowing easy insertion in its housing. This is often useful when installing steering head outer races in the frame.

Bearing types and markings

- Plain shell bearings, ball bearings, needle roller bearings and tapered roller bearings will all be found on motorcycles **(see illustrations 5.14 and 5.15)**. The ball and roller types are usually caged between an inner and outer race, but uncaged variations may be found.

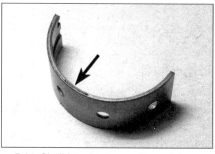

5.14 Shell bearings are either plain or grooved. They are usually identified by colour code (arrow)

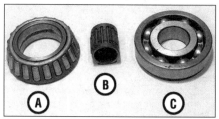

5.15 Tapered roller bearing (A), needle roller bearing (B) and ball journal bearing (C)

- Shell bearings (often called inserts) are usually found at the crankshaft main and connecting rod big-end where they are good at coping with high loads. They are made of a phosphor-bronze material and are impregnated with self-lubricating properties.
- Ball bearings and needle roller bearings consist of a steel inner and outer race with the balls or rollers between the races. They require constant lubrication by oil or grease and are good at coping with axial loads. Taper roller bearings consist of rollers set in a tapered cage set on the inner race; the outer race is separate. They are good at coping with axial loads and prevent movement along the shaft - a typical application is in the steering head.
- Bearing manufacturers produce bearings to ISO size standards and stamp one face of the bearing to indicate its internal and external diameter, load capacity and type **(see illustration 5.16)**.
- Metal bushes are usually of phosphor-bronze material. Rubber bushes are used in suspension mounting eyes. Fibre bushes have also been used in suspension pivots.

5.16 Typical bearing marking

Bearing fault finding

- If a bearing outer race has spun in its housing, the housing material will be damaged. You can use a bearing locking compound to bond the outer race in place if damage is not too severe.
- Shell bearings will fail due to damage of their working surface, as a result of lack of lubrication, corrosion or abrasive particles in the oil **(see illustration 5.17)**. Small particles of dirt in the oil may embed in the bearing material whereas larger particles will score the bearing and shaft journal. If a number of short journeys are made, insufficient heat will be generated to drive off condensation which has built up on the bearings.

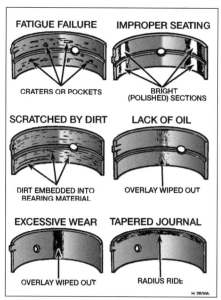

5.17 Typical bearing failures

- Ball and roller bearings will fail due to lack of lubrication or damage to the balls or rollers. Tapered-roller bearings can be damaged by overloading them. Unless the bearing is sealed on both sides, wash it in paraffin (kerosene) to remove all old grease then allow it to dry. Make a visual inspection looking to dented balls or rollers, damaged cages and worn or pitted races **(see illustration 5.18)**.
- A ball bearing can be checked for wear by listening to it when spun. Apply a film of light oil to the bearing and hold it close to the ear - hold the outer race with one hand and spin the inner

5.18 Example of ball journal bearing with damaged balls and cages

5.19 Hold outer race and listen to inner race when spun

race with the other hand **(see illustration 5.19)**. The bearing should be almost silent when spun; if it grates or rattles it is worn.

6 Oil seals

Oil seal removal and installation

- Oil seals should be renewed every time a component is dismantled. This is because the seal lips will become set to the sealing surface and will not necessarily reseal.
- Oil seals can be prised out of position using a large flat-bladed screwdriver **(see illustration 6.1)**. In the case of crankcase seals, check first that the seal is not lipped on the inside, preventing its removal with the crankcases joined.

6.1 Prise out oil seals with a large flat-bladed screwdriver

- New seals are usually installed with their marked face (containing the seal reference code) outwards and the spring side towards the fluid being retained. In certain cases, such as a two-stroke engine crankshaft seal, a double lipped seal may be used due to there being fluid or gas on each side of the joint.

Tools and Workshop Tips

- Use a bearing driver or socket which bears only on the outer hard edge of the seal to install it in the casing - tapping on the inner edge will damage the sealing lip.

Oil seal types and markings

- Oil seals are usually of the single-lipped type. Double-lipped seals are found where a liquid or gas is on both sides of the joint.
- Oil seals can harden and lose their sealing ability if the motorcycle has been in storage for a long period - renewal is the only solution.
- Oil seal manufacturers also conform to the ISO markings for seal size - these are moulded into the outer face of the seal **(see illustration 6.2)**.

6.2 These oil seal markings indicate inside diameter, outside diameter and seal thickness

7 Gaskets and sealants

Types of gasket and sealant

- Gaskets are used to seal the mating surfaces between components and keep lubricants, fluids, vacuum or pressure contained within the assembly. Aluminium gaskets are sometimes found at the cylinder joints, but most gaskets are paper-based. If the mating surfaces of the components being joined are undamaged the gasket can be installed dry, although a dab of sealant or grease will be useful to hold it in place during assembly.
- RTV (Room Temperature Vulcanising) silicone rubber sealants cure when exposed to moisture in the atmosphere. These sealants are good at filling pits or irregular gasket faces, but will tend to be forced out of the joint under very high torque. They can be used to replace a paper gasket, but first make sure that the width of the paper gasket is not essential to the shimming of internal components. RTV sealants should not be used on components containing petrol (gasoline).
- Non-hardening, semi-hardening and hard setting liquid gasket compounds can be used with a gasket or between a metal-to-metal joint. Select the sealant to suit the application: universal non-hardening sealant can be used on virtually all joints; semi-hardening on joint faces which are rough or damaged; hard setting sealant on joints which require a permanent bond and are subjected to high temperature and pressure. **Note:** *Check first if the paper gasket has a bead of sealant impregnated in its surface before applying additional sealant.*
- When choosing a sealant, make sure it is suitable for the application, particularly if being applied in a high-temperature area or in the vicinity of fuel. Certain manufacturers produce sealants in either clear, silver or black colours to match the finish of the engine. This has a particular application on motorcycles where much of the engine is exposed.
- Do not over-apply sealant. That which is squeezed out on the outside of the joint can be wiped off, whereas an excess of sealant on the inside can break off and clog oilways.

Breaking a sealed joint

- Age, heat, pressure and the use of hard setting sealant can cause two components to stick together so tightly that they are difficult to separate using finger pressure alone. Do not resort to using levers unless there is a pry point provided for this purpose **(see illustration 7.1)** or else the gasket surfaces will be damaged.
- Use a soft-faced hammer **(see illustration 7.2)** or a wood block and conventional hammer to strike the component near the mating surface. Avoid hammering against cast extremities since they may break off. If this method fails, try using a wood wedge between the two components.

Caution: If the joint will not separate, double-check that you have removed all the fasteners.

7.1 If a pry point is provided, apply gently pressure with a flat-bladed screwdriver

7.2 Tap around the joint with a soft-faced mallet if necessary - don't strike cooling fins

Removal of old gasket and sealant

- Paper gaskets will most likely come away complete, leaving only a few traces stuck on the sealing faces of the components. It is imperative that all traces are removed to ensure correct sealing of the new gasket.
- Very carefully scrape all traces of gasket away making sure that the sealing surfaces are not gouged or scored by the scraper **(see illustrations 7.3, 7.4 and 7.5)**. Stubborn deposits can be removed by spraying with an aerosol gasket remover. Final preparation of

Most components have one or two hollow locating dowels between the two gasket faces. If a dowel cannot be removed, do not resort to gripping it with pliers - it will almost certainly be distorted. Install a close-fitting socket or Phillips screwdriver into the dowel and then grip the outer edge of the dowel to free it.

7.3 Paper gaskets can be scraped off with a gasket scraper tool . . .

7.4 . . . a knife blade . . .

7.5 . . . or a household scraper

REF•18 Tools and Workshop Tips

7.6 Fine abrasive paper is wrapped around a flat file to clean up the gasket face

7.7 A kitchen scourer can be used on stubborn deposits

the gasket surface can be made with very fine abrasive paper or a plastic kitchen scourer (see illustrations 7.6 and 7.7).
● Old sealant can be scraped or peeled off components, depending on the type originally used. Note that gasket removal compounds are available to avoid scraping the components clean; make sure the gasket remover suits the type of sealant used.

8 Chains

Breaking and joining final drive chains

● Drive chains for all but small bikes are continuous and do not have a clip-type connecting link. The chain must be broken using a chain breaker tool and the new chain securely riveted together using a new soft rivet-type link. Never use a clip-type connecting link instead of a rivet-type link, except in an emergency. Various chain breaking and riveting tools are available, either as separate tools or combined as illustrated in the accompanying photographs - read the instructions supplied with the tool carefully.

> **Warning: The need to rivet the new link pins correctly cannot be overstressed - loss of control of the motorcycle is very likely to result if the chain breaks in use.**

● Rotate the chain and look for the soft link. The soft link pins look like they have been

8.1 Tighten the chain breaker to push the pin out of the link . . .

8.2 . . . withdraw the pin, remove the tool . . .

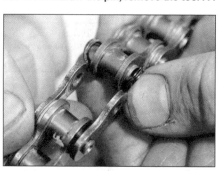

8.3 . . . and separate the chain link

deeply centre-punched instead of peened over like all the other pins (see illustration 8.9) and its sideplate may be a different colour. Position the soft link midway between the sprockets and assemble the chain breaker tool over one of the soft link pins (see illustration 8.1). Operate the tool to push the pin out through the chain (see illustration 8.2). On an O-ring chain, remove the O-rings (see illustration 8.3). Carry out the same procedure on the other soft link pin.

> **Caution: Certain soft link pins (particularly on the larger chains) may require their ends to be filed or ground off before they can be pressed out using the tool.**

● Check that you have the correct size and strength (standard or heavy duty) new soft link - do not reuse the old link. Look for the size marking on the chain sideplates (see illustration 8.10).
● Position the chain ends so that they are engaged over the rear sprocket. On an O-ring

8.4 Insert the new soft link, with O-rings, through the chain ends . . .

8.5 . . . install the O-rings over the pin ends . . .

8.6 . . . followed by the sideplate

chain, install a new O-ring over each pin of the link and insert the link through the two chain ends (see illustration 8.4). Install a new O-ring over the end of each pin, followed by the sideplate (with the chain manufacturer's marking facing outwards) (see illustrations 8.5 and 8.6). On an unsealed chain, insert the link through the two chain ends, then install the sideplate with the chain manufacturer's marking facing outwards.
● Note that it may not be possible to install the sideplate using finger pressure alone. If using a joining tool, assemble it so that the plates of the tool clamp the link and press the sideplate over the pins (see illustration 8.7). Otherwise, use two small sockets placed over

8.7 Push the sideplate into position using a clamp

Tools and Workshop Tips

8.8 Assemble the chain riveting tool over one pin at a time and tighten it fully

8.9 Pin end correctly riveted (A), pin end unriveted (B)

the rivet ends and two pieces of the wood between a G-clamp. Operate the clamp to press the sideplate over the pins.
● Assemble the joining tool over one pin (following the maker's instructions) and tighten the tool down to spread the pin end securely **(see illustrations 8.8 and 8.9)**. Do the same on the other pin.

 Warning: Check that the pin ends are secure and that there is no danger of the sideplate coming loose. If the pin ends are cracked the soft link must be renewed.

Final drive chain sizing

● Chains are sized using a three digit number, followed by a suffix to denote the chain type **(see illustration 8.10)**. Chain type is either standard or heavy duty (thicker sideplates), and also unsealed or O-ring/X-ring type.
● The first digit of the number relates to the pitch of the chain, ie the distance from the centre of one pin to the centre of the next pin **(see illustration 8.11)**. Pitch is expressed in eighths of an inch, as follows:

8.10 Typical chain size and type marking

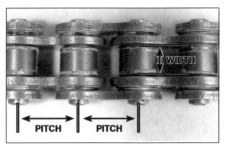

8.11 Chain dimensions

| Sizes commencing with a 4 (eg 428) have a pitch of 1/2 inch (12.7 mm) |
| Sizes commencing with a 5 (eg 520) have a pitch of 5/8 inch (15.9 mm) |
| Sizes commencing with a 6 (eg 630) have a pitch of 3/4 inch (19.1 mm) |

● The second and third digits of the chain size relate to the width of the rollers, again in imperial units, eg the 525 shown has 5/16 inch (7.94 mm) rollers **(see illustration 8.11)**.

9 Hoses

Clamping to prevent flow

● Small-bore flexible hoses can be clamped to prevent fluid flow whilst a component is worked on. Whichever method is used, ensure that the hose material is not permanently distorted or damaged by the clamp.
a) A brake hose clamp available from auto accessory shops **(see illustration 9.1)**.
b) A wingnut type hose clamp **(see illustration 9.2)**.

9.1 Hoses can be clamped with an automotive brake hose clamp . . .

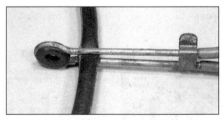

9.2 . . . a wingnut type hose clamp . . .

c) Two sockets placed each side of the hose and held with straight-jawed self-locking grips **(see illustration 9.3)**.
d) Thick card each side of the hose held between straight-jawed self-locking grips **(see illustration 9.4)**.

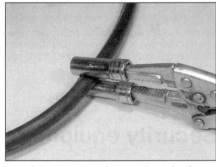

9.3 . . . two sockets and a pair of self-locking grips . . .

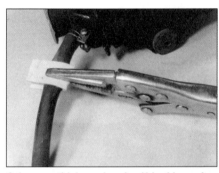

9.4 . . . or thick card and self-locking grips

Freeing and fitting hoses

● Always make sure the hose clamp is moved well clear of the hose end. Grip the hose with your hand and rotate it whilst pulling it off the union. If the hose has hardened due to age and will not move, slit it with a sharp knife and peel its ends off the union **(see illustration 9.5)**.
● Resist the temptation to use grease or soap on the unions to aid installation; although it helps the hose slip over the union it will equally aid the escape of fluid from the joint. It is preferable to soften the hose ends in hot water and wet the inside surface of the hose with water or a fluid which will evaporate.

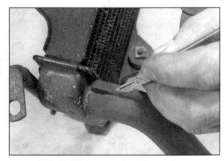

9.5 Cutting a coolant hose free with a sharp knife

REF•20 Security

Introduction

In less time than it takes to read this introduction, a thief could steal your motorcycle. Returning only to find your bike has gone is one of the worst feelings in the world. Even if the motorcycle is insured against theft, once you've got over the initial shock, you will have the inconvenience of dealing with the police and your insurance company.

The motorcycle is an easy target for the professional thief and the joyrider alike and the official figures on motorcycle theft make for depressing reading; on average a motorcycle is stolen every 16 minutes in the UK!

Motorcycle thefts fall into two categories, those stolen 'to order' and those taken by opportunists. The thief stealing to order will be on the look out for a specific make and model and will go to extraordinary lengths to obtain that motorcycle. The opportunist thief on the other hand will look for easy targets which can be stolen with the minimum of effort and risk.

Whilst it is never going to be possible to make your machine 100% secure, it is estimated that around half of all stolen motorcycles are taken by opportunist thieves. Remember that the opportunist thief is always on the look out for the easy option: if there are two similar motorcycles parked side-by-side, they will target the one with the lowest level of security. By taking a few precautions, you can reduce the chances of your motorcycle being stolen.

Security equipment

There are many specialised motorcycle security devices available and the following text summarises their applications and their good and bad points.

Once you have decided on the type of security equipment which best suits your needs, we recommended that you read one of the many equipment tests regularly carried out by the motorcycle press. These tests compare the products from all the major manufacturers and give impartial ratings on their effectiveness, value-for-money and ease of use.

No one item of security equipment can provide complete protection. It is highly recommended that two or more of the items described below are combined to increase the security of your motorcycle (a lock and chain plus an alarm system is just about ideal). The more security measures fitted to the bike, the less likely it is to be stolen.

Ensure the lock and chain you buy is of good quality and long enough to shackle your bike to a solid object

Lock and chain

Pros: *Very flexible to use; can be used to secure the motorcycle to almost any immovable object. On some locks and chains, the lock can be used on its own as a disc lock (see below).*

Cons: *Can be very heavy and awkward to carry on the motorcycle, although some types will be supplied with a carry bag which can be strapped to the pillion seat.*

● Heavy-duty chains and locks are an excellent security measure **(see illustration 1)**. Whenever the motorcycle is parked, use the lock and chain to secure the machine to a solid, immovable object such as a post or railings. This will prevent the machine from being ridden away or being lifted into the back of a van.

● When fitting the chain, always ensure the chain is routed around the motorcycle frame or swingarm **(see illustrations 2 and 3)**. Never merely pass the chain around one of the wheel rims; a thief may unbolt the wheel and lift the rest of the machine into a van, leaving you with just the wheel! Try to avoid having excess chain free, thus making it difficult to use cutting tools, and keep the chain and lock off the ground to prevent thieves attacking it with a cold chisel. Position the lock so that its lock barrel is facing downwards; this will make it harder for the thief to attack the lock mechanism.

Pass the chain through the bike's frame, rather than just through a wheel . . .

. . . and loop it around a solid object

Security

U-locks

Pros: *Highly effective deterrent which can be used to secure the bike to a post or railings. Most U-locks come with a carrier which allows the lock to be easily carried on the bike.*

Cons: *Not as flexible to use as a lock and chain.*

- These are solid locks which are similar in use to a lock and chain. U-locks are lighter than a lock and chain but not so flexible to use. The length and shape of the lock shackle limit the objects to which the bike can be secured **(see illustration 4)**.

Disc locks

Pros: *Small, light and very easy to carry; most can be stored underneath the seat.*

Cons: *Does not prevent the motorcycle being lifted into a van. Can be very embarrassing if you forget to remove the lock before attempting to ride off!*

- Disc locks are designed to be attached to the front brake disc. The lock passes through one of the holes in the disc and prevents the wheel rotating by jamming against the fork/brake caliper **(see illustration 5)**. Some are equipped with an alarm siren which sounds if the disc lock is moved; this not only acts as a theft deterrent but also as a handy reminder if you try to move the bike with the lock still fitted.

- Combining the disc lock with a length of cable which can be looped around a post or railings provides an additional measure of security **(see illustration 6)**.

Alarms and immobilisers

Pros: *Once installed it is completely hassle-free to use. If the system is 'Thatcham' or 'Sold Secure-approved', insurance companies may give you a discount.*

Cons: *Can be expensive to buy and complex to install. No system will prevent the motorcycle from being lifted into a van and taken away.*

- Electronic alarms and immobilisers are available to suit a variety of budgets. There are three different types of system available: pure alarms, pure immobilisers, and the more expensive systems which are combined alarm/immobilisers **(see illustration 7)**.
- An alarm system is designed to emit an audible warning if the motorcycle is being tampered with.
- An immobiliser prevents the motorcycle being started and ridden away by disabling its electrical systems.
- When purchasing an alarm/immobiliser system, check the cost of installing the system unless you are able to do it yourself. If the motorcycle is not used regularly, another consideration is the current drain of the system. All alarm/immobiliser systems are powered by the motorcycle's battery; purchasing a system with a very low current drain could prevent the battery losing its charge whilst the motorcycle is not being used.

U-locks can be used to secure the bike to a solid object – ensure you purchase one which is long enough

A typical disc lock attached through one of the holes in the disc

A disc lock combined with a security cable provides additional protection

A typical alarm/immobiliser system

REF•22 Security

Indelible markings can be applied to most areas of the bike – always apply the manufacturer's sticker to warn off thieves

Chemically-etched code numbers can be applied to main body panels . . .

. . . again, always ensure that the kit manufacturer's sticker is applied in a prominent position

Security marking kits

Pros: *Very cheap and effective deterrent. Many insurance companies will give you a discount on your insurance premium if a recognised security marking kit is used on your motorcycle.*

Cons: *Does not prevent the motorcycle being stolen by joyriders.*

● There are many different types of security marking kits available. The idea is to mark as many parts of the motorcycle as possible with a unique security number **(see illustrations 8, 9 and 10)**. A form will be included with the kit to register your personal details and those of the motorcycle with the kit manufacturer. This register is made available to the police to help them trace the rightful owner of any motorcycle or components which they recover should all other forms of identification have been removed. Always apply the warning stickers provided with the kit to deter thieves.

Ground anchors, wheel clamps and security posts

Pros: *An excellent form of security which will deter all but the most determined of thieves.*

Cons: *Awkward to install and can be expensive.*

● Whilst the motorcycle is at home, it is a good idea to attach it securely to the floor or a solid wall, even if it is kept in a securely locked garage. Various types of ground anchors, security posts and wheel clamps are available for this purpose **(see illustration 11)**. These security devices are either bolted to a solid concrete or brick structure or can be cemented into the ground.

Permanent ground anchors provide an excellent level of security when the bike is at home

Security at home

A high percentage of motorcycle thefts are from the owner's home. Here are some things to consider whenever your motorcycle is at home:

✔ Where possible, always keep the motorcycle in a securely locked garage. Never rely solely on the standard lock on the garage door, these are usual hopelessly inadequate. Fit an additional locking mechanism to the door and consider having the garage alarmed. A security light, activated by a movement sensor, is also a good investment.

✔ Always secure the motorcycle to the ground or a wall, even if it is inside a securely locked garage.

✔ Do not regularly leave the motorcycle outside your home, try to keep it out of sight wherever possible. If a garage is not available, fit a motorcycle cover over the bike to disguise its true identity.

✔ It is not uncommon for thieves to follow a motorcyclist home to find out where the bike is kept. They will then return at a later date. Be aware of this whenever you are returning home on your motorcycle. If you suspect you are being followed, do not return home, instead ride to a garage or shop and stop as a precaution.

✔ When selling a motorcycle, do not provide your home address or the location where the bike is normally kept. Arrange to meet the buyer at a location away from your home. Thieves have been known to pose as potential buyers to find out where motorcycles are kept and then return later to steal them.

Security away from the home

As well as fitting security equipment to your motorcycle here are a few general rules to follow whenever you park your motorcycle.

✔ Park in a busy, public place.
✔ Use car parks which incorporate security features, such as CCTV.

✔ At night, park in a well-lit area, preferably directly underneath a street light.
✔ Engage the steering lock.
✔ Secure the motorcycle to a solid, immovable object such as a post or railings with an additional lock. If this is not possible, secure the bike to a friend's motorcycle. Some public parking places provide security loops for motorcycles.
✔ Never leave your helmet or luggage attached to the motorcycle. Take them with you at all times.

Lubricants and fluids

A wide range of lubricants, fluids and cleaning agents is available for motor-cycles. This is a guide as to what is available, its applications and properties.

Four-stroke engine oil

● Engine oil is without doubt the most important component of any four-stroke engine. Modern motorcycle engines place a lot of demands on their oil and choosing the right type is essential. Using an unsuitable oil will lead to an increased rate of engine wear and could result in serious engine damage. Before purchasing oil, always check the recommended oil specification given by the manufacturer. The manufacturer will state a recommended 'type or classification' and also a specific 'viscosity' range for engine oil.

● The oil 'type or classification' is identified by its API (American Petroleum Institute) rating. The API rating will be in the form of two letters, e.g. SG. The S identifies the oil as being suitable for use in a petrol (gasoline) engine (S stands for spark ignition) and the second letter, ranging from A to J, identifies the oil's performance rating. The later this letter, the higher the specification of the oil; for example API SG oil exceeds the requirements of API SF oil. **Note:** *On some oils there may also be a second rating consisting of another two letters, the first letter being C, e.g. API SF/CD. This rating indicates the oil is also suitable for use in a diesel engines (the C stands for compression ignition) and is thus of no relevance for motorcycle use.*

● The 'viscosity' of the oil is identified by its SAE (Society of Automotive Engineers) rating. All modern engines require multigrade oils and the SAE rating will consist of two numbers, the first followed by a W, e.g. 10W/40. The first number indicates the viscosity rating of the oil at low temperatures (W stands for winter – tested at –20°C) and the second number represents the viscosity of the oil at high temperatures (tested at 100°C). The lower the number, the thinner the oil. For example an oil with an SAE 10W/40 rating will give better cold starting and running than an SAE 15W/40 oil.

● As well as ensuring the 'type' and 'viscosity' of the oil match the recommendations, another consideration to make when buying engine oil is whether to purchase a standard mineral-based oil, a semi-synthetic oil (also known as a synthetic blend or synthetic-based oil) or a fully-synthetic oil. Although all oils will have a similar rating and viscosity, their cost will vary considerably; mineral-based oils are the cheapest, the fully-synthetic oils the most expensive with the semi-synthetic oils falling somewhere in-between. This decision is very much up to the owner, but it should be noted that modern synthetic oils have far better lubricating and cleaning qualities than traditional mineral-based oils and tend to retain these properties for far longer. Bearing in mind the operating conditions inside a modern, high-revving motorcycle engine it is highly recommended that a fully synthetic oil is used. The extra expense at each service could save you money in the long term by preventing premature engine wear.

● As a final note always ensure that the oil is specifically designed for use in motorcycle engines. Engine oils designed primarily for use in car engines sometimes contain additives or friction modifiers which could cause clutch slip on a motorcycle fitted with a wet-clutch.

Two-stroke engine oil

● Modern two-stroke engines, with their high power outputs, place high demands on their oil. If engine seizure is to be avoided it is essential that a high-quality oil is used. Two-stroke oils differ hugely from four-stroke oils. The oil lubricates only the crankshaft and piston(s) (the transmission has its own lubricating oil) and is used on a total-loss basis where it is burnt completely during the combustion process.

● The Japanese have recently introduced a classification system for two-stroke oils, the JASO rating. This rating is in the form of two letters, either FA, FB or FC – FA is the lowest classification and FC the highest. Ensure the oil being used meets or exceeds the recommended rating specified by the manufacturer.

● As well as ensuring the oil rating matches the recommendation, another consideration to make when buying engine oil is whether to purchase a standard mineral-based oil, a semi-synthetic oil (also known as a synthetic blend or synthetic-based oil) or a fully-synthetic oil. The cost of each type of oil varies considerably; mineral-based oils are the cheapest, the fully-synthetic oils the most expensive with the semi-synthetic oils falling somewhere in-between. This decision is very much up to the owner, but it should be noted that modern synthetic oils have far better lubricating properties and burn cleaner than traditional mineral-based oils. It is therefore recommended that a fully synthetic oil is used. The extra expense could save you money in the long term by preventing premature engine wear, engine performance will be improved, carbon deposits and exhaust smoke will be reduced.

REF•24 Lubricants and fluids

● Always ensure that the oil is specifically designed for use in an injector system. Many high quality two-stroke oils are designed for competition use and need to be pre-mixed with fuel. These oils are of a much higher viscosity and are not designed to flow through the injector pumps used on road-going two-stroke motorcycles.

Transmission (gear) oil

● On a two-stroke engine, the transmission and clutch are lubricated by their own separate oil bath which must be changed in accordance with the Maintenance Schedule.
● Although the engine and transmission units of most four-strokes use a common lubrication supply, there are some exceptions where the engine and gearbox have separate oil reservoirs and a dry clutch is used.
● Motorcycle manufacturers will either recommend a monograde transmission oil or a four-stroke multigrade engine oil to lubricate the transmission.
● Transmission oils, or gear oils as they are often called, are designed specifically for use in transmission systems. The viscosity of these oils is represented by an SAE number, but the scale of measurement applied is different to that used to grade engine oils. As a rough guide a SAE90 gear oil will be of the same viscosity as an SAE50 engine oil.

Shaft drive oil

● On models equipped with shaft final drive, the shaft drive gears are will have their own oil supply. The manufacturer will state a recommended 'type or classification' and also a specific 'viscosity' range in the same manner as for four-stroke engine oil.
● Gear oil classification is given by the number which follows the API GL (GL standing for gear lubricant) rating, the higher the number, the higher the specification of the oil, e.g. API GL5 oil is a higher specification than API GL4 oil. Ensure the oil meets or exceeds the classification specified and is of the correct viscosity. The viscosity of gear oils is also represented by an SAE number but the scale of measurement used is different to that used to grade engine oils. As a rough guide an SAE90 gear oil will be of the same viscosity as an SAE50 engine oil.
● If the use of an EP (Extreme Pressure) gear oil is specified, ensure the oil purchased is suitable.

Fork oil and suspension fluid

● Conventional telescopic front forks are hydraulic and require fork oil to work. To ensure the forks function correctly, the fork oil must be changed in accordance with the Maintenance Schedule.
● Fork oil is available in a variety of viscosities, identified by their SAE rating; fork oil ratings vary from light (SAE 5) to heavy (SAE 30). When purchasing fork oil, ensure the viscosity rating matches that specified by the manufacturer.
● Some lubricant manufacturers also produce a range of high-quality suspension fluids which are very similar to fork oil but are designed mainly for competition use. These fluids may have a different viscosity rating system which is not to be confused with the SAE rating of normal fork oil. Refer to the manufacturer's instructions if in any doubt.

Brake and clutch fluid

● All disc brake systems and some clutch systems are hydraulically operated. To ensure correct operation, the hydraulic fluid must be changed in accordance with the Maintenance Schedule.
● Brake and clutch fluid is classified by its DOT rating with most motorcycle manufacturers specifying DOT 3 or 4 fluid. Both fluid types are glycol-based and can be mixed together without adverse effect; DOT 4 fluid exceeds the requirements of DOT 3 fluid. Although it is safe to use DOT 4 fluid in a system designed for use with DOT 3 fluid, never use DOT 3 fluid in a system which specifies the use of DOT 4 as this will adversely affect the system's performance. The type required for the system will be marked on the fluid reservoir cap.
● Some manufacturers also produce a DOT 5 hydraulic fluid. DOT 5 hydraulic fluid is silicone-based and is not compatible with the glycol-based DOT 3 and 4 fluids. Never mix DOT 5 fluid with DOT 3 or 4 fluid as this will seriously affect the performance of the hydraulic system.

Coolant/antifreeze

● When purchasing coolant/antifreeze, always ensure it is suitable for use in an aluminium engine and contains corrosion inhibitors to prevent possible blockages of the internal coolant passages of the system. As a general rule, most coolants are designed to be used neat and should not be diluted whereas antifreeze can be mixed with distilled water to provide a coolant solution of the required strength. Refer to the manufacturer's instructions on the bottle.
● Ensure the coolant is changed in accordance with the Maintenance Schedule.

Chain lube

● Chain lube is an aerosol-type spray lubricant specifically designed for use on motorcycle final drive chains. Chain lube has two functions, to minimise friction between the final drive chain and sprockets and to prevent corrosion of the chain. Regular use of a good-quality chain lube will extend the life of the drive chain and sprockets and thus maximise the power being transmitted from the transmission to the rear wheel.
● When using chain lube, always allow some time for the solvents in the lube to evaporate before riding the motorcycle. This will minimise the amount of lube which will

Lubricants and fluids

'fling' off from the chain when the motorcycle is used. If the motorcycle is equipped with an 'O-ring' chain, ensure the chain lube is labelled as being suitable for use on 'O-ring' chains.

Degreasers and solvents

● There are many different types of solvents and degreasers available to remove the grime and grease which accumulate around the motorcycle during normal use. Degreasers and solvents are usually available as an aerosol-type spray or as a liquid which you apply with a brush. Always closely follow the manufacturer's instructions and wear eye protection during use. Be aware that many solvents are flammable and may give off noxious fumes; take adequate precautions when using them (see Safety First!).

● For general cleaning, use one of the many solvents or degreasers available from most motorcycle accessory shops. These solvents are usually applied then left for a certain time before being washed off with water.

Brake cleaner is a solvent specifically designed to remove all traces of oil, grease and dust from braking system components. Brake cleaner is designed to evaporate quickly and leaves behind no residue.

Carburettor cleaner is an aerosol-type solvent specifically designed to clear carburettor blockages and break down the hard deposits and gum often found inside carburettors during overhaul.

Contact cleaner is an aerosol-type solvent designed for cleaning electrical components. The cleaner will remove all traces of oil and dirt from components such as switch contacts or fouled spark plugs and then dry, leaving behind no residue.

Gasket remover is an aerosol-type solvent designed for removing stubborn gaskets from engine components during overhaul. Gasket remover will minimise the amount of scraping required to remove the gasket and therefore reduce the risk of damage to the mating surface.

Spray lubricants

● Aerosol-based spray lubricants are widely available and are excellent for lubricating lever pivots and exposed cables and switches. Try to use a lubricant which is of the dry-film type as the fluid evaporates, leaving behind a dry-film of lubricant. Lubricants which leave behind an oily residue will attract dust and dirt which will increase the rate of wear of the cable/lever.

● Most lubricants also act as a moisture dispersant and a penetrating fluid. This means they can also be used to 'dry out' electrical components such as wiring connectors or switches as well as helping to free seized fasteners.

Greases

● Grease is used to lubricate many of the pivot-points. A good-quality multi-purpose grease is suitable for most applications but some manufacturers will specify the use of specialist greases for use on components such as swingarm and suspension linkage bushes. These specialist greases can be purchased from most motorcycle (or car) accessory shops; commonly specified types include molybdenum disulphide grease, lithium-based grease, graphite-based grease, silicone-based grease and high-temperature copper-based grease.

Gasket sealing compounds

● Gasket sealing compounds can be used in conjunction with gaskets, to improve their sealing capabilities, or on their own to seal metal-to-metal joints. Depending on their type, sealing compounds either set hard or stay relatively soft and pliable.

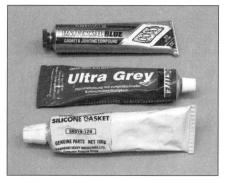

● When purchasing a gasket sealing compound, ensure that it is designed specifically for use on an internal combustion engine. General multi-purpose sealants available from DIY stores may appear visibly similar but they are not designed to withstand the extreme heat or contact with fuel and oil encountered when used on an engine (see 'Tools and Workshop Tips' for further information).

Thread locking compound

● Thread locking compounds are used to secure certain threaded fasteners in position to prevent them from loosening due to vibration. Thread locking compounds can be purchased from most motorcycle (and car) accessory shops. Ensure the threads of the both components are completely clean and dry before sparingly applying the locking compound (see 'Tools and Workshop Tips' for further information).

Fuel additives

● Fuel additives which protect and clean the fuel system components are widely available. These additives are designed to remove all traces of deposits that build up on the carburettors/injectors and prevent wear, helping the fuel system to operate more efficiently. If a fuel additive is being used, check that it is suitable for use with your motorcycle, especially if your motorcycle is equipped with a catalytic converter.

● Octane boosters are also available. These additives are designed to improve the performance of highly-tuned engines being run on normal pump-fuel and are of no real use on standard motorcycles.

Conversion Factors

Length (distance)
Inches (in)	x 25.4	= Millimetres (mm)	x 0.0394	=	Inches (in)
Feet (ft)	x 0.305	= Metres (m)	x 3.281	=	Feet (ft)
Miles	x 1.609	= Kilometres (km)	x 0.621	=	Miles

Volume (capacity)
Cubic inches (cu in; in^3)	x 16.387	= Cubic centimetres (cc; cm^3)	x 0.061	=	Cubic inches (cu in; in^3)
Imperial pints (Imp pt)	x 0.568	= Litres (l)	x 1.76	=	Imperial pints (Imp pt)
Imperial quarts (Imp qt)	x 1.137	= Litres (l)	x 0.88	=	Imperial quarts (Imp qt)
Imperial quarts (Imp qt)	x 1.201	= US quarts (US qt)	x 0.833	=	Imperial quarts (Imp qt)
US quarts (US qt)	x 0.946	= Litres (l)	x 1.057	=	US quarts (US qt)
Imperial gallons (Imp gal)	x 4.546	= Litres (l)	x 0.22	=	Imperial gallons (Imp gal)
Imperial gallons (Imp gal)	x 1.201	= US gallons (US gal)	x 0.833	=	Imperial gallons (Imp gal)
US gallons (US gal)	x 3.785	= Litres (l)	x 0.264	=	US gallons (US gal)

Mass (weight)
Ounces (oz)	x 28.35	= Grams (g)	x 0.035	=	Ounces (oz)
Pounds (lb)	x 0.454	= Kilograms (kg)	x 2.205	=	Pounds (lb)

Force
Ounces-force (ozf; oz)	x 0.278	= Newtons (N)	x 3.6	=	Ounces-force (ozf; oz)
Pounds-force (lbf; lb)	x 4.448	= Newtons (N)	x 0.225	=	Pounds-force (lbf; lb)
Newtons (N)	x 0.1	= Kilograms-force (kgf; kg)	x 9.81	=	Newtons (N)

Pressure
Pounds-force per square inch (psi; lbf/in^2; lb/in^2)	x 0.070	= Kilograms-force per square centimetre (kgf/cm^2; kg/cm^2)	x 14.223	=	Pounds-force per square inch (psi; lbf/in^2; lb/in^2)
Pounds-force per square inch (psi; lbf/in^2; lb/in^2)	x 0.068	= Atmospheres (atm)	x 14.696	=	Pounds-force per square inch (psi; lbf/in^2; lb/in^2)
Pounds-force per square inch (psi; lbf/in^2; lb/in^2)	x 0.069	= Bars	x 14.5	=	Pounds-force per square inch (psi; lbf/in^2; lb/in^2)
Pounds-force per square inch (psi; lbf/in^2; lb/in^2)	x 6.895	= Kilopascals (kPa)	x 0.145	=	Pounds-force per square inch (psi; lbf/in^2; lb/in^2)
Kilopascals (kPa)	x 0.01	= Kilograms-force per square centimetre (kgf/cm^2; kg/cm^2)	x 98.1	=	Kilopascals (kPa)
Millibar (mbar)	x 100	= Pascals (Pa)	x 0.01	=	Millibar (mbar)
Millibar (mbar)	x 0.0145	= Pounds-force per square inch (psi; lbf/in^2; lb/in^2)	x 68.947	=	Millibar (mbar)
Millibar (mbar)	x 0.75	= Millimetres of mercury (mmHg)	x 1.333	=	Millibar (mbar)
Millibar (mbar)	x 0.401	= Inches of water (inH$_2$O)	x 2.491	=	Millibar (mbar)
Millimetres of mercury (mmHg)	x 0.535	= Inches of water (inH$_2$O)	x 1.868	=	Millimetres of mercury (mmHg)
Inches of water (inH$_2$O)	x 0.036	= Pounds-force per square inch (psi; lbf/in^2; lb/in^2)	x 27.68	=	Inches of water (inH$_2$O)

Torque (moment of force)
Pounds-force inches (lbf in; lb in)	x 1.152	= Kilograms-force centimetre (kgf cm; kg cm)	x 0.868	=	Pounds-force inches (lbf in; lb in)
Pounds-force inches (lbf in; lb in)	x 0.113	= Newton metres (Nm)	x 8.85	=	Pounds-force inches (lbf in; lb in)
Pounds-force inches (lbf in; lb in)	x 0.083	= Pounds-force feet (lbf ft; lb ft)	x 12	=	Pounds-force inches (lbf in; lb in)
Pounds-force feet (lbf ft; lb ft)	x 0.138	= Kilograms-force metres (kgf m; kg m)	x 7.233	=	Pounds-force feet (lbf ft; lb ft)
Pounds-force feet (lbf ft; lb ft)	x 1.356	= Newton metres (Nm)	x 0.738	=	Pounds-force feet (lbf ft; lb ft)
Newton metres (Nm)	x 0.102	= Kilograms-force metres (kgf m; kg m)	x 9.804	=	Newton metres (Nm)

Power
Horsepower (hp)	x 745.7	= Watts (W)	x 0.0013	=	Horsepower (hp)

Velocity (speed)
Miles per hour (miles/hr; mph)	x 1.609	= Kilometres per hour (km/hr; kph)	x 0.621	=	Miles per hour (miles/hr; mph)

Fuel consumption*
Miles per gallon (mpg)	x 0.354	= Kilometres per litre (km/l)	x 2.825	=	Miles per gallon (mpg)

Temperature

Degrees Fahrenheit = (°C x 1.8) + 32   Degrees Celsius (Degrees Centigrade; °C) = (°F - 32) x 0.56

It is common practice to convert from miles per gallon (mpg) to litres/100 kilometres (l/100km), where mpg x l/100 km = 282

MOT Test Checks

About the MOT Test

In the UK, all vehicles more than three years old are subject to an annual test to ensure that they meet minimum safety requirements. A current test certificate must be issued before a machine can be used on public roads, and is required before a road fund licence can be issued. Riding without a current test certificate will also invalidate your insurance.

For most owners, the MOT test is an annual cause for anxiety, and this is largely due to owners not being sure what needs to be checked prior to submitting the motorcycle for testing. The simple answer is that a fully roadworthy motorcycle will have no difficulty in passing the test.

This is a guide to getting your motorcycle through the MOT test. Obviously it will not be possible to examine the motorcycle to the same standard as the professional MOT tester, particularly in view of the equipment required for some of the checks. However, working through the following procedures will enable you to identify any problem areas before submitting the motorcycle for the test.

It has only been possible to summarise the test requirements here, based on the regulations in force at the time of printing. Test standards are becoming increasingly stringent, although there are some exemptions for older vehicles. More information about the MOT test can be obtained from the TSO publications, *How Safe is your Motorcycle* and *The MOT Inspection Manual for Motorcycle Testing*.

Many of the checks require that one of the wheels is raised off the ground. If the motorcycle doesn't have a centre stand, note that an auxiliary stand will be required. Additionally, the help of an assistant may prove useful.

Certain exceptions apply to machines under 50 cc, machines without a lighting system, and Classic bikes - if in doubt about any of the requirements listed below seek confirmation from an MOT tester prior to submitting the motorcycle for the test.

Check that the frame number is clearly visible.

> **HAYNES HiNT** *If a component is in borderline condition, the tester has discretion in deciding whether to pass or fail it. If the motorcycle presented is clean and evidently well cared for, the tester may be more inclined to pass a borderline component than if the motorcycle is scruffy and apparently neglected.*

Electrical System

Lights, turn signals, horn and reflector

✔ With the ignition on, check the operation of the following electrical components. **Note:** *The electrical components on certain small-capacity machines are powered by the generator, requiring that the engine is run for this check.*

 a) Headlight and tail light. Check that both illuminate in the low and high beam switch positions.
 b) Position lights. Check that the front position (or sidelight) and tail light illuminate in this switch position.
 c) Turn signals. Check that all flash at the correct rate, and that the warning light(s) function correctly. Check that the turn signal switch works correctly.
 d) Hazard warning system (where fitted). Check that all four turn signals flash in this switch position.
 e) Brake stop light. Check that the light comes on when the front and rear brakes are independently applied. Models first used on or after 1st April 1986 must have a brake light switch on each brake.
 f) Horn. Check that the sound is continuous and of reasonable volume.

✔ Check that there is a red reflector on the rear of the machine, either mounted separately or as part of the tail light lens.
✔ Check the condition of the headlight, tail light and turn signal lenses.

Headlight beam height

✔ The MOT tester will perform a headlight beam height check using specialised beam setting equipment **(see illustration 1)**. This equipment will not be available to the home mechanic, but if you suspect that the headlight is incorrectly set or may have been maladjusted in the past, you can perform a rough test as follows.

✔ Position the bike in a straight line facing a brick wall. The bike must be off its stand, upright and with a rider seated. Measure the height from the ground to the centre of the headlight and mark a horizontal line on the wall at this height. Position the motorcycle 3.8 metres from the wall and draw a vertical line up the wall central to the centreline of the motorcycle. Switch to dipped beam and check that the beam pattern falls slightly lower than the horizontal line and to the left of the vertical line **(see illustration 2)**.

Headlight beam height checking equipment

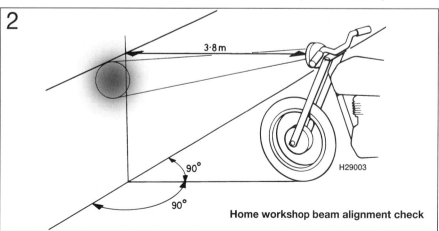

Home workshop beam alignment check

REF•28 MOT Test Checks

Exhaust System and Final Drive

Exhaust

✔ Check that the exhaust mountings are secure and that the system does not foul any of the rear suspension components.
✔ Start the motorcycle. When the revs are increased, check that the exhaust is neither holed nor leaking from any of its joints. On a linked system, check that the collector box is not leaking due to corrosion.

✔ Note that the exhaust decibel level ("loudness" of the exhaust) is assessed at the discretion of the tester. If the motorcycle was first used on or after 1st January 1985 the silencer must carry the BSAU 193 stamp, or a marking relating to its make and model, or be of OE (original equipment) manufacture. If the silencer is marked NOT FOR ROAD USE, RACING USE ONLY or similar, it will fail the MOT.

Final drive

✔ On chain or belt drive machines, check that the chain/belt is in good condition and does not have excessive slack. Also check that the sprocket is securely mounted on the rear wheel hub. Check that the chain/belt guard is in place.
✔ On shaft drive bikes, check for oil leaking from the drive unit and fouling the rear tyre.

Steering and Suspension

Steering

✔ With the front wheel raised off the ground, rotate the steering from lock to lock. The handlebar or switches must not contact the fuel tank or be close enough to trap the rider's hand. Problems can be caused by damaged lock stops on the lower yoke and frame, or by the fitting of non-standard handlebars.
✔ When performing the lock to lock check, also ensure that the steering moves freely without drag or notchiness. Steering movement can be impaired by poorly routed cables, or by overtight head bearings or worn bearings. The tester will perform a check of the steering head bearing lower race by mounting the front wheel on a surface plate, then performing a lock to lock check with the weight of the machine on the lower bearing (see illustration 3).
✔ Grasp the fork sliders (lower legs) and attempt to push and pull on the forks (see

Front wheel mounted on a surface plate for steering head bearing lower race check

illustration 4). Any play in the steering head bearings will be felt. Note that in extreme cases, wear of the front fork bushes can be misinterpreted for head bearing play.
✔ Check that the handlebars are securely mounted.
✔ Check that the handlebar grip rubbers are secure. They should by bonded to the bar left end and to the throttle cable pulley on the right end.

Front suspension

✔ With the motorcycle off the stand, hold the front brake on and pump the front forks up and down (see illustration 5). Check that they are adequately damped.

Checking the steering head bearings for freeplay

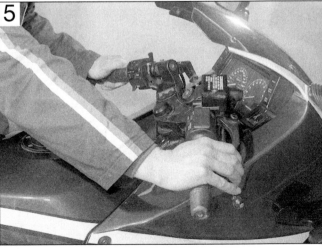

Hold the front brake on and pump the front forks up and down to check operation

MOT Test Checks

Inspect the area around the fork dust seal for oil leakage (arrow)

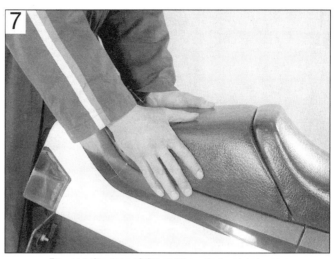
Bounce the rear of the motorcycle to check rear suspension operation

Checking for rear suspension linkage play

✔ Inspect the area above and around the front fork oil seals **(see illustration 6)**. There should be no sign of oil on the fork tube (stanchion) nor leaking down the slider (lower leg). On models so equipped, check that there is no oil leaking from the anti-dive units.
✔ On models with swingarm front suspension, check that there is no freeplay in the linkage when moved from side to side.

Rear suspension

✔ With the motorcycle off the stand and an assistant supporting the motorcycle by its handlebars, bounce the rear suspension **(see illustration 7)**. Check that the suspension components do not foul on any of the cycle parts and check that the shock absorber(s) provide adequate damping.
✔ Visually inspect the shock absorber(s) and check that there is no sign of oil leakage from its damper. This is somewhat restricted on certain single shock models due to the location of the shock absorber.
✔ With the rear wheel raised off the ground, grasp the wheel at the highest point and attempt to pull it up **(see illustration 8)**. Any play in the swingarm pivot or suspension linkage bearings will be felt as movement.
Note: *Do not confuse play with actual suspension movement.* Failure to lubricate suspension linkage bearings can lead to bearing failure **(see illustration 9)**.
✔ With the rear wheel raised off the ground, grasp the swingarm ends and attempt to move the swingarm from side to side and forwards and backwards - any play indicates wear of the swingarm pivot bearings **(see illustration 10)**.

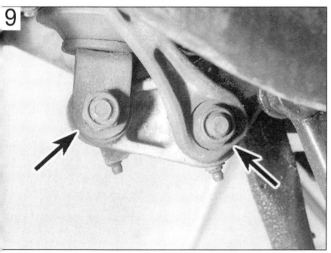
Worn suspension linkage pivots (arrows) are usually the cause of play in the rear suspension

Grasp the swingarm at the ends to check for play in its pivot bearings

REF•30 MOT Test Checks

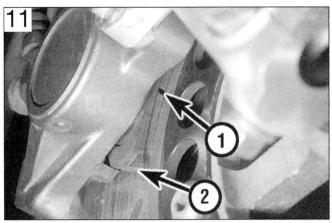

Brake pad wear can usually be viewed without removing the caliper. Most pads have wear indicator grooves (1) and some also have indicator tangs (2)

On drum brakes, check the angle of the operating lever with the brake fully applied. Most drum brakes have a wear indicator pointer and scale.

Brakes, Wheels and Tyres

Brakes

✔ With the wheel raised off the ground, apply the brake then free it off, and check that the wheel is about to revolve freely without brake drag.
✔ On disc brakes, examine the disc itself. Check that it is securely mounted and not cracked.
✔ On disc brakes, view the pad material through the caliper mouth and check that the pads are not worn down beyond the limit **(see illustration 11)**.
✔ On drum brakes, check that when the brake is applied the angle between the operating lever and cable or rod is not too great **(see illustration 12)**. Check also that the operating lever doesn't foul any other components.
✔ On disc brakes, examine the flexible hoses from top to bottom. Have an assistant hold the brake on so that the fluid in the hose is under pressure, and check that there is no sign of fluid leakage, bulges or cracking. If there are any metal brake pipes or unions, check that these are free from corrosion and damage. Where a brake-linked anti-dive system is fitted, check the hoses to the anti-dive in a similar manner.
✔ Check that the rear brake torque arm is secure and that its fasteners are secured by self-locking nuts or castellated nuts with split-pins or R-pins **(see illustration 13)**.
✔ On models with ABS, check that the self-check warning light in the instrument panel works.
✔ The MOT tester will perform a test of the motorcycle's braking efficiency based on a calculation of rider and motorcycle weight. Although this cannot be carried out at home, you can at least ensure that the braking systems are properly maintained. For hydraulic disc brakes, check the fluid level, lever/pedal feel (bleed of air if its spongy) and pad material. For drum brakes, check adjustment, cable or rod operation and shoe lining thickness.

Wheels and tyres

✔ Check the wheel condition. Cast wheels should be free from cracks and if of the built-up design, all fasteners should be secure. Spoked wheels should be checked for broken, corroded, loose or bent spokes.
✔ With the wheel raised off the ground, spin the wheel and visually check that the tyre and wheel run true. Check that the tyre does not foul the suspension or mudguards.
✔ With the wheel raised off the ground grasp the wheel and attempt to move it about the axle (spindle) **(see illustration 14)**. Any play felt here indicates wheel bearing failure.

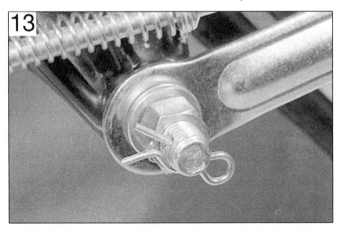

Brake torque arm must be properly secured at both ends

Check for wheel bearing play by trying to move the wheel about the axle (spindle)

MOT Test Checks

Checking the tyre tread depth

Tyre direction of rotation arrow can be found on tyre sidewall

Castellated type wheel axle (spindle) nut must be secured by a split pin or R-pin

Two straightedges are used to check wheel alignment

✔ Check the tyre tread depth, tread condition and sidewall condition **(see Illustration 15)**.
✔ Check the tyre type. Front and rear tyre types must be compatible and be suitable for road use. Tyres marked NOT FOR ROAD USE, COMPETITION USE ONLY or similar, will fail the MOT.

✔ If the tyre sidewall carries a direction of rotation arrow, this must be pointing in the direction of normal wheel rotation **(see illustration 16)**.
✔ Check that the wheel axle (spindle) nuts (where applicable) are properly secured. A self-locking nut or castellated nut with a split-pin or R-pin can be used **(see illustration 17)**.
✔ Wheel alignment is checked with the motorcycle off the stand and a rider seated. With the front wheel pointing straight ahead, two perfectly straight lengths of metal or wood and placed against the sidewalls of both tyres **(see illustration 18)**. The gap each side of the front tyre must be equidistant on both sides. Incorrect wheel alignment may be due to a cocked rear wheel (often as the result of poor chain adjustment) or in extreme cases, a bent frame.

General checks and condition

✔ Check the security of all major fasteners, bodypanels, seat, fairings (where fitted) and mudguards.

✔ Check that the rider and pillion footrests, handlebar levers and brake pedal are securely mounted.

✔ Check for corrosion on the frame or any load-bearing components. If severe, this may affect the structure, particularly under stress.

Sidecars

A motorcycle fitted with a sidecar requires additional checks relating to the stability of the machine and security of attachment and swivel joints, plus specific wheel alignment (toe-in) requirements. Additionally, tyre and lighting requirements differ from conventional motorcycle use. Owners are advised to check MOT test requirements with an official test centre.

REF•32 Storage

Preparing for storage

Before you start

If repairs or an overhaul is needed, see that this is carried out now rather than left until you want to ride the bike again.

Give the bike a good wash and scrub all dirt from its underside. Make sure the bike dries completely before preparing for storage.

Engine

● Remove the spark plug(s) and lubricate the cylinder bores with approximately a teaspoon of motor oil using a spout-type oil can **(see illustration 1)**. Reinstall the spark plug(s). Crank the engine over a couple of times to coat the piston rings and bores with oil. If the bike has a kickstart, use this to turn the engine over. If not, flick the kill switch to the OFF position and crank the engine over on the starter **(see illustration 2)**. If the nature on the ignition system prevents the starter operating with the kill switch in the OFF position, remove the spark plugs and fit them back in their caps; ensure that the plugs are earthed (grounded) against the cylinder head when the starter is operated **(see illustration 3)**.

 Warning: It is important that the plugs are earthed (grounded) away from the spark plug holes otherwise there is a risk of atomised fuel from the cylinders igniting.

 On a single cylinder four-stroke engine, you can seal the combustion chamber completely by positioning the piston at TDC on the compression stroke.

● Drain the carburettor(s) otherwise there is a risk of jets becoming blocked by gum deposits from the fuel **(see illustration 4)**.

● If the bike is going into long-term storage consider adding a fuel stabiliser to the fuel in the tank. If the tank is drained completely corrosion of its internal surfaces may occur if left unprotected for a long period. The tank can be treated with a rust preventative especially for this purpose. Alternatively remove the tank and pour half a litre of motor oil into it, install the filler cap and shake the tank to coat its internals with oil before draining off the excess. The same effect can also be achieved by spraying WD40 or a similar water-dispersant around the inside of the tank via its flexible nozzle.

● Make sure the cooling system contains the correct mix of antifreeze. Antifreeze also contains important corrosion inhibitors.

● The air intakes and exhaust can be sealed off by covering or plugging the openings. Ensure that you do not seal in any condensation; run the engine until it is hot

Squirt a drop of motor oil into each cylinder

Flick the kill switch to OFF . . .

. . . and ensure that the metal bodies of the plugs (arrows) are earthed against the cylinder head

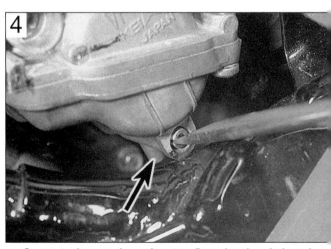

Connect a hose to the carburettor float chamber drain stub (arrow) and unscrew the drain screw

Storage REF•33

Exhausts can be sealed off with a plastic bag

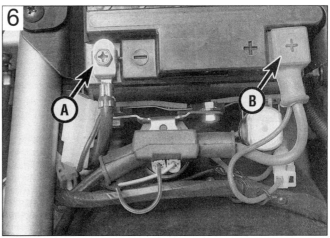

Disconnect the negative lead (A) first, followed by the positive lead (B)

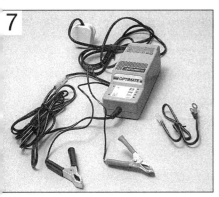

Use a suitable battery charger - this kit also assess battery condition

then switch off and allow to cool. Tape a piece of thick plastic over the silencer end(s) **(see illustration 5)**. Note that some advocate pouring a tablespoon of motor oil into the silencer(s) before sealing them off.

Battery

● Remove it from the bike - in extreme cases of cold the battery may freeze and crack its case **(see illustration 6)**.

● Check the electrolyte level and top up if necessary (conventional refillable batteries). Clean the terminals.
● Store the battery off the motorcycle and away from any sources of fire. Position a wooden block under the battery if it is to sit on the ground.
● Give the battery a trickle charge for a few hours every month **(see illustration 7)**.

Tyres

● Place the bike on its centrestand or an auxiliary stand which will support the motorcycle in an upright position. Position wood blocks under the tyres to keep them off the ground and to provide insulation from damp. If the bike is being put into long-term storage, ideally both tyres should be off the ground; not only will this protect the tyres, but will also ensure that no load is placed on the steering head or wheel bearings.
● Deflate each tyre by 5 to 10 psi, no more or the beads may unseat from the rim, making subsequent inflation difficult on tubeless tyres.

Pivots and controls

● Lubricate all lever, pedal, stand and footrest pivot points. If grease nipples are fitted to the rear suspension components, apply lubricant to the pivots.
● Lubricate all control cables.

Cycle components

● Apply a wax protectant to all painted and plastic components. Wipe off any excess, but don't polish to a shine. Where fitted, clean the screen with soap and water.
● Coat metal parts with Vaseline (petroleum jelly). When applying this to the fork tubes, do not compress the forks otherwise the seals will rot from contact with the Vaseline.
● Apply a vinyl cleaner to the seat.

Storage conditions

● Aim to store the bike in a shed or garage which does not leak and is free from damp.
● Drape an old blanket or bedspread over the bike to protect it from dust and direct contact with sunlight (which will fade paint). This also hides the bike from prying eyes. Beware of tight-fitting plastic covers which may allow condensation to form and settle on the bike.

Getting back on the road

Engine and transmission

● Change the oil and replace the oil filter. If this was done prior to storage, check that the oil hasn't emulsified - a thick whitish substance which occurs through condensation.
● Remove the spark plugs. Using a spout-type oil can, squirt a few drops of oil into the cylinder(s). This will provide initial lubrication as the piston rings and bores comes back into contact. Service the spark plugs, or fit new ones, and install them in the engine.

● Check that the clutch isn't stuck on. The plates can stick together if left standing for some time, preventing clutch operation. Engage a gear and try rocking the bike back and forth with the clutch lever held against the handlebar. If this doesn't work on cable-operated clutches, hold the clutch lever back against the handlebar with a strong elastic band or cable tie for a couple of hours **(see illustration 8)**.
● If the air intakes or silencer end(s) were blocked off, remove the bung or cover used.
● If the fuel tank was coated with a rust

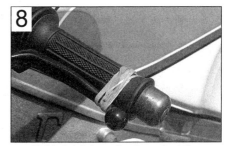

Hold clutch lever back against the handlebar with elastic bands or a cable tie

Storage

preventative, oil or a stabiliser added to the fuel, drain and flush the tank and dispose of the fuel sensibly. If no action was taken with the fuel tank prior to storage, it is advised that the old fuel is disposed of since it will go off over a period of time. Refill the fuel tank with fresh fuel.

Frame and running gear

- Oil all pivot points and cables.
- Check the tyre pressures. They will definitely need inflating if pressures were reduced for storage.
- Lubricate the final drive chain (where applicable).
- Remove any protective coating applied to the fork tubes (stanchions) since this may well destroy the fork seals. If the fork tubes weren't protected and have picked up rust spots, remove them with very fine abrasive paper and refinish with metal polish.
- Check that both brakes operate correctly. Apply each brake hard and check that it's not possible to move the motorcycle forwards, then check that the brake frees off again once released. Brake caliper pistons can stick due to corrosion around the piston head, or on the sliding caliper types, due to corrosion of the slider pins. If the brake doesn't free after repeated operation, take the caliper off for examination. Similarly drum brakes can stick due to a seized operating cam, cable or rod linkage.
- If the motorcycle has been in long-term storage, renew the brake fluid and clutch fluid (where applicable).
- Depending on where the bike has been stored, the wiring, cables and hoses may have been nibbled by rodents. Make a visual check and investigate disturbed wiring loom tape.

Battery

- If the battery has been previously removal and given top up charges it can simply be reconnected. Remember to connect the positive cable first and the negative cable last.
- On conventional refillable batteries, if the battery has not received any attention, remove it from the motorcycle and check its electrolyte level. Top up if necessary then charge the battery. If the battery fails to hold a charge and a visual checks show heavy white sulphation of the plates, the battery is probably defective and must be renewed. This is particularly likely if the battery is old. Confirm battery condition with a specific gravity check.
- On sealed (MF) batteries, if the battery has not received any attention, remove it from the motorcycle and charge it according to the information on the battery case - if the battery fails to hold a charge it must be renewed.

Starting procedure

- If a kickstart is fitted, turn the engine over a couple of times with the ignition OFF to distribute oil around the engine. If no kickstart is fitted, flick the engine kill switch OFF and the ignition ON and crank the engine over a couple of times to work oil around the upper cylinder components. If the nature of the ignition system is such that the starter won't work with the kill switch OFF, remove the spark plugs, fit them back into their caps and earth (ground) their bodies on the cylinder head. Reinstall the spark plugs afterwards.
- Switch the kill switch to RUN, operate the choke and start the engine. If the engine won't start don't continue cranking the engine - not only will this flatten the battery, but the starter motor will overheat. Switch the ignition off and try again later. If the engine refuses to start, go through the fault finding procedures in this manual. **Note:** *If the bike has been in storage for a long time, old fuel or a carburettor blockage may be the problem. Gum deposits in carburettors can block jets - if a carburettor cleaner doesn't prove successful the carburettors must be dismantled for cleaning.*
- Once the engine has started, check that the lights, turn signals and horn work properly.
- Treat the bike gently for the first ride and check all fluid levels on completion. Settle the bike back into the maintenance schedule.

Fault Finding REF•35

This Section provides an easy reference-guide to the more common faults that are likely to afflict your machine. Obviously, the opportunities are almost limitless for faults to occur as a result of obscure failures, and to try and cover all eventualities would require a book. Indeed, a number have been written on the subject.

Successful troubleshooting is not a mysterious 'black art' but the application of a bit of knowledge combined with a systematic and logical approach to the problem. Approach any troubleshooting by first accurately identifying the symptom and then checking through the list of possible causes, starting with the simplest or most obvious and progressing in stages to the most complex.

Take nothing for granted, but above all apply liberal quantities of common sense.

The main symptom of a fault is given in the text as a major heading below which are listed the various systems or areas which may contain the fault. Details of each possible cause for a fault and the remedial action to be taken are given, in brief, in the paragraphs below each heading. Further information should be sought in the relevant Chapter.

1 Engine doesn't start or is difficult to start
- [ ] Starter motor doesn't rotate
- [ ] Starter motor rotates but engine does not turn over
- [ ] Starter works but engine won't turn over (seized)
- [ ] No fuel flow
- [ ] Engine flooded
- [ ] No spark or weak spark
- [ ] Compression low
- [ ] Stalls after starting
- [ ] Rough idle

2 Poor running at low speed
- [ ] Spark weak
- [ ] Fuel/air mixture incorrect
- [ ] Compression low
- [ ] Poor acceleration

3 Poor running or no power at high speed
- [ ] Firing incorrect
- [ ] Fuel/air mixture incorrect
- [ ] Compression low
- [ ] Knocking or pinging
- [ ] Miscellaneous causes

4 Overheating
- [ ] Engine overheats
- [ ] Firing incorrect
- [ ] Fuel/air mixture incorrect
- [ ] Compression too high
- [ ] Engine load excessive
- [ ] Lubrication inadequate
- [ ] Miscellaneous causes

5 Clutch problems
- [ ] Clutch slipping
- [ ] Clutch not disengaging completely

6 Gearchanging problems
- [ ] Doesn't go into gear, or lever doesn't return
- [ ] Jumps out of gear
- [ ] Overshifts

7 Abnormal engine noise
- [ ] Knocking or pinking
- [ ] Piston slap or rattling
- [ ] Valve noise
- [ ] Other noise

8 Abnormal driveline noise
- [ ] Clutch noise
- [ ] Transmission noise
- [ ] Final drive noise

9 Abnormal frame and suspension noise
- [ ] Front end noise
- [ ] Shock absorber noise
- [ ] Brake noise

10 Excessive exhaust smoke
- [ ] White smoke
- [ ] Black smoke
- [ ] Brown smoke

11 Poor handling or stability
- [ ] Handlebar hard to turn
- [ ] Handlebar shakes or vibrates excessively
- [ ] Handlebar pulls to one side
- [ ] Poor shock absorbing qualities

12 Braking problems
- [ ] Brakes are spongy, don't hold
- [ ] Brake lever or pedal pulsates
- [ ] Brakes drag

13 Electrical problems
- [ ] Battery dead or weak
- [ ] Battery overcharged

1 Engine doesn't start or is difficult to start

Starter motor doesn't rotate

- [ ] Engine kill switch OFF.
- [ ] Fuse blown. Check main fuse and ignition circuit fuse (Chapter 9).
- [ ] Battery voltage low. Check and recharge battery (Chapter 9).
- [ ] Starter motor defective. Make sure the wiring to the starter is secure. Make sure the starter relay clicks when the start button is pushed. If the relay clicks, then the fault is in the wiring or motor.
- [ ] Starter relay faulty. Check it according to the procedure in Chapter 9.
- [ ] Starter switch not contacting. The contacts could be wet, corroded or dirty. Disassemble and clean the switch (Chapter 9).
- [ ] Wiring open or shorted. Check all wiring connections and harnesses to make sure that they are dry, tight and not corroded. Also check for broken or frayed wires that can cause a short to ground (earth) (see wiring diagram, Chapter 9).
- [ ] Ignition (main) switch defective. Check the switch according to the procedure in Chapter 9. Replace the switch with a new one if it is defective.
- [ ] Engine kill switch defective. Check for wet, dirty or corroded contacts. Clean or replace the switch as necessary (Chapter 9).
- [ ] Faulty neutral, side stand or clutch switch. Check the wiring to each switch and the switch itself according to the procedures in Chapter 9.

Starter motor rotates but engine does not turn over

- [ ] Starter motor clutch defective. Inspect and repair or replace (Chapter 2).
- [ ] Damaged idler or starter gears. Inspect and replace the damaged parts (Chapter 2).

Starter works but engine won't turn over (seized)

- [ ] Seized engine caused by one or more internally damaged components. Failure due to wear, abuse or lack of lubrication. Damage can include seized valves, followers, camshafts, pistons, crankshaft, connecting rod bearings, or transmission gears or bearings. Refer to Chapter 2 for engine disassembly.

No fuel flow

- [ ] No fuel in tank.
- [ ] Fuel tank breather hose obstructed.
- [ ] Fuel tap strainer or in-line filter clogged. Remove the tap and clean it and check the filter (Chapters 1 and 4).
- [ ] Fuel line clogged. Pull the fuel line loose and carefully blow through it.
- [ ] Float needle valve clogged. For all of the valves to be clogged, either a very bad batch of fuel with an unusual additive has been used, or some other foreign material has entered the tank. Many times after a machine has been stored for many months without running, the fuel turns to a varnish-like liquid and forms deposits on the inlet needle valves and jets. The carburettors should be removed and overhauled if draining the float chambers doesn't solve the problem.
- [ ] Fuel pump faulty. Check the fuel pump flow and renew the pump if necessary (Chapter 4).

Engine flooded

- [ ] Fuel level (float height) too high. Check as described in Chapter 4.
- [ ] Float needle valve worn or stuck open. A piece of dirt, rust or other debris can cause the valve to seat improperly, causing excess fuel to be admitted to the float chamber. In this case, the float chamber should be cleaned and the needle valve and seat inspected. If the needle and seat are worn, then the leaking will persist and the parts should be replaced with new ones (Chapter 4).
- [ ] Starting technique incorrect. Under normal circumstances (i.e., if all the carburettor functions are sound) the machine should start with little or no throttle. When the engine is cold, the choke should be operated and the engine started without opening the throttle. When the engine is at operating temperature, only a very slight amount of throttle should be necessary. If the engine is flooded, turn the fuel tap OFF or disconnect the vacuum hose (according to model - see Chapter 4) and hold the throttle open while cranking the engine. This will allow additional air to reach the cylinders. Remember to turn the fuel tap back ON or attach the vacuum hose.

No spark or weak spark

- [ ] Ignition switch OFF.
- [ ] Engine kill switch turned to the OFF position.
- [ ] Battery voltage low. Check and recharge the battery as necessary (Chapter 9).
- [ ] Spark plugs dirty, defective or worn out. Locate reason for fouled plugs using spark plug condition chart and follow the plug maintenance procedures (Chapter 1).
- [ ] Spark plug caps or secondary (HT) wiring faulty. Check condition. Renew either or both components if cracks or deterioration are evident (Chapter 5).
- [ ] Spark plug caps not making good contact. Make sure that the plug caps fit snugly over the plug ends.
- [ ] Ignition control unit defective. Check the unit (Chapter 5).
- [ ] Pick-up coil defective. Check the unit (Chapter 5).
- [ ] Ignition HT coils defective. Check the coils (Chapter 5).
- [ ] Ignition or kill switch shorted. This is usually caused by water, corrosion, damage or excessive wear. The switches can be disassembled and cleaned with electrical contact cleaner. If cleaning does not help, renew the switches (Chapter 9).
- [ ] Wiring shorted or broken between:
    a) *Ignition (main) switch and engine kill switch (or blown fuse)*
    b) *Ignition control unit and engine kill switch*
    c) *Ignition control unit and ignition HT coils*
    d) *Ignition HT coils and spark plugs*
    e) *Ignition control unit and pick-up coil*
- [ ] Make sure that all wiring connections are clean, dry and tight. Look for chafed and broken wires (Chapters 5 and 9).

1 Engine doesn't start or is difficult to start (continued)

Compression low

- [ ] Spark plugs loose. Remove the plugs and inspect their threads. Reinstall and tighten to the specified torque (Chapter 1).
- [ ] Cylinder head not sufficiently tightened down. If the cylinder head is suspected of being loose, then there's a chance that the gasket or head is damaged if the problem has persisted for any length of time. The head nuts should be tightened to the proper torque in the correct sequence (Chapter 2).
- [ ] Improper valve clearance. This means that the valve is not closing completely and compression pressure is leaking past the valve. Check and adjust the valve clearances (Chapter 1).
- [ ] Cylinder and/or piston worn. Excessive wear will cause compression pressure to leak past the rings. This is usually accompanied by worn rings as well. A top-end overhaul is necessary (Chapter 2).
- [ ] Piston rings worn, weak, broken, or sticking. Broken or sticking piston rings usually indicate a lubrication or carburation problem that causes excess carbon deposits or seizures to form on the pistons and rings. Top-end overhaul is necessary (Chapter 2).
- [ ] Piston ring-to-groove clearance excessive. This is caused by excessive wear of the piston ring lands. Piston renewal is necessary (Chapter 2).
- [ ] Cylinder head gasket damaged. If the head is allowed to become loose, or if excessive carbon build-up on the piston crown and combustion chamber causes extremely high compression, the head gasket may leak. Retorquing the head is not always sufficient to restore the seal, so gasket renewal is necessary (Chapter 2).
- [ ] Cylinder head warped. This is caused by overheating or improperly tightened head nuts. Machine shop resurfacing or head renewal is necessary (Chapter 2).
- [ ] Valve spring broken or weak. Caused by component failure or wear; the springs must be renewed (Chapter 2).
- [ ] Valve not seating properly. This is caused by a bent valve (from over-revving or improper valve adjustment), burned valve or seat (improper carburation) or an accumulation of carbon deposits on the seat (from carburation or lubrication problems). The valves must be cleaned and/or replaced and the seats serviced if possible (Chapter 2).

Stalls after starting

- [ ] Improper choke action. Make sure the choke linkage shaft is getting a full stroke and staying in the out position (Chapter 4).
- [ ] Ignition malfunction (Chapter 5).
- [ ] Carburettor malfunction (Chapter 4).
- [ ] Fuel contaminated. The fuel can be contaminated with either dirt or water, or can change chemically if the machine is allowed to sit for several months or more. Drain the tank and float chambers (Chapter 4).
- [ ] Intake air leak. Check for loose carburettor-to-intake manifold connections, loose or missing vacuum gauge adapter screws or hoses, or loose carburettor tops (Chapter 4).
- [ ] Engine idle speed incorrect. Turn idle adjusting screw until the engine idles at the specified rpm (Chapter 1).

Rough idle

- [ ] Ignition malfunction (Chapter 5).
- [ ] Idle speed incorrect (Chapter 1).
- [ ] Carburettors not synchronised. Adjust carburettors with vacuum gauge or manometer set as described in Chapter 1.
- [ ] Carburettor malfunction (Chapter 4).
- [ ] Fuel contaminated. The fuel can be contaminated with either dirt or water, or can change chemically if the machine is allowed to sit for several months or more. Drain the tank and float chambers (Chapter 4).
- [ ] Intake air leak. Check for loose carburettor-to-intake manifold connections, loose or missing vacuum gauge adapter screws or hoses, or loose carburettor tops (Chapter 4).
- [ ] Air filter clogged. Renew the air filter element (Chapter 1).

Fault Finding

2 Poor running at low speeds

Spark weak

- [ ] Battery voltage low. Check and recharge battery (Chapter 9).
- [ ] Spark plugs fouled, defective or worn out. Refer to Chapter 1 for spark plug maintenance.
- [ ] Spark plug cap or HT wiring defective. Refer to Chapters 1 and 5 for details on the ignition system.
- [ ] Spark plug caps not making contact. Make sure they are securely pushed on to the plugs.
- [ ] Incorrect spark plugs. Wrong type, heat range or cap configuration. Check and install correct plugs listed in Chapter 1.
- [ ] Ignition control unit defective (Chapter 5).
- [ ] Pick-up coil defective (Chapter 5).
- [ ] Ignition HT coils defective (Chapter 5).

Fuel/air mixture incorrect

- [ ] Pilot screws out of adjustment (Chapter 4).
- [ ] Pilot jet or air passage clogged. Remove and overhaul the carburettors (Chapter 4).
- [ ] Air bleed holes clogged. Remove carburettor and blow out all passages (Chapter 4).
- [ ] Air filter clogged, poorly sealed or missing (Chapter 1).
- [ ] Air filter housing poorly sealed. Look for cracks, holes or loose clamps and renew or repair defective parts.
- [ ] Fuel level too high or too low. Check the level (Chapter 4).
- [ ] Fuel tank breather hose obstructed.
- [ ] Carburettor intake manifolds loose. Check for cracks, breaks, tears or loose clamps. Renew the rubber intake manifold joints if split or perished.

Compression low

- [ ] Spark plugs loose. Remove the plugs and inspect their threads. Reinstall and tighten to the specified torque (Chapter 1).
- [ ] Cylinder head not sufficiently tightened down. If the cylinder head is suspected of being loose, then there's a chance that the gasket and head are damaged if the problem has persisted for any length of time. The head nuts should be tightened to the proper torque in the correct sequence (Chapter 2).
- [ ] Improper valve clearance. This means that the valve is not closing completely and compression pressure is leaking past the valve. Check and adjust the valve clearances (Chapter 1).
- [ ] Cylinder and/or piston worn. Excessive wear will cause compression pressure to leak past the rings. This is usually accompanied by worn rings as well. A top-end overhaul is necessary (Chapter 2).
- [ ] Piston rings worn, weak, broken, or sticking. Broken or sticking piston rings usually indicate a lubrication or carburation problem that causes excess carbon deposits or seizures to form on the pistons and rings. Top-end overhaul is necessary (Chapter 2).
- [ ] Piston ring-to-groove clearance excessive. This is caused by excessive wear of the piston ring lands. Piston renewal is necessary (Chapter 2).
- [ ] Cylinder head gasket damaged. If the head is allowed to become loose, or if excessive carbon build-up on the piston crown and combustion chamber causes extremely high compression, the head gasket may leak. Retorquing the head is not always sufficient to restore the seal, so gasket renewal is necessary (Chapter 2).
- [ ] Cylinder head warped. This is caused by overheating or improperly tightened head nuts. Machine shop resurfacing or head renewal is necessary (Chapter 2).
- [ ] Valve spring broken or weak. Caused by component failure or wear; the springs must be renewed (Chapter 2).
- [ ] Valve not seating properly. This is caused by a bent valve (from over-revving or improper valve adjustment), burned valve or seat (improper carburation) or an accumulation of carbon deposits on the seat (from carburation, lubrication problems). The valves must be cleaned and/or replaced and the seats serviced if possible (Chapter 2).

Poor acceleration

- [ ] Carburettors leaking or dirty. Overhaul the carburettors (Chapter 4).
- [ ] Timing not advancing. The pick-up coil or the ignition control unit may be defective. If so, they must be renewed, as they can't be repaired.
- [ ] Carburettors not synchronised. Adjust them with a vacuum gauge set or manometer (Chapter 1).
- [ ] Engine oil viscosity too high. Using a heavier oil than that recommended in Chapter 1 can damage the oil pump or lubrication system and cause drag on the engine.
- [ ] Brakes dragging. Usually caused by debris which has entered the brake piston seals, or from a warped disc or bent axle. Repair as necessary (Chapter 7).
- [ ] Fuel pump flow rate insufficient. Check the pump (Chapter 4).

Fault Finding REF•39

3 Poor running or no power at high speed

Firing incorrect
- [ ] Air filter restricted. Clean or renew filter (Chapter 1).
- [ ] Spark plugs fouled, defective or worn out. See Chapter 1 for spark plug maintenance.
- [ ] Spark plug caps or HT wiring defective. See Chapters 1 and 5 for details of the ignition system.
- [ ] Spark plug caps not in good contact (Chapter 5).
- [ ] Incorrect spark plugs. Wrong type, heat range or cap configuration. Check and install correct plugs listed in Chapter 1.
- [ ] Ignition control unit defective (Chapter 5).
- [ ] Ignition HT coils defective (Chapter 5).

Fuel/air mixture incorrect
- [ ] Main jet clogged. Dirt, water or other contaminants can clog the main jets. Clean the fuel tap strainer, the float chamber area, and the jets and carburettor orifices (Chapter 4). Renew the in-line fuel filter (Chapter 1).
- [ ] Main jet wrong size. The standard jetting is for sea level atmospheric pressure and oxygen content.
- [ ] Throttle shaft-to-carburettor body clearance excessive. Refer to Chapter 4 for inspection and part renewal procedures.
- [ ] Air bleed holes clogged. Remove and overhaul carburettors (Chapter 4).
- [ ] Air filter clogged, poorly sealed, or missing (Chapter 1).
- [ ] Air filter housing poorly sealed. Look for cracks, holes or loose clamps, and renew or repair defective parts.
- [ ] Fuel level too high or too low. Check the level (Chapter 4).
- [ ] Fuel tank breather hose obstructed.
- [ ] Carburettor intake manifolds loose. Check for cracks, breaks, tears or loose clamps. Renew the rubber intake manifolds if they are split or perished (Chapter 4).

Compression low
- [ ] Spark plugs loose. Remove the plugs and inspect their threads. Reinstall and tighten to the specified torque (Chapter 1).
- [ ] Cylinder head not sufficiently tightened down. If the cylinder head is suspected of being loose, then there's a chance that the gasket and head are damaged if the problem has persisted for any length of time. The head nuts should be tightened to the proper torque in the correct sequence (Chapter 2).
- [ ] Improper valve clearance. This means that the valve is not closing completely and compression pressure is leaking past the valve. Check and adjust the valve clearances (Chapter 1).
- [ ] Cylinder and/or piston worn. Excessive wear will cause compression pressure to leak past the rings. This is usually accompanied by worn rings as well. A top-end overhaul is necessary (Chapter 2).
- [ ] Piston rings worn, weak, broken, or sticking. Broken or sticking piston rings usually indicate a lubrication or carburation problem that causes excess carbon deposits or seizures to form on the pistons and rings. Top-end overhaul is necessary (Chapter 2).
- [ ] Piston ring-to-groove clearance excessive. This is caused by excessive wear of the piston ring lands. Piston renewal is necessary (Chapter 2).
- [ ] Cylinder head gasket damaged. If the head is allowed to become loose, or if excessive carbon build-up on the piston crown and combustion chamber causes extremely high compression, the head gasket may leak. Retorquing the head is not always sufficient to restore the seal, so gasket renewal is necessary (Chapter 2).
- [ ] Cylinder head warped. This is caused by overheating or improperly tightened head nuts. Machine shop resurfacing or head renewal is necessary (Chapter 2).
- [ ] Valve spring broken or weak. Caused by component failure or wear; the springs must be renewed (Chapter 2).
- [ ] Valve not seating properly. This is caused by a bent valve (from over-revving or improper valve adjustment), burned valve or seat (improper carburation) or an accumulation of carbon deposits on the seat (from carburation or lubrication problems). The valves must be cleaned and/or renewed and the seats serviced if possible (Chapter 2).

Knocking or pinking
- [ ] Carbon build-up in combustion chamber. Use of a fuel additive that will dissolve the adhesive bonding the carbon particles to the crown and chamber is the easiest way to remove the build-up. Otherwise, the cylinder head will have to be removed and decarbonised (Chapter 2).
- [ ] Incorrect or poor quality fuel. Old or improper grades of fuel can cause detonation. This causes the piston to rattle, thus the knocking or pinking sound. Drain old fuel and always use the recommended fuel grade.
- [ ] Spark plug heat range incorrect. Uncontrolled detonation indicates the plug heat range is too hot. The plug in effect becomes a glow plug, raising cylinder temperatures. Install the proper heat range plug (Chapter 1).
- [ ] Improper air/fuel mixture. This will cause the cylinders to run hot, which leads to detonation. Clogged jets or an air leak can cause this imbalance. See Chapter 4.

Miscellaneous causes
- [ ] Throttle valve doesn't open fully. Adjust the throttle grip freeplay (Chapter 1).
- [ ] Clutch slipping. May be caused by loose or worn clutch components. Refer to Chapter 2 for clutch overhaul procedures.
- [ ] Engine oil viscosity too high. Using a heavier oil than the one recommended in Chapter 1 can damage the oil pump or lubrication system and cause drag on the engine.
- [ ] Brakes dragging. Usually caused by debris which has entered the brake piston seals, or from a warped disc or bent axle. Repair as necessary.
- [ ] Fuel pump flow rate insufficient. Check the pump (Chapter 4).

Fault Finding

4 Overheating

Engine overheats
- [ ] Coolant level low. Check and add coolant (Chapter 1).
- [ ] Leak in cooling system. Check cooling system hoses and radiator for leaks and other damage. Repair or renew parts as necessary (Chapter 3).
- [ ] Thermostat sticking open or closed. Check and renew as described in Chapter 3.
- [ ] Faulty radiator cap. Remove the cap and have it pressure tested.
- [ ] Coolant passages clogged. Have the entire system drained and flushed, then refill with fresh coolant.
- [ ] Water pump defective. Remove the pump and check the components (Chapter 3).
- [ ] Clogged radiator fins. Clean them by blowing compressed air through the fins from the backside.
- [ ] Cooling fan or fan switch fault (Chapter 3).

Firing incorrect
- [ ] Spark plugs fouled, defective or worn out. See Chapter 1 for spark plug maintenance.
- [ ] Incorrect spark plugs.
- [ ] Ignition control unit defective (Chapter 5).
- [ ] Faulty ignition HT coils (Chapter 5).

Fuel/air mixture incorrect
- [ ] Main jet clogged. Dirt, water and other contaminants can clog the main jets. Clean the fuel tap strainer, the float chamber area and the jets and carburettor orifices (Chapter 4). Renew the in-line fuel filter (Chapter 1).
- [ ] Main jet wrong size. The standard jetting is for sea level atmospheric pressure and oxygen content.
- [ ] Air filter clogged, poorly sealed or missing (Chapter 1).
- [ ] Air filter housing poorly sealed. Look for cracks, holes or loose clamps and renew or repair.
- [ ] Fuel level too low. Check the level (Chapter 4).
- [ ] Fuel tank breather hose obstructed.
- [ ] Carburettor intake manifolds loose. Check for cracks, breaks, tears or loose clamps. Renew the rubber intake manifold joints if split or perished.

Compression too high
- [ ] Carbon build-up in combustion chamber. Use of a fuel additive that will dissolve the adhesive bonding the carbon particles to the piston crown and chamber is the easiest way to remove the build-up. Otherwise, the cylinder head will have to be removed and decarbonised (Chapter 2).
- [ ] Improperly machined head surface or installation of incorrect gasket during engine assembly.

Engine load excessive
- [ ] Clutch slipping. Can be caused by damaged, loose or worn clutch components. Refer to Chapter 2 for overhaul procedures.
- [ ] Engine oil level too high. The addition of too much oil will cause pressurisation of the crankcase and inefficient engine operation. Check Specifications and drain to proper level (Chapter 1).
- [ ] Engine oil viscosity too high. Using a heavier oil than the one recommended in Chapter 1 can damage the oil pump or lubrication system as well as cause drag on the engine.
- [ ] Brakes dragging. Usually caused by debris which has entered the brake piston seals, or from a warped disc or bent axle. Repair as necessary.

Lubrication inadequate
- [ ] Engine oil level too low. Friction caused by intermittent lack of lubrication or from oil that is overworked can cause overheating. The oil provides a definite cooling function in the engine. Check the oil level (Chapter 1).
- [ ] Poor quality engine oil or incorrect viscosity or type. Oil is rated not only according to viscosity but also according to type. Some oils are not rated high enough for use in this engine. Check the Specifications section and change to the correct oil (Chapter 1).

Miscellaneous causes
- [ ] Modification to exhaust system. Most aftermarket exhaust systems cause the engine to run leaner, which make them run hotter. When installing an accessory exhaust system, always rejet the carburettors.

5 Clutch problems

Clutch slipping
- [ ] Incorrectly adjusted cable (see Chapter 1).
- [ ] Friction plates worn or warped. Overhaul the clutch assembly (Chapter 2).
- [ ] Plain plates warped (Chapter 2).
- [ ] Clutch release mechanism defective. Renew any defective parts (Chapter 2).
- [ ] Clutch centre or housing unevenly worn. This causes improper engagement of the plates. Renew the damaged or worn parts (Chapter 2).

Clutch not disengaging completely
- [ ] Incorrectly adjusted cable (see Chapter 1).
- [ ] Clutch plates warped or damaged. This will cause clutch drag, which in turn will cause the machine to creep. Overhaul the clutch assembly (Chapter 2).
- [ ] Clutch diaphragm spring broken or weak (Chapter 2).
- [ ] Engine oil deteriorated. Old, thin, worn out oil will not provide proper lubrication for the plates, causing the clutch to drag. Change the oil and filter (Chapter 1).
- [ ] Engine oil viscosity too high. Using a heavier oil than recommended in Chapter 1 can cause the plates to stick together, putting a drag on the engine. Change to the correct weight oil (Chapter 1).
- [ ] Clutch needle bearing seized on input shaft. Lack of lubrication, severe wear or damage can cause the guide to seize on the shaft. Overhaul of the clutch, and perhaps transmission, may be necessary to repair the damage (Chapter 2).
- [ ] Clutch release mechanism defective. Overhaul it (Chapter 2).
- [ ] Loose clutch centre nut. Causes housing and centre misalignment putting a drag on the engine. Engagement adjustment continually varies. Overhaul the clutch assembly (Chapter 2).

Fault Finding

6 Gearchanging problems

Doesn't go into gear or lever doesn't return

- [ ] Clutch not disengaging. See above.
- [ ] Selector fork(s) bent or seized. Often caused by dropping the machine or from lack of lubrication. Overhaul the transmission (Chapter 2).
- [ ] Gear(s) stuck on shaft. Most often caused by a lack of lubrication or excessive wear in transmission gears and bushes. Overhaul the transmission (Chapter 2).
- [ ] Selector drum binding. Caused by lubrication failure or excessive wear. Renew the drum and bearing (Chapter 2).
- [ ] Gearchange lever return spring weak or broken (Chapter 2).
- [ ] Gearchange lever broken. Splines stripped out of lever or shaft, caused by allowing the lever to get loose or from dropping the machine. Renew necessary parts (Chapter 2).
- [ ] Gearchange mechanism stopper arm broken or worn. Full engagement and rotary movement of selector drum results. Renew the arm (Chapter 2).
- [ ] Stopper arm spring broken. Allows arm to float, causing sporadic gearchange operation. Renew spring (Chapter 2).

Jumps out of gear

- [ ] Selector fork(s) worn. Overhaul the transmission (Chapter 2).
- [ ] Gear groove(s) worn. Overhaul the transmission (Chapter 2).
- [ ] Gear dogs or dog slots worn or damaged. The gears should be inspected and renewed. No attempt should be made to service the worn parts.

Overselects

- [ ] Stopper arm spring weak or broken (Chapter 2).
- [ ] Gearchange shaft return spring post broken or distorted (Chapter 2).

7 Abnormal engine noise

Knocking or pinking

- [ ] Carbon build-up in combustion chamber. Use of a fuel additive that will dissolve the adhesive bonding the carbon particles to the piston crown and chamber is the easiest way to remove the build-up. Otherwise, the cylinder head will have to be removed and decarbonised (Chapter 2).
- [ ] Incorrect or poor quality fuel. Old or improper fuel can cause detonation. This causes the pistons to rattle, thus the knocking or pinking sound. Drain the old fuel and always use the recommended grade fuel (Chapter 4).
- [ ] Spark plug heat range incorrect. Uncontrolled detonation indicates that the plug heat range is too hot. The plug in effect becomes a glow plug, raising cylinder temperatures. Install the proper heat range plug (Chapter 1).
- [ ] Improper air/fuel mixture. This will cause the cylinders to run hot and lead to detonation. Clogged jets or an air leak can cause this imbalance. See Chapter 4.

Piston slap or rattling

- [ ] Cylinder-to-piston clearance excessive. Caused by improper assembly. Inspect and overhaul top-end parts (Chapter 2).
- [ ] Connecting rod bent. Caused by over-revving, trying to start a badly flooded engine or from ingesting a foreign object into the combustion chamber. Renew the damaged parts (Chapter 2).
- [ ] Piston pin or piston pin bore worn or seized from wear or lack of lubrication. Renew damaged parts (Chapter 2).
- [ ] Piston ring(s) worn, broken or sticking. Overhaul the top-end (Chapter 2).
- [ ] Piston seizure damage. Usually from lack of lubrication or overheating. Renew the pistons and crankcases, as necessary (Chapter 2).
- [ ] Connecting rod upper or lower end clearance excessive. Caused by excessive wear or lack of lubrication. Renew worn parts.

Valve noise

- [ ] Incorrect valve clearances. Adjust the clearances by referring to Chapter 1.
- [ ] Valve spring broken or weak. Check and renew weak valve springs (Chapter 2).
- [ ] Camshaft or cylinder head worn or damaged. Lack of lubrication at high rpm is usually the cause of damage. Insufficient oil or failure to change the oil at the recommended intervals are the chief causes. Since there are no replaceable bearings in the head, the head itself will have to be renewed if there is excessive wear or damage (Chapter 2).

Other noise

- [ ] Cylinder head gasket leaking.
- [ ] Exhaust pipe leaking at cylinder head connection. Caused by improper fit of pipe(s) or loose exhaust flange. All exhaust fasteners should be tightened evenly and carefully. Failure to do this will lead to a leak.
- [ ] Crankshaft runout excessive. Caused by a bent crankshaft (from over-revving) or damage from an upper cylinder component failure. Can also be attributed to dropping the machine on either of the crankshaft ends.
- [ ] Engine mounting bolts loose. Tighten all engine mount bolts (Chapter 2).
- [ ] Crankshaft bearings worn (Chapter 2).
- [ ] Camshaft drive gear assembly defective. Renew according to the procedure in Chapter 2.

8 Abnormal driveline noise

Clutch noise
- [ ] Clutch housing/friction plate clearance excessive (Chapter 2).
- [ ] Loose or damaged clutch pressure plate and/or bolts (Chapter 2).

Transmission noise
- [ ] Bearings/bushes worn. Also includes the possibility that the shafts are worn. Overhaul the transmission (Chapter 2).
- [ ] Gears worn or chipped (Chapter 2).
- [ ] Metal chips jammed in gear teeth. Probably pieces from a broken clutch, gear or selector mechanism that were picked up by the gears. This will cause early bearing failure (Chapter 2).
- [ ] Engine oil level too low. Causes a howl from transmission. Also affects engine power and clutch operation (Chapter 1).

Final drive noise
- [ ] Chain not adjusted properly (Chapter 1).
- [ ] Front or rear sprocket loose. Tighten fasteners (Chapter 6).
- [ ] Sprockets worn. Renew sprockets (Chapter 6).
- [ ] Rear sprocket warped. Renew sprockets (Chapter 6).
- [ ] Loose or worn rear wheel or sprocket coupling bearings. Check and renew as needed (Chapter 7).

9 Abnormal frame and suspension noise

Front end noise
- [ ] Low fluid level or improper viscosity oil in forks. This can sound like spurting and is usually accompanied by irregular fork action (Chapter 6).
- [ ] Spring weak or broken. Makes a clicking or scraping sound. Fork oil, when drained, will have a lot of metal particles in it (Chapter 6).
- [ ] Steering head bearings loose or damaged. Clicks when braking. Check and adjust or renew as necessary (Chapters 1 and 6).
- [ ] Fork yokes loose. Make sure all clamp pinch bolts are tightened to the specified torque (Chapter 6).
- [ ] Fork tube bent. Good possibility if machine has been dropped. Replace tube with a new one (Chapter 6).
- [ ] Front axle bolt or axle clamp bolt loose. Tighten them to the specified torque (Chapter 7).
- [ ] Loose or worn wheel bearings. Check and renew as needed (Chapter 7).

Shock absorber noise
- [ ] Fluid level incorrect. Indicates a leak caused by defective seal. Shock will be covered with oil. Renew shock or seek advice on repair from a Yamaha dealer (Chapter 6).
- [ ] Defective shock absorber with internal damage. This is in the body of the shock and can't be remedied. The shock must be replaced with a new one (Chapter 6).
- [ ] Bent or damaged shock body. Replace the shock with a new one (Chapter 6).
- [ ] Loose or worn suspension linkage components. Check and renew as necessary (Chapter 6).

Brake noise
- [ ] Squeal caused by pad shim not installed or positioned correctly (where fitted) (Chapter 7).
- [ ] Squeal caused by dust on brake pads. Usually found in combination with glazed pads. Clean using brake cleaning solvent (Chapter 7).
- [ ] Contamination of brake pads. Oil, brake fluid or dirt causing brake to chatter or squeal. Clean or renew pads (Chapter 7).
- [ ] Pads glazed. Caused by excessive heat from prolonged use or from contamination. Do not use sandpaper, emery cloth, carborundum cloth or any other abrasive to roughen the pad surfaces as abrasives will stay in the pad material and damage the disc. A very fine flat file can be used, but pad renewal is suggested as a cure (Chapter 7).
- [ ] Disc warped. Can cause a chattering, clicking or intermittent squeal. Usually accompanied by a pulsating lever and uneven braking. Renew the disc (Chapter 7).
- [ ] Loose or worn wheel bearings. Check and renew as needed (Chapter 7).

Fault Finding REF•43

10 Excessive exhaust smoke

White smoke

- ☐ Piston oil ring worn. The ring may be broken or damaged, causing oil from the crankcase to be pulled past the piston into the combustion chamber. Replace the rings with new ones (Chapter 2).
- ☐ Cylinders worn, cracked, or scored. Caused by overheating or oil starvation. Check the cylinder bores, lubrication system and cooling system (see Chapters 2 and 3).
- ☐ Valve oil seal damaged or worn. Replace oil seals with new ones (Chapter 2).
- ☐ Valve guide worn. Perform a complete valve job (Chapter 2).
- ☐ Engine oil level too high, which causes the oil to be forced past the rings. Drain oil to the proper level (Chapter 1).
- ☐ Head gasket broken between oil return and cylinder. Causes oil to be pulled into the combustion chamber. Replace the head gasket and check the head for warpage (Chapter 2).
- ☐ Abnormal crankcase pressurisation, which forces oil past the rings. Clogged breather is usually the cause.

Black smoke

- ☐ Air filter clogged. Clean or replace the element (Chapter 1).
- ☐ Main jet too large or loose. Compare the jet size to the Specifications (Chapter 4).
- ☐ Choke cable or linkage shaft stuck, causing fuel to be pulled through choke circuit (Chapter 4).
- ☐ Fuel level too high. Check and adjust the float height(s) as necessary (Chapter 4).
- ☐ Float needle valve held off needle seat. Clean the float chambers and fuel line and replace the needles and seats if necessary (Chapter 4).

Brown smoke

- ☐ Main jet too small or clogged. Lean condition caused by wrong size main jet or by a restricted orifice. Clean float chambers and jets and compare jet size to Specifications (Chapter 4).
- ☐ Fuel flow insufficient - float needle valve stuck closed due to chemical reaction with old fuel; fuel level incorrect; restricted fuel line; faulty fuel pump (Chapter 4).
- ☐ Carburettor intake manifold clamps loose (Chapter 4).
- ☐ Air filter poorly sealed or not installed (Chapter 1).

11 Poor handling or stability

Handlebar hard to turn

- ☐ Steering head bearing adjuster nut too tight. Check adjustment as described in Chapter 1.
- ☐ Bearings damaged. Roughness can be felt as the bars are turned from side-to-side. Renew bearings and races (Chapter 6).
- ☐ Races dented or worn. Denting results from wear in only one position (e.g., straight ahead), from a collision or hitting a pothole or from dropping the machine. Renew races and bearings (Chapter 6)
- ☐ Steering stem lubrication inadequate. Causes are grease getting hard from age or being washed out by high pressure car washes. Disassemble steering head and repack bearings (Chapter 6).
- ☐ Steering stem bent. Caused by a collision, hitting a pothole or by dropping the machine. Renew damaged part. Don't try to straighten the steering stem (Chapter 6).
- ☐ Front tyre air pressure too low (Chapter 1).

Handlebar shakes or vibrates excessively

- ☐ Tyres worn or out of balance (Chapter 7).
- ☐ Swingarm bearings worn. Renew worn bearings (Chapter 6).
- ☐ Wheel rim(s) warped or damaged. Inspect wheels for runout (Chapter 7).
- ☐ Wheel bearings worn. Worn front or rear wheel bearings can cause poor tracking. Worn front bearings will cause wobble (Chapter 7).
- ☐ Handlebar clamp bolts loose (Chapter 6).
- ☐ Fork yoke bolts loose. Tighten them to the specified torque (Chapter 6).
- ☐ Engine mounting bolts loose. Will cause excessive vibration with increased engine rpm (Chapter 2).

Handlebar pulls to one side

- ☐ Frame bent. Definitely suspect this if the machine has been dropped. May or may not be accompanied by cracking near the bend. Renew the frame (Chapter 6).
- ☐ Wheels out of alignment. Caused by improper location of axle spacers or from bent steering stem or frame (Chapter 6).
- ☐ Swingarm bent or twisted. Caused by age (metal fatigue) or impact damage. Renew the arm (Chapter 6).
- ☐ Steering stem bent. Caused by impact damage or by dropping the motorcycle. Renew the steering stem (Chapter 6).
- ☐ Fork tube bent. Disassemble the forks and renew the damaged parts (Chapter 6).
- ☐ Fork oil level uneven. Check and add or drain as necessary (Chapter 6).

Poor shock absorbing qualities

- ☐ Too hard:
   a) Fork oil level excessive (Chapter 6).
   b) Fork oil viscosity too high. Use a lighter oil (see the Specifications in Chapter 6).
   c) Fork tube bent. Causes a harsh, sticking feeling (Chapter 6).
   d) Fork internal damage (Chapter 6).
   e) Shock shaft or body bent or damaged (Chapter 6).
   f) Shock internal damage.
   g) Tyre pressure too high (Chapter 1).
- ☐ Too soft:
   a) Fork or shock oil insufficient and/or leaking (Chapter 6).
   b) Fork oil level too low (Chapter 6).
   c) Fork oil viscosity too light (Chapter 6).
   d) Fork springs weak or broken (Chapter 6).
   e) Shock internal damage or leakage (Chapter 6).

12 Braking problems

Brakes are spongy, don't hold
- [ ] Air in brake line. Caused by inattention to master cylinder fluid level or by leakage. Locate problem and bleed brakes (Chapter 7).
- [ ] Pad or disc worn (Chapters 1 and 7).
- [ ] Brake fluid leak. See paragraph 1.
- [ ] Contaminated pads. Caused by contamination with oil, grease, brake fluid, etc. Clean or renew pads. Clean disc thoroughly with brake cleaner (Chapter 7).
- [ ] Brake fluid deteriorated. Fluid is old or contaminated. Drain system, replenish with new fluid and bleed the system (Chapter 7).
- [ ] Master cylinder internal parts worn or damaged causing fluid to bypass (Chapter 7).
- [ ] Master cylinder bore scratched by foreign material or broken spring. Repair or renew master cylinder (Chapter 7).
- [ ] Disc warped. Renew disc (Chapter 7).

Brake lever or pedal pulsates
- [ ] Disc warped. Renew disc (Chapter 7).
- [ ] Axle bent. Renew axle (Chapter 7).
- [ ] Brake caliper bolts loose (Chapter 7).
- [ ] Wheel warped or otherwise damaged (Chapter 7).
- [ ] Wheel bearings damaged or worn (Chapter 7).

Brakes drag
- [ ] Master cylinder piston seized. Caused by wear or damage to piston or cylinder bore (Chapter 7).
- [ ] Lever balky or stuck. Check pivot and lubricate (Chapter 7).
- [ ] Brake caliper piston seized in bore. Caused by wear or ingestion of dirt past deteriorated seal (Chapter 7).
- [ ] Brake pad damaged. Pad material separated from backing plate. Usually caused by faulty manufacturing process or from contact with chemicals. Renew pads (Chapter 7).
- [ ] Pads improperly installed (Chapter 7).

13 Electrical problems

Battery dead or weak
- [ ] Battery faulty. Caused by sulphated plates which are shorted through sedimentation. Also, broken battery terminal making only occasional contact (Chapter 9).
- [ ] Battery cables making poor contact (Chapter 9).
- [ ] Load excessive. Caused by addition of high wattage lights or other electrical accessories.
- [ ] Ignition (main) switch defective. Switch either grounds (earths) internally or fails to shut off system. Renew the switch (Chapter 9).
- [ ] Regulator/rectifier defective (Chapter 9).
- [ ] Alternator stator coil open or shorted (Chapter 9).
- [ ] Wiring faulty. Wiring grounded (earthed) or connections loose in ignition, charging or lighting circuits (Chapter 9).

Battery overcharged
- [ ] Regulator/rectifier defective. Overcharging is noticed when battery gets excessively warm (Chapter 9).
- [ ] Battery defective. Renew battery (Chapter 9).
- [ ] Battery amperage too low, wrong type or size. Install manufacturer's specified amp-hour battery to handle charging load (Chapter 9).

Fault Finding Equipment

Checking engine compression

- Low compression will result in exhaust smoke, heavy oil consumption, poor starting and poor performance. A compression test will provide useful information about an engine's condition and if performed regularly, can give warning of trouble before any other symptoms become apparent.
- A compression gauge will be required, along with an adapter to suit the spark plug hole thread size. Note that the screw-in type gauge/adapter set up is preferable to the rubber cone type.
- Before carrying out the test, first check the valve clearances as described in Chapter 1.

1 Run the engine until it reaches normal operating temperature, then stop it and remove the spark plug(s), taking care not to scald your hands on the hot components.
2 Install the gauge adapter and compression gauge in No. 1 cylinder spark plug hole **(see illustration 1)**.

Screw the compression gauge adapter into the spark plug hole, then screw the gauge into the adapter

3 On kickstart-equipped motorcycles, make sure the ignition switch is OFF, then open the throttle fully and kick the engine over a couple of times until the gauge reading stabilises.
4 On motorcycles with electric start only, the procedure will differ depending on the nature of the ignition system. Flick the engine kill switch (engine stop switch) to OFF and turn the ignition switch ON; open the throttle fully and crank the engine over on the starter motor for a couple of revolutions until the gauge reading stabilises. If the starter will not operate with the kill switch OFF, turn the ignition switch OFF and refer to the next paragraph.
5 Install the plugs back in their caps and arrange the plug electrodes so that their metal bodies are earthed (grounded) against the cylinder head; this is essential to prevent damage to the ignition system **(see illustration 2)**. Position the plugs well away from the plug holes otherwise there is a risk

All spark plugs must be earthed (grounded) against the cylinder head

of atomised fuel escaping from the plug holes and igniting. As a safety precaution, cover the cylinder head cover with rag and disconnect the fuel pump wiring connector (see Chapter 4). Turn the ignition switch and kill switch ON, open the throttle fully and crank the engine over on the starter motor for a couple of revolutions until the gauge reading stabilises.
6 After one or two revolutions the pressure should build up to a maximum figure and then stabilise. Take a note of this reading and on multi-cylinder engines repeat the test on the remaining cylinders.
7 The correct pressures are given in Chapter 1 Specifications. If the results fall within the specified range and on multi-cylinder engines all are relatively equal, the engine is in good condition. If there is a marked difference between the readings, or if the readings are lower than specified, inspection of the top-end components will be required.
8 Low compression pressure may be due to worn cylinder bores, pistons or rings, failure of the cylinder head gasket, worn valve seals, or poor valve seating.
9 To distinguish between cylinder/piston wear and valve leakage, pour a small quantity of oil into the bore to temporarily seal the piston rings, then repeat the compression tests **(see illustration 3)**. If the readings show

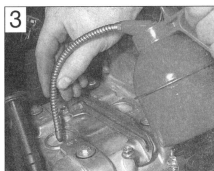

Bores can be temporarily sealed with a squirt of motor oil

a noticeable increase in pressure this confirms that the cylinder bore, piston, or rings are worn. If, however, no change is indicated, the cylinder head gasket or valves should be examined.
10 High compression pressure indicates excessive carbon build-up in the combustion chamber and on the piston crown. If this is the case the cylinder head should be removed and the deposits removed. Note that excessive carbon build-up is less likely with the use of modern fuels.

Checking battery open-circuit voltage

⚠️ *Warning: The gases produced by the battery are explosive - never smoke or create any sparks in the vicinity of the battery. Never allow the electrolyte to contact your skin or clothing - if it does, wash it off and seek immediate medical attention.*

REF•46 Fault Finding Equipment

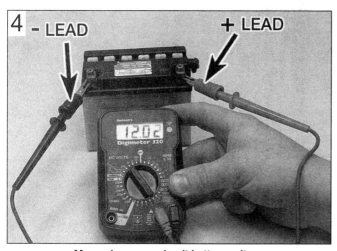
Measuring open-circuit battery voltage

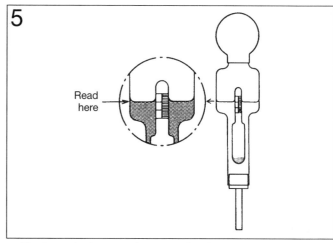
Float-type hydrometer for measuring battery specific gravity

- Before any electrical fault is investigated the battery should be checked.
- You'll need a dc voltmeter or multimeter to check battery voltage. Check that the leads are inserted in the correct terminals on the meter, red lead to positive (+ve), black lead to negative (-ve). Incorrect connections can damage the meter.
- A sound fully-charged 12 volt battery should produce between 12.3 and 12.6 volts across its terminals (12.8 volts for a maintenance-free battery). On machines with a 6 volt battery, voltage should be between 6.1 and 6.3 volts.

1 Set a multimeter to the 0 to 20 volts dc range and connect its probes across the battery terminals. Connect the meter's positive (+ve) probe, usually red, to the battery positive (+ve) terminal, followed by the meter's negative (-ve) probe, usually black, to the battery negative terminal (-ve) **(see illustration 4)**.
2 If battery voltage is low (below 10 volts on a 12 volt battery or below 4 volts on a six volt battery), charge the battery and test the voltage again. If the battery repeatedly goes flat, investigate the motorcycle's charging system.

Checking battery specific gravity (SG)

⚠️ *Warning: The gases produced by the battery are explosive - never smoke or create any sparks in the vicinity of the battery. Never allow the electrolyte to contact your skin or clothing - if it does, wash it off and seek immediate medical attention.*

- The specific gravity check gives an indication of a battery's state of charge.
- A hydrometer is used for measuring specific gravity. Make sure you purchase one which has a small enough hose to insert in the aperture of a motorcycle battery.
- Specific gravity is simply a measure of the electrolyte's density compared with that of water. Water has an SG of 1.000 and fully-charged battery electrolyte is about 26% heavier, at 1.260.
- Specific gravity checks are not possible on maintenance-free batteries. Testing the open-circuit voltage is the only means of determining their state of charge.

1 To measure SG, remove the battery from the motorcycle and remove the first cell cap. Draw

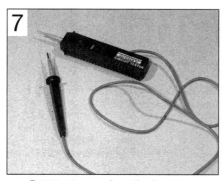
Digital multimeter can be used for all electrical tests

some electrolyte into the hydrometer and note the reading **(see illustration 5)**. Return the electrolyte to the cell and install the cap.
2 The reading should be in the region of 1.260 to 1.280. If SG is below 1.200 the battery needs charging. Note that SG will vary with temperature; it should be measured at 20°C (68°F). Add 0.007 to the reading for every 10°C above 20°C, and subtract 0.007 from the reading for every 10°C below 20°C. Add 0.004 to the reading for every 10°F above 68°F, and subtract 0.004 from the reading for every 10°F below 68°F.
3 When the check is complete, rinse the hydrometer thoroughly with clean water.

Checking for continuity

- The term continuity describes the uninterrupted flow of electricity through an electrical circuit. A continuity check will determine whether an **open-circuit** situation exists.
- Continuity can be checked with an ohmmeter, multimeter, continuity tester or battery and bulb test circuit **(see illustrations 6, 7 and 8)**.

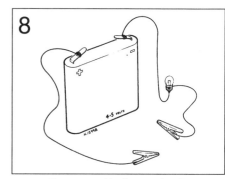
Battery and bulb test circuit

Battery-powered continuity tester

Fault Finding Equipment   REF•47

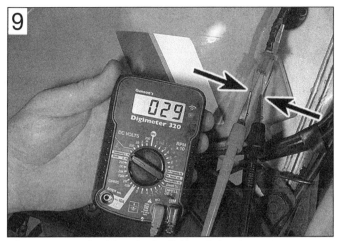
Continuity check of front brake light switch using a meter - note split pins used to access connector terminals

Continuity check of rear brake light switch using a continuity tester

- All of these instruments are self-powered by a battery, therefore the checks are made with the ignition OFF.
- As a safety precaution, always disconnect the battery negative (-ve) lead before making checks, particularly if ignition switch checks are being made.
- If using a meter, select the appropriate ohms scale and check that the meter reads infinity (∞). Touch the meter probes together and check that meter reads zero; where necessary adjust the meter so that it reads zero.
- After using a meter, always switch it OFF to conserve its battery.

Switch checks

1 If a switch is at fault, trace its wiring up to the wiring connectors. Separate the wire connectors and inspect them for security and condition. A build-up of dirt or corrosion here will most likely be the cause of the problem - clean up and apply a water dispersant such as WD40.
2 If using a test meter, set the meter to the ohms x 10 scale and connect its probes across the wires from the switch **(see illustration 9)**. Simple ON/OFF type switches, such as brake light switches, only have two wires whereas combination switches, like the ignition switch, have many internal links. Study the wiring diagram to ensure that you are connecting across the correct pair of wires. Continuity (low or no measurable resistance - 0 ohms) should be indicated with the switch ON and no continuity (high resistance) with it OFF.
3 Note that the polarity of the test probes doesn't matter for continuity checks, although care should be taken to follow specific test procedures if a diode or solid-state component is being checked.
4 A continuity tester or battery and bulb circuit can be used in the same way. Connect its probes as described above **(see illustration 10)**. The light should come on to indicate continuity in the ON switch position, but should extinguish in the OFF position.

Wiring checks

- Many electrical faults are caused by damaged wiring, often due to incorrect routing or chaffing on frame components.
- Loose, wet or corroded wire connectors can also be the cause of electrical problems, especially in exposed locations.

1 A continuity check can be made on a single length of wire by disconnecting it at each end and connecting a meter or continuity tester across both ends of the wire **(see illustration 11)**.
2 Continuity (low or no resistance - 0 ohms) should be indicated if the wire is good. If no continuity (high resistance) is shown, suspect a broken wire.

Checking for voltage

- A voltage check can determine whether current is reaching a component.
- Voltage can be checked with a dc voltmeter, multimeter set on the dc volts scale, test light or buzzer **(see illustrations 12 and 13)**. A meter has the advantage of being able to measure actual voltage.
- When using a meter, check that its leads are inserted in the correct terminals on the meter, red to positive (+ve), black to negative (-ve). Incorrect connections can damage the meter.
- A voltmeter (or multimeter set to the dc volts scale) should always be connected in parallel (across the load). Connecting it in series will destroy the meter.
- Voltage checks are made with the ignition ON.

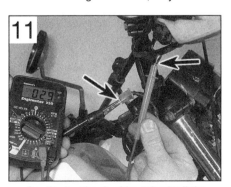
Continuity check of front brake light switch sub-harness

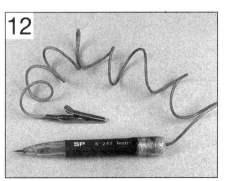
A simple test light can be used for voltage checks

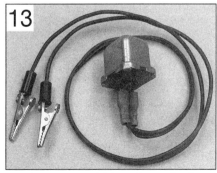
A buzzer is useful for voltage checks

REF•48 Fault Finding Equipment

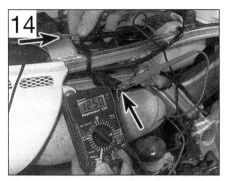

Checking for voltage at the rear brake light power supply wire using a meter . . .

1 First identify the relevant wiring circuit by referring to the wiring diagram at the end of this manual. If other electrical components share the same power supply (ie are fed from the same fuse), take note whether they are working correctly - this is useful information in deciding where to start checking the circuit.
2 If using a meter, check first that the meter leads are plugged into the correct terminals on the meter (see above). Set the meter to the dc volts function, at a range suitable for the battery voltage. Connect the meter red probe (+ve) to the power supply wire and the black probe to a good metal earth (ground) on the motorcycle's frame or directly to the battery negative (-ve) terminal **(see illustration 14)**. Battery voltage should be shown on the meter

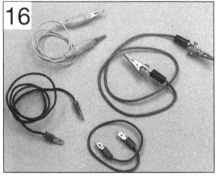

A selection of jumper wires for making earth (ground) checks

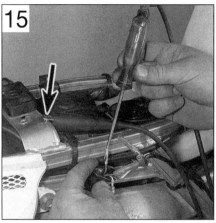

. . . or a test light - note the earth connection to the frame (arrow)

with the ignition switched ON.
3 If using a test light or buzzer, connect its positive (+ve) probe to the power supply terminal and its negative (-ve) probe to a good earth (ground) on the motorcycle's frame or directly to the battery negative (-ve) terminal **(see illustration 15)**. With the ignition ON, the test light should illuminate or the buzzer sound.
4 If no voltage is indicated, work back towards the fuse continuing to check for voltage. When you reach a point where there is voltage, you know the problem lies between that point and your last check point.

Checking the earth (ground)

● Earth connections are made either directly to the engine or frame (such as sensors, neutral switch etc. which only have a positive feed) or by a separate wire into the earth circuit of the wiring harness. Alternatively a short earth wire is sometimes run directly from the component to the motorcycle's frame.
● Corrosion is often the cause of a poor earth connection.
● If total failure is experienced, check the security of the main earth lead from the negative (-ve) terminal of the battery and also the main earth (ground) point on the wiring harness. If corroded, dismantle the connection and clean all surfaces back to bare metal.
1 To check the earth on a component, use an insulated jumper wire to temporarily bypass its earth connection **(see illustration 16)**. Connect one end of the jumper wire between the earth terminal or metal body of the component and the other end to the motorcycle's frame.
2 If the circuit works with the jumper wire installed, the original earth circuit is faulty. Check the wiring for open-circuits or poor connections. Clean up direct earth connections, removing all traces of corrosion and remake the joint. Apply petroleum jelly to the joint to prevent future corrosion.

Tracing a short-circuit

● A short-circuit occurs where current shorts to earth (ground) bypassing the circuit components. This usually results in a blown fuse.

● A short-circuit is most likely to occur where the insulation has worn through due to wiring chafing on a component, allowing a direct path to earth (ground) on the frame.

1 Remove any bodypanels necessary to access the circuit wiring.
2 Check that all electrical switches in the circuit are OFF, then remove the circuit fuse and connect a test light, buzzer or voltmeter (set to the dc scale) across the fuse terminals. No voltage should be shown.
3 Move the wiring from side to side whilst observing the test light or meter. When the test light comes on, buzzer sounds or meter shows voltage, you have found the cause of the short. It will usually shown up as damaged or burned insulation.
4 Note that the same test can be performed on each component in the circuit, even the switch.

Technical Terms Explained REF•49

A

ABS (Anti-lock braking system) A system, usually electronically controlled, that senses incipient wheel lockup during braking and relieves hydraulic pressure at wheel which is about to skid.
Aftermarket Components suitable for the motorcycle, but not produced by the motorcycle manufacturer.
Allen key A hexagonal wrench which fits into a recessed hexagonal hole.
Alternating current (ac) Current produced by an alternator. Requires converting to direct current by a rectifier for charging purposes.
Alternator Converts mechanical energy from the engine into electrical energy to charge the battery and power the electrical system.
Ampere (amp) A unit of measurement for the flow of electrical current. Current = Volts ÷ Ohms.
Ampere-hour (Ah) Measure of battery capacity.
Angle-tightening A torque expressed in degrees. Often follows a conventional tightening torque for cylinder head or main bearing fasteners **(see illustration)**.

Angle-tightening cylinder head bolts

Antifreeze A substance (usually ethylene glycol) mixed with water, and added to the cooling system, to prevent freezing of the coolant in winter. Antifreeze also contains chemicals to inhibit corrosion and the formation of rust and other deposits that would tend to clog the radiator and coolant passages and reduce cooling efficiency.
Anti-dive System attached to the fork lower leg (slider) to prevent fork dive when braking hard.
Anti-seize compound A coating that reduces the risk of seizing on fasteners that are subjected to high temperatures, such as exhaust clamp bolts and nuts.
API American Petroleum Institute. A quality standard for 4-stroke motor oils.
Asbestos A natural fibrous mineral with great heat resistance, commonly used in the composition of brake friction materials. Asbestos is a health hazard and the dust created by brake systems should never be inhaled or ingested.
ATF Automatic Transmission Fluid. Often used in front forks.
ATU Automatic Timing Unit. Mechanical device for advancing the ignition timing on early engines.
ATV All Terrain Vehicle. Often called a Quad.
Axial play Side-to-side movement.
Axle A shaft on which a wheel revolves. Also known as a spindle.

B

Backlash The amount of movement between meshed components when one component is held still. Usually applies to gear teeth.
Ball bearing A bearing consisting of a hardened inner and outer race with hardened steel balls between the two races.
Bearings Used between two working surfaces to prevent wear of the components and a build-up of heat. Four types of bearing are commonly used on motorcycles: plain shell bearings, ball bearings, tapered roller bearings and needle roller bearings.
Bevel gears Used to turn the drive through 90°. Typical applications are shaft final drive and camshaft drive **(see illustration)**.

Bevel gears are used to turn the drive through 90°

BHP Brake Horsepower. The British measurement for engine power output. Power output is now usually expressed in kilowatts (kW).
Bias-belted tyre Similar construction to radial tyre, but with outer belt running at an angle to the wheel rim.
Big-end bearing The bearing in the end of the connecting rod that's attached to the crankshaft.
Bleeding The process of removing air from an hydraulic system via a bleed nipple or bleed screw.
Bottom-end A description of an engine's crankcase components and all components contained there-in.
BTDC Before Top Dead Centre in terms of piston position. Ignition timing is often expressed in terms of degrees or millimetres BTDC.
Bush A cylindrical metal or rubber component used between two moving parts.
Burr Rough edge left on a component after machining or as a result of excessive wear.

C

Cam chain The chain which takes drive from the crankshaft to the camshaft(s).
Canister The main component in an evaporative emission control system (California market only); contains activated charcoal granules to trap vapours from the fuel system rather than allowing them to vent to the atmosphere.
Castellated Resembling the parapets along the top of a castle wall. For example, a castellated wheel axle or spindle nut.
Catalytic converter A device in the exhaust system of some machines which converts certain pollutants in the exhaust gases into less harmful substances.
Charging system Description of the components which charge the battery, ie the alternator, rectifer and regulator.
Circlip A ring-shaped clip used to prevent endwise movement of cylindrical parts and shafts. An internal circlip is installed in a groove in a housing; an external circlip fits into a groove on the outside of a cylindrical piece such as a shaft. Also known as a snap-ring.
Clearance The amount of space between two parts. For example, between a piston and a cylinder, between a bearing and a journal, etc.
Coil spring A spiral of elastic steel found in various sizes throughout a vehicle, for example as a springing medium in the suspension and in the valve train.
Compression Reduction in volume, and increase in pressure and temperature, of a gas, caused by squeezing it into a smaller space.
Compression damping Controls the speed the suspension compresses when hitting a bump.
Compression ratio The relationship between cylinder volume when the piston is at top dead centre and cylinder volume when the piston is at bottom dead centre.
Continuity The uninterrupted path in the flow of electricity. Little or no measurable resistance.
Continuity tester Self-powered bleeper or test light which indicates continuity.
Cp Candlepower. Bulb rating commonly found on US motorcycles.
Crossply tyre Tyre plies arranged in a criss-cross pattern. Usually four or six plies used, hence 4PR or 6PR in tyre size codes.
Cush drive Rubber damper segments fitted between the rear wheel and final drive sprocket to absorb transmission shocks **(see illustration)**.

Cush drive rubbers dampen out transmission shocks

D

Degree disc Calibrated disc for measuring piston position. Expressed in degrees.
Dial gauge Clock-type gauge with adapters for measuring runout and piston position. Expressed in mm or inches.
Diaphragm The rubber membrane in a master cylinder or carburettor which seals the upper chamber.
Diaphragm spring A single sprung plate often used in clutches.
Direct current (dc) Current produced by a dc generator.

REF•50 Technical Terms Explained

Decarbonisation The process of removing carbon deposits - typically from the combustion chamber, valves and exhaust port/system.
Detonation Destructive and damaging explosion of fuel/air mixture in combustion chamber instead of controlled burning.
Diode An electrical valve which only allows current to flow in one direction. Commonly used in rectifiers and starter interlock systems.
Disc valve (or rotary valve) A induction system used on some two-stroke engines.
Double-overhead camshaft (DOHC) An engine that uses two overhead camshafts, one for the intake valves and one for the exhaust valves.
Drivebelt A toothed belt used to transmit drive to the rear wheel on some motorcycles. A drivebelt has also been used to drive the camshafts. Drivebelts are usually made of Kevlar.
Driveshaft Any shaft used to transmit motion. Commonly used when referring to the final driveshaft on shaft drive motorcycles.

E

Earth return The return path of an electrical circuit, utilising the motorcycle's frame.
ECU (Electronic Control Unit) A computer which controls (for instance) an ignition system, or an anti-lock braking system.
EGO Exhaust Gas Oxygen sensor. Sometimes called a Lambda sensor.
Electrolyte The fluid in a lead-acid battery.
EMS (Engine Management System) A computer controlled system which manages the fuel injection and the ignition systems in an integrated fashion.
Endfloat The amount of lengthways movement between two parts. As applied to a crankshaft, the distance that the crankshaft can move side-to-side in the crankcase.
Endless chain A chain having no joining link. Common use for cam chains and final drive chains.
EP (Extreme Pressure) Oil type used in locations where high loads are applied, such as between gear teeth.
Evaporative emission control system Describes a charcoal filled canister which stores fuel vapours from the tank rather than allowing them to vent to the atmosphere. Usually only fitted to California models and referred to as an EVAP system.
Expansion chamber Section of two-stroke engine exhaust system so designed to improve engine efficiency and boost power.

F

Feeler blade or gauge A thin strip or blade of hardened steel, ground to an exact thickness, used to check or measure clearances between parts.
Final drive Description of the drive from the transmission to the rear wheel. Usually by chain or shaft, but sometimes by belt.
Firing order The order in which the engine cylinders fire, or deliver their power strokes, beginning with the number one cylinder.
Flooding Term used to describe a high fuel level in the carburettor float chambers, leading to fuel overflow. Also refers to excess fuel in the combustion chamber due to incorrect starting technique.

Free length The no-load state of a component when measured. Clutch, valve and fork spring lengths are measured at rest, without any preload.
Freeplay The amount of travel before any action takes place. The looseness in a linkage, or an assembly of parts, between the initial application of force and actual movement. For example, the distance the rear brake pedal moves before the rear brake is actuated.
Fuel injection The fuel/air mixture is metered electronically and directed into the engine intake ports (indirect injection) or into the cylinders (direct injection). Sensors supply information on engine speed and conditions.
Fuel/air mixture The charge of fuel and air going into the engine. See **Stoichiometric ratio**.
Fuse An electrical device which protects a circuit against accidental overload. The typical fuse contains a soft piece of metal which is calibrated to melt at a predetermined current flow (expressed as amps) and break the circuit.

G

Gap The distance the spark must travel in jumping from the centre electrode to the side electrode in a spark plug. Also refers to the distance between the ignition rotor and the pickup coil in an electronic ignition system.
Gasket Any thin, soft material - usually cork, cardboard, asbestos or soft metal - installed between two metal surfaces to ensure a good seal. For instance, the cylinder head gasket seals the joint between the block and the cylinder head.
Gauge An instrument panel display used to monitor engine conditions. A gauge with a movable pointer on a dial or a fixed scale is an analogue gauge. A gauge with a numerical readout is called a digital gauge.
Gear ratios The drive ratio of a pair of gears in a gearbox, calculated on their number of teeth.
Glaze-busting see **Honing**
Grinding Process for renovating the valve face and valve seat contact area in the cylinder head.
Gudgeon pin The shaft which connects the connecting rod small-end with the piston. Often called a piston pin or wrist pin.

H

Helical gears Gear teeth are slightly curved and produce less gear noise that straight-cut gears. Often used for primary drives.

Installing a Helicoil thread insert in a cylinder head

Helicoil A thread insert repair system. Commonly used as a repair for stripped spark plug threads **(see illustration)**.
Honing A process used to break down the glaze on a cylinder bore (also called glaze-busting). Can also be carried out to roughen a rebored cylinder to aid ring bedding-in.
HT (High Tension) Description of the electrical circuit from the secondary winding of the ignition coil to the spark plug.
Hydraulic A liquid filled system used to transmit pressure from one component to another. Common uses on motorcycles are brakes and clutches.
Hydrometer An instrument for measuring the specific gravity of a lead-acid battery.
Hygroscopic Water absorbing. In motorcycle applications, braking efficiency will be reduced if DOT 3 or 4 hydraulic fluid absorbs water from the air - care must be taken to keep new brake fluid in tightly sealed containers.

I

lbf ft Pounds-force feet. An imperial unit of torque. Sometimes written as ft-lbs.
lbf in Pound-force inch. An imperial unit of torque, applied to components where a very low torque is required. Sometimes written as in-lbs.
IC Abbreviation for Integrated Circuit.
Ignition advance Means of increasing the timing of the spark at higher engine speeds. Done by mechanical means (ATU) on early engines or electronically by the ignition control unit on later engines.
Ignition timing The moment at which the spark plug fires, expressed in the number of crankshaft degrees before the piston reaches the top of its stroke, or in the number of millimetres before the piston reaches the top of its stroke.
Infinity (∞) Description of an open-circuit electrical state, where no continuity exists.
Inverted forks (upside down forks) The sliders or lower legs are held in the yokes and the fork tubes or stanchions are connected to the wheel axle (spindle). Less unsprung weight and stiffer construction than conventional forks.

J

JASO Quality standard for 2-stroke oils.
Joule The unit of electrical energy.
Journal The bearing surface of a shaft.

K

Kickstart Mechanical means of turning the engine over for starting purposes. Only usually fitted to mopeds, small capacity motorcycles and off-road motorcycles.
Kill switch Handebar-mounted switch for emergency ignition cut-out. Cuts the ignition circuit on all models, and additionally prevent starter motor operation on others.
km Symbol for kilometre.
kmh Abbreviation for kilometres per hour.

L

Lambda (λ) sensor A sensor fitted in the exhaust system to measure the exhaust gas oxygen content (excess air factor).

Technical Terms Explained REF•51

Lapping see **Grinding**.
LCD Abbreviation for Liquid Crystal Display.
LED Abbreviation for Light Emitting Diode.
Liner A steel cylinder liner inserted in a aluminium alloy cylinder block.
Locknut A nut used to lock an adjustment nut, or other threaded component, in place.
Lockstops The lugs on the lower triple clamp (yoke) which abut those on the frame, preventing handlebar-to-fuel tank contact.
Lockwasher A form of washer designed to prevent an attaching nut from working loose.
LT Low Tension Description of the electrical circuit from the power supply to the primary winding of the ignition coil.

M

Main bearings The bearings between the crankshaft and crankcase.
Maintenance-free (MF) battery A sealed battery which cannot be topped up.
Manometer Mercury-filled calibrated tubes used to measure intake tract vacuum. Used to synchronise carburettors on multi-cylinder engines.
Micrometer A precision measuring instrument that measures component outside diameters **(see illustration)**.

Tappet shims are measured with a micrometer

MON (Motor Octane Number) A measure of a fuel's resistance to knock.
Monograde oil An oil with a single viscosity, eg SAE80W.
Monoshock A single suspension unit linking the swingarm or suspension linkage to the frame.
mph Abbreviation for miles per hour.
Multigrade oil Having a wide viscosity range (eg 10W40). The W stands for Winter, thus the viscosity ranges from SAE10 when cold to SAE40 when hot.
Multimeter An electrical test instrument with the capability to measure voltage, current and resistance. Some meters also incorporate a continuity tester and buzzer.

N

Needle roller bearing Inner race of caged needle rollers and hardened outer race. Examples of uncaged needle rollers can be found on some engines. Commonly used in rear suspension applications and in two-stroke engines.
Nm Newton metres.
NOx Oxides of Nitrogen. A common toxic pollutant emitted by petrol engines at higher temperatures.

O

Octane The measure of a fuel's resistance to knock.
OE (Original Equipment) Relates to components fitted to a motorcycle as standard or replacement parts supplied by the motorcycle manufacturer.
Ohm The unit of electrical resistance. Ohms = Volts ÷ Current.
Ohmmeter An instrument for measuring electrical resistance.
Oil cooler System for diverting engine oil outside of the engine to a radiator for cooling purposes.
Oil injection A system of two-stroke engine lubrication where oil is pump-fed to the engine in accordance with throttle position.
Open-circuit An electrical condition where there is a break in the flow of electricity - no continuity (high resistance).
O-ring A type of sealing ring made of a special rubber-like material; in use, the O-ring is compressed into a groove to provide the sealing action.
Oversize (OS) Term used for piston and ring size options fitted to a rebored cylinder.
Overhead cam (sohc) engine An engine with single camshaft located on top of the cylinder head.
Overhead valve (ohv) engine An engine with the valves located in the cylinder head, but with the camshaft located in the engine block or crankcase.
Oxygen sensor A device installed in the exhaust system which senses the oxygen content in the exhaust and converts this information into an electric current. Also called a Lambda sensor.

P

Plastigauge A thin strip of plastic thread, available in different sizes, used for measuring clearances. For example, a strip of Plastigauge is laid across a bearing journal. The parts are assembled and dismantled; the width of the crushed strip indicates the clearance between journal and bearing.
Polarity Either negative or positive earth (ground), determined by which battery lead is connected to the frame (earth return). Modern motorcycles are usually negative earth.
Pre-ignition A situation where the fuel/air mixture ignites before the spark plug fires. Often due to a hot spot in the combustion chamber caused by carbon build-up. Engine has a tendency to 'run-on'.
Pre-load (suspension) The amount a spring is compressed when in the unloaded state. Preload can be applied by gas, spacer or mechanical adjuster.
Premix The method of engine lubrication on older two-stroke engines. Engine oil is mixed with the petrol in the fuel tank in a specific ratio. The fuel/oil mix is sometimes referred to as "petroil".
Primary drive Description of the drive from the crankshaft to the clutch. Usually by gear or chain.
PS Pfedestärke - a German interpretation of BHP.
PSI Pounds-force per square inch. Imperial measurement of tyre pressure and cylinder pressure measurement.
PTFE Polytetrafluroethylene. A low friction substance.

Pulse secondary air injection system A process of promoting the burning of excess fuel present in the exhaust gases by routing fresh air into the exhaust ports.

Q

Quartz halogen bulb Tungsten filament surrounded by a halogen gas. Typically used for the headlight **(see illustration)**.

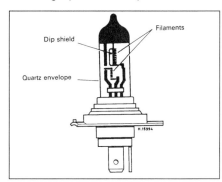

Quartz halogen headlight bulb construction

R

Rack-and-pinion A pinion gear on the end of a shaft that mates with a rack (think of a geared wheel opened up and laid flat). Sometimes used in clutch operating systems.
Radial play Up and down movement about a shaft.
Radial ply tyres Tyre plies run across the tyre (from bead to bead) and around the circumference of the tyre. Less resistant to tread distortion than other tyre types.
Radiator A liquid-to-air heat transfer device designed to reduce the temperature of the coolant in a liquid cooled engine.
Rake A feature of steering geometry - the angle of the steering head in relation to the vertical **(see illustration)**.

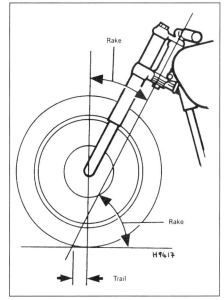

Steering geometry

Technical Terms Explained

Rebore Providing a new working surface to the cylinder bore by boring out the old surface. Necessitates the use of oversize piston and rings.

Rebound damping A means of controlling the oscillation of a suspension unit spring after it has been compressed. Resists the spring's natural tendency to bounce back after being compressed.

Rectifier Device for converting the ac output of an alternator into dc for battery charging.

Reed valve An induction system commonly used on two-stroke engines.

Regulator Device for maintaining the charging voltage from the generator or alternator within a specified range.

Relay A electrical device used to switch heavy current on and off by using a low current auxiliary circuit.

Resistance Measured in ohms. An electrical component's ability to pass electrical current.

RON (Research Octane Number) A measure of a fuel's resistance to knock.

rpm revolutions per minute.

Runout The amount of wobble (in-and-out movement) of a wheel or shaft as it's rotated. The amount a shaft rotates 'out-of-true'. The out-of-round condition of a rotating part.

S

SAE (Society of Automotive Engineers) A standard for the viscosity of a fluid.

Sealant A liquid or paste used to prevent leakage at a joint. Sometimes used in conjunction with a gasket.

Service limit Term for the point where a component is no longer useable and must be renewed.

Shaft drive A method of transmitting drive from the transmission to the rear wheel.

Shell bearings Plain bearings consisting of two shell halves. Most often used as big-end and main bearings in a four-stroke engine. Often called bearing inserts.

Shim Thin spacer, commonly used to adjust the clearance or relative positions between two parts. For example, shims inserted into or under tappets or followers to control valve clearances. Clearance is adjusted by changing the thickness of the shim.

Short-circuit An electrical condition where current shorts to earth (ground) bypassing the circuit components.

Skimming Process to correct warpage or repair a damaged surface, eg on brake discs or drums.

Slide-hammer A special puller that screws into or hooks onto a component such as a shaft or bearing; a heavy sliding handle on the shaft bottoms against the end of the shaft to knock the component free.

Small-end bearing The bearing in the upper end of the connecting rod at its joint with the gudgeon pin.

Spalling Damage to camshaft lobes or bearing journals shown as pitting of the working surface.

Specific gravity (SG) The state of charge of the electrolyte in a lead-acid battery. A measure of the electrolyte's density compared with water.

Straight-cut gears Common type gear used on gearbox shafts and for oil pump and water pump drives.

Stanchion The inner sliding part of the front forks, held by the yokes. Often called a fork tube.

Stoichiometric ratio The optimum chemical air/fuel ratio for a petrol engine, said to be 14.7 parts of air to 1 part of fuel.

Sulphuric acid The liquid (electrolyte) used in a lead-acid battery. Poisonous and extremely corrosive.

Surface grinding (lapping) Process to correct a warped gasket face, commonly used on cylinder heads.

T

Tapered-roller bearing Tapered inner race of caged needle rollers and separate tapered outer race. Examples of taper roller bearings can be found on steering heads.

Tappet A cylindrical component which transmits motion from the cam to the valve stem, either directly or via a pushrod and rocker arm. Also called a cam follower.

TCS Traction Control System. An electronically-controlled system which senses wheel spin and reduces engine speed accordingly.

TDC Top Dead Centre denotes that the piston is at its highest point in the cylinder.

Thread-locking compound Solution applied to fastener threads to prevent slackening. Select type to suit application.

Thrust washer A washer positioned between two moving components on a shaft. For example, between gear pinions on gearshaft.

Timing chain See **Cam Chain.**

Timing light Stroboscopic lamp for carrying out ignition timing checks with the engine running.

Top-end A description of an engine's cylinder block, head and valve gear components.

Torque Turning or twisting force about a shaft.

Torque setting A prescribed tightness specified by the motorcycle manufacturer to ensure that the bolt or nut is secured correctly. Undertightening can result in the bolt or nut coming loose or a surface not being sealed. Overtightening can result in stripped threads, distortion or damage to the component being retained.

Torx key A six-point wrench.

Tracer A stripe of a second colour applied to a wire insulator to distinguish that wire from another one with the same colour insulator. For example, Br/W is often used to denote a brown insulator with a white tracer.

Trail A feature of steering geometry. Distance from the steering head axis to the tyre's central contact point.

Triple clamps The cast components which extend from the steering head and support the fork stanchions or tubes. Often called fork yokes.

Turbocharger A centrifugal device, driven by exhaust gases, that pressurises the intake air. Normally used to increase the power output from a given engine displacement.

TWI Abbreviation for Tyre Wear Indicator. Indicates the location of the tread depth indicator bars on tyres.

U

Universal joint or U-joint (UJ) A double-pivoted connection for transmitting power from a driving to a driven shaft through an angle. Typically found in shaft drive assemblies.

Unsprung weight Anything not supported by the bike's suspension (ie the wheel, tyres, brakes, final drive and bottom (moving) part of the suspension).

V

Vacuum gauges Clock-type gauges for measuring intake tract vacuum. Used for carburettor synchronisation on multi-cylinder engines.

Valve A device through which the flow of liquid, gas or vacuum may be stopped, started or regulated by a moveable part that opens, shuts or partially obstructs one or more ports or passageways. The intake and exhaust valves in the cylinder head are of the poppet type.

Valve clearance The clearance between the valve tip (the end of the valve stem) and the rocker arm or tappet/follower. The valve clearance is measured when the valve is closed. The correct clearance is important - if too small the valve won't close fully and will burn out, whereas if too large noisy operation will result.

Valve lift The amount a valve is lifted off its seat by the camshaft lobe.

Valve timing The exact setting for the opening and closing of the valves in relation to piston position.

Vernier caliper A precision measuring instrument that measures inside and outside dimensions. Not quite as accurate as a micrometer, but more convenient.

VIN Vehicle Identification Number. Term for the bike's engine and frame numbers.

Viscosity The thickness of a liquid or its resistance to flow.

Volt A unit for expressing electrical "pressure" in a circuit. Volts = current x ohms.

W

Water pump A mechanically-driven device for moving coolant around the engine.

Watt A unit for expressing electrical power. Watts = volts x current.

Wear limit see **Service limit**

Wet liner A liquid-cooled engine design where the pistons run in liners which are directly surrounded by coolant **(see illustration)**.

Wet liner arrangement

Wheelbase Distance from the centre of the front wheel to the centre of the rear wheel.

Wiring harness or loom Describes the electrical wires running the length of the motorcycle and enclosed in tape or plastic sheathing. Wiring coming off the main harness is usually referred to as a sub harness.

Woodruff key A key of semi-circular or square section used to locate a gear to a shaft. Often used to locate the alternator rotor on the crankshaft.

Wrist pin Another name for gudgeon or piston pin.

Index

Note: *References throughout this index are in the form - "Chapter number" • "Page number"*

A

Aim (headlight) – 1•26
Air filter – 1•10, 4•4
Air/fuel mixture adjustment – 4•5
Air induction system
 check – 1•11
 function, disassembly, reassembly – 4•21
Alignment (wheel) – 7•13
Alternator – 9•23

B

Battery
 charging – 9•3
 check – 1•15, 9•3
 removal/installation – 9•3
 specifications – 9•1
Bearings
 connecting rod – 2•42, 2•46
 general advice – REF•14
 main – 2•42, 2•44
 steering head –1•19, 1•23, 6•12
 swingarm – 1•19, 1•23, 6•17
 wheel – 1•15, 7•16
Bleeding (brakes) – 7•12
Body panels – 8•3
Brake
 bleeding – 7•12
 calipers – 7•3, 7•8
 discs – 7•4, 7•10
 fluid change – 1•23
 fluid top-up – 0•14
 hoses and unions – 1•26, 7•12
 master cylinder – 7•5, 7•10
 pads – 1•13, 7•2, 7•7
 seal renewal – 1•23
 specifications – 7•1
 system check – 1•14
Brake light
 bulbs – 9•6
 circuit check – 9•5
 switches – 1•14, 9•9
Brake lever
 pivot lubrication – 1•18
 removal/installation – 6•5
 span adjuster – 1•14
Brake pedal
 height adjustment – 1•14
 pivot lubrication – 1•18
 removal/installation – 6•2
Bulbs
 headlight – 9•5
 sidelight – 9•6
 tail/brake light – 9•6
 turn signals – 9•8
 warning lights (instruments) – 9•11
 wattages – 9•2

C

Cables
 choke – 1•18, 4•15
 clutch – 1•14, 1•18, 2•27
 EXUP – 1•11, 4•17
 throttle – 1•16, 1•18, 4•14
Calipers (brake) – 7•3, 7•8
Cam chain, tensioner, guides/blades – 2•11
Camshafts – 2•13
Carburettors
 cleaning – 4•7
 dismantling – 4•7
 fuel level check – 4•12
 general overhaul advice – 4•5
 idle speed – 1•9
 inspection – 4•7
 joining – 4•11
 reassembly – 4•12
 removal/installation – 4•5
 separation – 4•11
 specifications – 4•1, 4•2
 synchronisation – 1•9
 throttle position sensor – 5•6
Chain
 cam – 2•11
 drive chain (final) – 0•16, 1•6, 6•18
 fitting advice – REF• 18
Charging (battery) – 9•3
Charging system checks – 9•23
Choke cable
 check/adjust – 1•18
 lubrication – 1•18
 removal/installation – 4•15
Clutch
 overhaul – 2•23
 specifications – 2•3
Clutch cable
 check/adjust – 1•14
 lubrication – 1•18
 removal/installation – 2•27
Clutch lever
 pivot lubrication – 1•18
 removal/installation – 6•4
Clutch switch – 9•14
Cockpit trim panels – 8•5
Coil (pick-up) – 5•4
Coils (HT) – 5•2
Compression test – 1•27, REF•44
Connecting rods – 2•44
Conversion factors – REF•26
Coolant
 top-up – 0•15
 change – 1•21
 reservoir – 3•2
Coolant temperature display/light – 3•4
Cooling fan, fan switch/relay – 3•2
Cooling system
 check – 1•16
 hoses, pipes and unions – 3•8
 radiator – 3•6
 thermostat – 3•5
 water pump – 3•7
Crankcases – 2•39
Crankshaft – 2•42
Cush drive (rear wheel) – 6•20
Cylinder bores – 2•41
Cylinder head
 cover – 2•10
 overhaul and valves – 2•18
 removal and installation – 2•17

Index

D

Dimensions (model) – 0•11
Disc (brake)
  front – 7•4
  rear – 7•10
Drive chain (final)
  adjustment – 1•6
  checks – 0•16, 1•6
  fitting advice – REF• 18
  lubrication – 1•6
  removal, inspection, cleaning, installation – 6•18

E

Electrical system
  alternator – 9•23
  battery – 1•15, 9•3
  brake light switches – 1•14, 9•9
  charging system – 9•23
  clutch switch – 9•14
  fault finding – 9•2
  fuses – 9•4
  handlebar switches – 9•12, 9•13
  headlights – 1•26, 9•5, 9•6
  horn – 9•15
  ignition (main) switch – 9•11
  instruments – 9•9, 9•10
  lighting system check – 9•5
  neutral switch – 9•13
  oil level sensor/relay – 9•15
  regulator/rectifier – 9•25
  relay assembly – 9•14
  sidestand switch – 9•14
  starter motor – 9•16, 9•19
  starter relay – 9•16
  tail/brake light – 9•6, 9•7
  turn signals – 9•8
  warning lights – 9•11
  wiring diagrams – 9•26
Engine
  bearings – 2•42
  cam chain, tensioner and blades/guides – 2•11
  camshafts – 2•13
  compression test – 1•27, REF•44
  connecting rods – 2•44
  crankcases – 2•39, 2•41
  crankshaft – 2•42
  cylinder bores – 2•41
  cylinder head – 2•17, 2•18
  followers – 2•13
  initial start-up – 2•58
  oil cooler – 2•9
  oil pressure test – 1•27
  oil pump – 2•34
  oil strainer and pressure relief valve – 2•33
  piston rings – 2•49
  pistons – 2•47
  removal/installation – 2•6
  running-in – 2•58
  specifications – 2•1
  sprocket – 1•6, 6•18
  starter clutch – 2•28
  starter idle/reduction gear – 2•28
  sump – 2•33
  valve clearance check/adjust – 1•23
  valve cover – 2•10
  valve overhaul – 2•18
Engine number – 0•9
Engine/transmission oil
  level check – 0•13
  filter change – 1•21
  oil change – 1•12
Exhaust system – 4•16
EXUP
  cable renewal – 4•17
  check – 1•11, 4•18
  removal/installation – 4•19

F

Fairing – 8•3
Fan and switch – 3•2
Fault finding – 5•2, 9•2, REF•35 et seq
Fault finding equipment – REF•45 et seq
Filter
  air – 1•10, 4•4
  fuel – 1•11, 1•23
  oil – 1•21
Final drive chain
  adjustment – 1•6
  checks – 0•16, 1•6
  fitting advice – REF• 18
  lubrication – 1•6
  removal, inspection, cleaning, installation – 6•18
Followers (camshaft) – 2•13
Footrests
  pivot lubrication – 1•18
  removal/installation – 6•2
Frame – 6•2
Frame number – 0•9
Front brake
  calipers – 7•3
  discs – 7•4
  fluid top-up – 0•14
  master cylinder – 7•5
  pads – 1•13, 7•2
Front brake lever
  pivot lubrication – 1•18
  removal/installation – 6•5
  span adjuster – 1•14
Front brake light switch – 1•14, 9•9
Front forks
  adjustment – 6•15
  check – 0•16, 1•18
  oil change – 1•27
  overhaul – 6•5
  removal/installation – 6•5
Front mudguard – 8•6
Front wheel
  bearings – 1•15, 7•16
  removal/installation – 7•14
Fuel level check – 4•12
Fuel level warning light and sensor – 4•21
Fuel pump and relay – 4•19
Fuel system
  carburettors – 1•9, 4•5 to 4•12
  check – 1•11
  filters – 1•11, 1•23
  hose renewal – 1•27
Fuel tank and tap – 4•3
Fuses
  check and renewal – 9•4
  ratings – 9•2

Index

G
Gaskets – REF•17
Gearbox
  selector drum and forks – 2•57
  shafts – 2•50, 2•51
  specifications – 2•4
Gearchange lever linkage
  pivot lubrication – 1•18
  removal/installation – 6•2
Gearchange mechanism – 2•30

H
Handlebar switches – 9•12, 9•13
Handlebars – 6•3
Headlights
  aim – 1•26
  bulbs – 9•5
  circuit check and relays – 9•5
  removal/installation – 9•6
Horn – 9•15
HT coils – 5•2

I
Idle speed – 1•9
Ignition (main) switch – 9•11
Ignition system
  check – 5•2
  HT coils – 5•2
  ignition control unit – 5•5
  pick-up coil – 5•4
  specifications – 5•1
  throttle position sensor – 5•6
  timing – 5•5
Instrument cluster – 9•9, 9•10, 9•11

L
Legal checks – 0•16
Lighting system checks – 9•5
Lower fairing – 8•3
Lubricants – 1•2, REF•23
Lubrication system (engine/transmission)
  oil change – 1•12
  oil cooler – 2•9
  oil filter change – 1•21
  oil level check – 0•13
  oil level sensor/relay – 9•15
  oil pressure check – 1•27
  oil pump – 2•34
  oil sump, strainer and pressure relief valve – 2•33

M
Main bearings – 2•42
Maintenance
  procedures – 1•6 et seq
  schedule – 1•3
Master cylinder (brake)
  front – 7•5
  rear – 7•10
Micrometer – REF•9
Mirrors – 8•5
Mixture adjustment – 4•5
Model codes – 0•9
MOT test checks – REF• 27
Mudguard (front) – 8•6

N
Neutral switch – 9•13

O
Oil (engine/transmission)
  change – 1•12
  level check – 0•13
  pressure check – 1•27
  recommendations – 1•2
Oil (front forks) – 1•27
Oil cooler – 2•9
Oil filter – 1•21
Oil level sensor/relay – 9•15
Oil pump – 2•34
Oil seals – REF•16
Oil sump, strainer and pressure relief valve – 2•33

P
Pads (brake)
  check – 1•13
  renewal – 7•2, 7•7
Performance data – 0•10
Pick-up coil – 5•4
Piston rings – 2•49
Pistons – 2•47
Plastigauge – REF•11
Pressure relief valve – 2•33
Pump (fuel) – 4•19
Pump (oil) – 2•34
Pump (water) – 3•7

R
Radiator – 3•6
Radiator pressure cap – 1•16, 3•2
Rear brake
  caliper – 7•8
  disc – 7•10
  fluid top-up – 0•14
  master cylinder – 7•10
  pads – 1•13, 7•7
Rear brake light switch – 1•14, 9•9
Rear brake pedal
  height adjustment – 1•14
  pivot lubrication – 1•18
  removal/installation – 6•2
Rear sprocket – 1•6, 6•18
Rear sprocket coupling bearing – 7•18
Rear sprocket coupling/damper – 6•20
Rear suspension linkage
  bearing lubrication – 1•23
  removal, inspection, installation – 6•14
Rear suspension unit (shock)
  adjustment – 6•16
  check – 0•16, 1•18
  removal, inspection, installation – 6•13
Rear view mirrors – 8•5
Rear wheel
  bearings – 1•15, 7•17
  removal/installation – 7•15
Regulator/rectifier – 9•25
Relay
  assembly – 9•14
  fan – 3•2
  fuel pump – 4•19
  headlights – 9•5

Index

oil level – 9•15
starter circuit cut-off – 9•14
starter motor – 9•16
turn signal – 9•8

S

Safety – 0•12, 0•16, 4•2, REF•4
Screen – 8•5
Sealants – REF•17
Seat cowling – 8•3
Seats – 8•2
Security advice – REF•20
Selector drum and forks – 2•57
Sidelights
  bulbs – 9•6
  circuit check – 9•5
Sidestand
  check – 1•15
  pivot lubrication – 1•18
  removal/installation – 6•3
  switch – 9•14
Spark plugs – 1•8
Specifications
  brakes – 7•1
  clutch – 2•3
  cooling system – 3•1
  electrical system – 9•1
  engine – 2•1
  final drive – 6•1
  fuel system – 4•1
  ignition system – 5•1
  maintenance – 1•2
  suspension – 6•1
  transmission – 2•4
  tyres – 7•1
  wheels – 7•1
Speedometer – 9•9, 9•10
Sprockets
  check – 1•6, 6•18
  renewal – 6•18
Starter circuit cut-off relay – 9•14
Starter clutch – 2•28
Starter idle/reduction gear – 2•28
Starter motor
  overhaul – 9•19
  removal/installation – 9•16
  specifications – 9•2
Starter relay – 9•16
Steering
  checks – 0•16
  head bearings check/adjust – 1•19
  head bearings inspection/renewal – 6•12
  head bearings lubrication – 1•23
  stem removal/installation – 6•10
Storage advice – REF•32
Sump – 2•33
Suspension
  adjustments – 6•15
  checks – 0•16, 1•18
  front forks – 1•27, 6•5
  rear shock and linkage – 1•23, 6•13, 6•14
  swingarm – 1•23, 6•16, 6•17
Swingarm
  bearing check – 1•19
  bearing lubrication – 1•23
  inspection and bearing renewal – 6•17
  removal/installation – 6•16

Switch
  brake light – 1•14, 9•9
  clutch – 9•14
  handlebar – 9•12, 9•13
  ignition (main) – 9•11
  neutral – 9•13
  sidestand – 9•14
Synchronisation (carburettors) – 1•9

T

Tachometer – 9•9, 9•10
Tail light
  bulbs – 9•6
  circuit check – 9•5
  removal/installation – 9•7
Tank (fuel) – 4•3, 4•4
Tap (fuel) – 4•3
Temperature (coolant) display/light – 3•4
Tensioner (cam chain) – 2•11
Timing (ignition) – 5•5
Timing (valve) – 2•16
Thermo unit – 3•4
Thermostat – 3•5
Throttle cables
  check/adjust – 1•16
  lubrication – 1•18
  removal/installation – 4•14
Throttle position sensor – 5•6
Tools – REF•2
Torque – REF• 13
Torque settings – 1•2, 2•4, 3•1, 4•1, 5•1, 6•1, 7•2, 9•2
Transmission
  selector drum and forks – 2•57
  shafts – 2•50, 2•51
  specifications – 2•4
Turn signal
  bulbs – 9•8
  circuit check and relay – 9•5, 9•8
  removal/installation – 9•8
Tyre
  general information and fitting – 7•18
  pressures and tread wear – 0•16
  specifications – 7•1

V

Valve clearance check/adjust – 1•23
Valve cover – 2•10
Valve overhaul – 2•18
Vernier caliper – REF•10
VIN (Vehicle Identification Number) – 0•9

W

Warning light bulbs – 9•11
Water pump – 3•7
Weight – 0•11
Wheels
  alignment – 7•13
  bearings – 1•15, 7•16
  check – 1•15
  inspection and repair – 7•13
  removal/installation – 7•14, 7•15
  specifications – 7•1
Windshield – 8•5
Wiring diagrams – 9•26

Haynes Motorcycle Manuals – The Complete List

Title	Book No
BMW	
BMW 2-valve Twins (70 - 96)	0249
BMW K100 & 75 2-valve Models (83 - 96)	1373
BMW R850 & R1100 4-valve Twins (93 - 97)	3466
BSA	
BSA Bantam (48 - 71)	0117
BSA Unit Singles (58 - 72)	0127
BSA Pre-unit Singles (54 - 61)	0326
BSA A7 & A10 Twins (47 - 62)	0121
BSA A50 & A65 Twins (62 - 73)	0155
DUCATI	
Ducati 600, 750 & 900 2-valve V-Twins (91 - 96)	3290
Ducati 748, 916 & 996 4-valve V-Twins (94 - 01)	3756
HARLEY-DAVIDSON	
Harley-Davidson Sportsters (70 - 01)	0702
Harley-Davidson Big Twins (70 - 99)	0703
HONDA	
Honda NB, ND, NP & NS50 Melody (81 - 85)	◊ 0622
Honda NE/NB50 Vision & SA50 Vision Met-in (85 - 95)	◊ 1278
Honda MB, MBX, MT & MTX50 (80 - 93)	0731
Honda C50, C70 & C90 (67 - 99)	0324
Honda XR80R & XR100R (85 - 96)	2218
Honda XL/XR 80, 100, 125, 185 & 200 2-valve Models (78 - 87)	0566
Honda H100 & H100S Singles (80 - 92)	◊ 0734
Honda CB/CD125T & CM125C Twins (77 - 88)	◊ 0571
Honda CG125 (76 - 00)	◊ 0433
Honda NS125 (86 - 93)	◊ 3056
Honda MBX/MTX125 & MTX200 (83 - 93)	◊ 1132
Honda CD/CM185 200T & CM250C 2-valve Twins (77 - 85)	0572
Honda XL/XR 250 & 500 (78 - 84)	0567
Honda XR250L, XR250R & XR400R (86 - 01)	2219
Honda CB250 & CB400N Super Dreams (78 - 84)	◊ 0540
Honda CR Motocross Bikes (86 - 01)	2222
Honda Elsinore 250 (73 - 75)	0217
Honda CBR400RR Fours (88 - 99)	3552
Honda VFR400 (NC30) & RVF400 (NC35) V-Fours (89 - 98)	3496
Honda CB500 (93 - 01)	3753
Honda CB400 & CB550 Fours (73 - 77)	0262
Honda CX/GL500 & 650 V-Twins (78 - 86)	0442
Honda CBX550 Four (82 - 86)	◊ 0940
Honda XL600R & XR600R (83 - 00)	2183
Honda XL600/650V Transalp & XRV750 Africa Twin (87 - 02)	3919
Honda CBR600F1 & 1000F Fours (87 - 96)	1730
Honda CBR600F2 & F3 Fours (91 - 98)	2070
Honda CBR600F4 (99 - 02)	3911
Honda CB600F Hornet (98 - 02)	3915
Honda NTV600/650/Deauville V-Twins (88 - 01)	3243
Honda Shadow VT600 & 750 (USA) (88 - 99)	2312
Honda CB750 sohc Four (69 - 79)	0131
Honda V45/65 Sabre & Magna (82 - 88)	0820
Honda VFR750 & 700 V-Fours (86 - 97)	2101
Honda VFR800 V-Fours (97 - 99)	3703
Honda VT1000 (FireStorm, Super Hawk) & XL1000V (Varadero) (97 - 02)	3744
Honda CB750 & CB900 dohc Fours (78 - 84)	0535
Honda CBR900RR FireBlade (92 - 99)	2161
Honda CBR1100XX Super Blackbird (97 - 02)	3901
Honda ST1100 Pan European V-Fours (90 - 01)	3384

Title	Book No
Honda Shadow VT1100 (USA) (85 - 98)	2313
Honda GL1000 Gold Wing (75 - 79)	0309
Honda GL1100 Gold Wing (79 - 81)	0669
Honda Gold Wing 1200 (USA) (84 - 87)	2199
Honda Gold Wing 1500 (USA) (88 - 00)	2225
KAWASAKI	
Kawasaki AE/AR 50 & 80 (81 - 95)	1007
Kawasaki KC, KE & KH100 (75 - 99)	1371
Kawasaki KMX125 & 200 (86 - 96)	◊ 3046
Kawasaki 250, 350 & 400 Triples (72 - 79)	0134
Kawasaki 400 & 440 Twins (74 - 81)	0281
Kawasaki 400, 500 & 550 Fours (79 - 91)	0910
Kawasaki EN450 & 500 Twins (Ltd/Vulcan) (85 - 93)	2053
Kawasaki EX & ER500 (GPZ500S & ER-5) Twins (87 - 99)	2052
Kawasaki ZX600 (Ninja ZX-6, ZZ-R600) Fours (90 - 00)	2146
Kawasaki ZX-6R Ninja Fours (95 - 98)	3541
Kawasaki ZX600 (GPZ600R, GPX600R, Ninja 600R & RX) & ZX750 (GPX750R, Ninja 750R) Fours (85 - 97)	1780
Kawasaki 650 Four (76 - 78)	0373
Kawasaki 750 Air-cooled Fours (80 - 91)	0574
Kawasaki ZR550 & 750 Zephyr Fours (90 - 97)	3382
Kawasaki ZX750 (Ninja ZX-7 & ZXR750) Fours (89 - 96)	2054
Kawasaki Ninja ZX-7R & ZX-9R (ZX750P, ZX900B/C/D/E) (94 - 00)	3721
Kawasaki 900 & 1000 Fours (73 - 77)	0222
Kawasaki ZX900, 1000 & 1100 Liquid-cooled Fours (83 - 97)	1681
MOTO GUZZI	
Moto Guzzi 750, 850 & 1000 V-Twins (74 - 78)	0339
MZ	
MZ ETZ Models (81 - 95)	◊ 1680
NORTON	
Norton 500, 600, 650 & 750 Twins (57 - 70)	0187
Norton Commando (68 - 77)	0125
PIAGGIO	
Piaggio (Vespa) Scooters (91 - 98)	3492
SUZUKI	
Suzuki GT, ZR & TS50 (77 - 90)	◊ 0799
Suzuki TS50X (84 - 00)	◊ 1599
Suzuki 100, 125, 185 & 250 Air-cooled Trail bikes (79 - 89)	0797
Suzuki GP100 & 125 Singles (78 - 93)	◊ 0576
Suzuki GS, GN, GZ & DR125 Singles (82 - 99)	◊ 0888
Suzuki GT250X7, GT200X5 & SB200 Twins (78 - 83)	◊ 0469
Suzuki GS/GSX250, 400 & 450 Twins (79 - 85)	0736
Suzuki GS500E Twin (89 - 97)	3238
Suzuki GS550 (77 - 82) & GS750 Fours (76 - 79)	0363
Suzuki GS/GSX550 4-valve Fours (83 - 88)	1133
Suzuki GSX-R600 & 750 (96 - 99)	3553
Suzuki GSF600 & 1200 Bandit Fours (95 - 01)	3367
Suzuki GS850 Fours (78 - 88)	0536
Suzuki GS1000 Four (77 - 79)	0484
Suzuki GSX-R750, GSX-R1100 (85 - 92), GSX600F, GSX750F, GSX1100F (Katana) Fours (88 - 96)	2055
Suzuki GS/GSX1000, 1100 & 1150 4-valve Fours (79 - 88)	0737
TRIUMPH	
Triumph 350 & 500 Unit Twins (58 - 73)	0137
Triumph Pre-Unit Twins (47 - 62)	0251
Triumph 650 & 750 2-valve Unit Twins (63 - 83)	0122
Triumph Trident & BSA Rocket 3 (69 - 75)	0136
Triumph Fuel Injected Triples (97 - 00)	3755
Triumph Triples & Fours (carburettor engines) (91 - 99)	2162

Title	Book No
VESPA	
Vespa P/PX125, 150 & 200 Scooters (78 - 95)	0707
Vespa Scooters (59 - 78)	0126
YAMAHA	
Yamaha DT50 & 80 Trail Bikes (78 - 95)	◊ 0800
Yamaha T50 & 80 Townmate (83 - 95)	◊ 1247
Yamaha YB100 Singles (73 - 91)	◊ 0474
Yamaha RS/RXS100 & 125 Singles (74 - 95)	0331
Yamaha RD & DT125LC (82 - 87)	◊ 0887
Yamaha TZR125 (87 - 93) & DT125R (88 - 95)	◊ 1655
Yamaha TY50, 80, 125 & 175 (74 - 84)	◊ 0464
Yamaha XT & SR125 (82 - 96)	1021
Yamaha Trail Bikes (81 - 00)	2350
Yamaha 250 & 350 Twins (70 - 79)	0040
Yamaha XS250, 360 & 400 sohc Twins (75 - 84)	0378
Yamaha RD250 & 350LC Twins (80 - 82)	0803
Yamaha RD350 YPVS Twins (83 - 95)	1158
Yamaha RD400 Twin (75 - 79)	0333
Yamaha XT, TT & SR500 Singles (75 - 83)	0342
Yamaha XZ550 Vision V-Twins (82 - 85)	0821
Yamaha FJ, FZ, XJ & YX600 Radian (84 - 92)	2100
Yamaha XJ600S (Diversion, Seca II) & XJ600N Fours (92 - 99)	2145
Yamaha YZF600R Thundercat & FZS600 Fazer (96 - 00)	3702
Yamaha YZF-R6 (98 - 02)	3900
Yamaha 650 Twins (70 - 83)	0341
Yamaha XJ650 & 750 Fours (80 - 84)	0738
Yamaha XS750 & 850 Triples (76 - 85)	0340
Yamaha TDM850, TRX850 & XTZ750 (89 - 99)	3540
Yamaha YZF750R & YZF1000R Thunderace (93 - 00)	3720
Yamaha FZR600, 750 & 1000 Fours (87 - 96)	2056
Yamaha XV V-Twins (81 - 96)	0802
Yamaha XJ900F Fours (83 - 94)	3239
Yamaha XJ900S Diversion (94 - 01)	3739
Yamaha YZF-R1 (98 - 01)	3754
Yamaha FJ1100 & 1200 Fours (84 - 96)	2057
ATVs	
Honda ATC70, 90, 110, 185 & 200 (71 - 85)	0565
Honda TRX300 Shaft Drive ATVs (88 - 00)	2125
Honda TRX300EX & TRX400EX ATVs (93 - 99)	2318
Kawasaki Bayou 220/300 & Prairie 300 ATVs (86 - 01)	2351
Polaris ATVs (85 to 97)	2302
Yamaha YT, YFM, YTM & YTZ ATVs (80 - 85)	1154
Yamaha YFS200 Blaster ATV (88 - 98)	2317
Yamaha YFB250 Timberwolf ATV (92 - 96)	2217
Yamaha YFM350 (ER and Big Bear) ATVs (87 - 99)	2126
Yamaha Warrior and Banshee ATVs (87 - 99)	2314
ATV Basics	10450
MOTORCYCLE TECHBOOKS	
Motorcycle Basics TechBook (2nd Edition)	3515
Motorcycle Electrical TechBook (3rd Edition)	3471
Motorcycle Fuel Systems TechBook	3514
Motorcycle Workshop Practice TechBook (2nd Edition)	3470

◊ = not available in the USA   **Bold type** = Superbike

The manuals on this page are available through good motorcycle dealers and accessory shops.
In case of difficulty, contact: **Haynes Publishing**
(UK) **+44 1963 442030**   (USA) **+1 805 4986703**
(FR) **+33 1 47 78 50 50**   (SV) **+46 18 124016**
(Australia/New Zealand) **+61 3 9763 8100**

Preserving Our Motoring Heritage

The Model J Duesenberg Derham Tourster. Only eight of these magnificent cars were ever built – this is the only example to be found outside the United States of America

Almost every car you've ever loved, loathed or desired is gathered under one roof at the Haynes Motor Museum. Over 300 immaculately presented cars and motorbikes represent every aspect of our motoring heritage, from elegant reminders of bygone days, such as the superb Model J Duesenberg to curiosities like the bug-eyed BMW Isetta. There are also many old friends and flames. Perhaps you remember the 1959 Ford Popular that you did your courting in? The magnificent 'Red Collection' is a spectacle of classic sports cars including AC, Alfa Romeo, Austin Healey, Ferrari, Lamborghini, Maserati, MG, Riley, Porsche and Triumph.

A Perfect Day Out

Each and every vehicle at the Haynes Motor Museum has played its part in the history and culture of Motoring. Today, they make a wonderful spectacle and a great day out for all the family. Bring the kids, bring Mum and Dad, but above all bring your camera to capture those golden memories for ever. You will also find an impressive array of motoring memorabilia, a comfortable 70 seat video cinema and one of the most extensive transport book shops in Britain. The Pit Stop Cafe serves everything from a cup of tea to wholesome, home-made meals or, if you prefer, you can enjoy the large picnic area nestled in the beautiful rural surroundings of Somerset.

John Haynes O.B.E., Founder and Chairman of the museum at the wheel of a Haynes Light 12.

The 1936 490cc sohc-engined International Norton – well known for its racing success

The Museum is situated on the A359 Yeovil to Frome road at Sparkford, just off the A303 in Somerset. It is about 40 miles south of Bristol, and 25 minutes drive from the M5 intersection at Taunton.
Open 9.30am - 5.30pm (10.00am - 4.00pm Winter) 7 days a week, *except Christmas Day, Boxing Day and New Years Day*
Special rates available for schools, coach parties and outings Charitable Trust No. 292048